ALSO BY DOUGLAS WALLER

Disciples: The World War II Missions of the CIA Directors
Who Fought for Wild Bill Donovan

Wild Bill Donovan: The Spymaster Who Created
the OSS and Modern American Espionage

A Question of Loyalty: Gen. Billy Mitchell and
the Court-Martial That Gripped the Nation

Big Red: The Three-Month Voyage of a Trident Nuclear Submarine

The Commandos: The Inside Story of America's Secret Soldiers

Air Warriors

THE INSIDE
STORY
OF THE MAKING
OF A
NAVY PILOT

Douglas Waller

SIMON & SCHUSTER PAPERBACKS
New York London Toronto Sydney New Delhi

Simon & Schuster Paperbacks
An Imprint of Simon & Schuster, Inc.
1230 Avenue of the Americas
New York, NY 10020

First Simon & Schuster trade paperback edition June 2019

SIMON & SCHUSTER PAPERBACKS and colophon are
registered trademarks of Simon & Schuster, Inc.

For information about special discounts for bulk purchases,
please contact Simon & Schuster Special Sales at 1-866-506-1949 or
business@simonandschuster.com.

The Simon & Schuster Speakers Bureau can bring authors
to your live event. For more information or to book an event,
contact the Simon & Schuster Speakers Bureau at 1-866-248-3049 or
visit our website at www.simonspeakers.com.

Designed by Karolina Harris

Manufactured in the United States of America

1   3   5   7   9   10   8   6   4   2

The Library of Congress has cataloged the hardcover edition as follows:
Waller, Douglas C.
Air warriors: the inside story of the making of a Navy Pilot/Douglas C. Waller
p.   cm.
1. United States. Navy—Aviation.
2. Air pilots, Military—Training of—United States. I. Title.
VG93.W34   1998
359.9'45'0973—dc21   98-9727
CIP

ISBN 978-0-6848-1430-8
ISBN 978-1-9821-2821-0 (pbk)

PHOTO CREDITS
1–5. Kevin Stephens, *All Hands* magazine;
6, 8–13, 15, 17–18, 20–22, 24–29, 32–35, 37–38, 41. Douglas C. Waller;
7, 14, 19–20, 30–31, 36, 39–40. Department of the Navy;
16. Bill Sigler;
23. Jonathan Wise

To Mom and Dad

NOTE TO READERS

Air Warriors was first published in 1998. Of course, in the more than twenty years that have passed, Naval aircraft and carrier equipment, technology, and organization have changed considerably. So have the military's social mores. But what has not changed is the physical and mental challenges young men and women face to become Naval aviators. The drive and determination I chronicled in the mid-1990s remains the same for these trainees today.

Douglas Waller
Raleigh, N.C.
2019

CONTENTS

PART IV FLIGHT OF THE HORNET

INTRODUCTION

DEAR GOD, please get me through this flight alive! I thought as I sucked down air as fast as I could from the oxygen mask pasted to my face. I was strapped into the back seat of Navy Lieutenant Kevin Nibblelink's cockpit during the dogfight, praying that I wouldn't throw up from the violent turns his T-45 jet trainer was taking or black out from the crushing Gs I was feeling. For twenty minutes we had been flying mock aerial combat in the bright sunny skies near the El Centro air station in Southern California. It seemed like twenty years. An instructor pilot, Nibblelink was battling one of his students, another Navy lieutenant named Bill Sigler, whose T-45 Goshawk raced just ahead of us.

So far, the dogfight for me had been a dizzying blur of earth, sky, and horizon jumbled up in no particular order. I was so confused I had no idea what had been happening during the engagement. Back on the ground, it would take me twelve hours of interviewing Sigler and Nibblelink to decipher the half hour dogfight. I bought a pad of large art paper to diagram each part of the battle. I ended up with thirty-seven pages of drawings.

Nibblelink flipped the radio switch down on the throttle that he gripped with his left hand so he could talk to me.

"Okay, you can go ahead and take the stick," he said over the radio. We were about to fly directly at Sigler, then bank hard to the left for the dogfight, and Nibblelink wanted me to take the stick that steered the aircraft. In other words, assume control of its flight. Be a fighter pilot for a few minutes, even though the only thing I had flown before this was a paper plane.

"You want me to take it?" I asked, my voice cracking, not believing what I'd just heard.

"Yeah, you take the stick," Nibblelink said nonchalantly as if he was giving me the wheel of a car for a Sunday drive. But this was more like the Daytona Speedway.

Terrified, I grabbed the stick between my legs as if it was the only thing holding me in the jet. To set up for the maneuver I had to bank the Goshawk hard to the right and aim it straight at Sigler, who was flying toward me in the opposite direction.

I moved the stick back slightly to push the nose up, then strained to pull it to the right. I felt like I was in a tug-of-war with the stick. The G force slammed at me as the jet tilted almost perpendicular to the ground during my right bank. It felt as if my children, when they were little, were hanging on my right arm as I struggled to keep a grip on the stick with the Gs dragging my limb down.

"Pull harder, pull harder, pull harder! Right at him! Right at him!" Nibblelink ordered as I arm-wrestled with the stick to keep the jet pointed up and toward Sigler.

In a flash, Sigler's jet whizzed past me on the left.

"Roll out here!" Nibblelink commanded in the next instant.

I shoved the stick back and to the left as hard as I could so the jet would reverse direction and break to the left in a tight circular turn.

"Roll against him!" Nibblelink shouted to get me to lift the Goshawk up. "Pull!"

I pulled, then pushed with all my might.

"Now, right there, feel the nibble of buffeting?" he asked.

Nibble? The damn plane seemed to me to be shuddering all over as it strained from the hard bank I was forcing it to fly.

"Now look back at him while you're turning," Nibblelink continued.

I darted my eyes quickly, afraid that I'd never get my head pointed back to the front if I turned it now. The jet was approaching five Gs.

"Do you see him over there, left eight o'clock, a little high?" Nibblelink asked quickly.

I didn't know where I was. I didn't know where eight o'clock was. And I didn't have the faintest idea where Sigler was.

"Very nice, you're fighting him," Nibblelink complimented.

No, I wasn't fighting. I was holding on for dear life.

And if I held these controls much longer, we'd crash and die. I was sure of it.

"You better take the stick!" I said to Nibblelink anxiously over our intercom radio.

Why's he giving up? Nibbs thought to himself. I had actually been flying a pretty decent turn up to that point and could have lined up to take a shot at Sigler. The jet was flying almost 20,000 feet above the ground, more than enough room for Nibblelink to grab the controls and stabilize it if I spun the Goshawk out of control.

"Okay, I got the stick," Nibblelink said reluctantly, wondering why I'd gotten cold feet.

I was just relieved that the aircraft was back in competent hands. Later on the ground, I would be embarrassed that I had wimped out so early. But sitting limp and exhausted in the cockpit seat, all I could think of was, How did I ever get myself into this project?

I have to admit that I started this book reluctantly. My research began several years after the Tailhook scandal, when hordes of drunken Naval aviators had assaulted women at a Las Vegas convention. I had set out to learn about the young men and, in a few cases, women who flew planes for the sea service. Who were these people who landed on and took off from aircraft carriers? Were they all Tom Cruises? Top Gun types? How were they made? How did they train? I didn't know and at first I questioned to myself whether I even cared to know. I had covered the Tailhook scandal as a magazine journalist and had come away not particularly impressed with Naval aviators. In September 1991, hundreds of them had gathered in Las Vegas, Nevada, for the annual Tailhook Association convention. Named for the hook that catches the arresting wire on the carrier flight deck and brings a landing plane to a stop, the Tailhook Association promoted Naval aviation and hosted an annual convention, which had become a sex and booze Mecca for aviators. That year, the young pilots had been in an ugly mood. The Air Force had run the Desert Storm air war over Iraq and along with the Army had hogged most of the glory. Because its admirals had scandalously mishandled aircraft procurement, the Navy's stealth attack fighter program had been canceled. Congress was prepared to allow women to serve in combat squadrons. Even gays were winning court approval to be recognized in uniform.

On the third floor of the Las Vegas Hilton, Navy squadrons each year set up hospitality suites that hosted stag films, strip shows, female leg shaving, and in one room a rhinoceros penis dispensing drinks. A "gauntlet" out in the hallway also became a tradition, where drunken aviators lined up on each side to grope and paw at women shoved down the hall. The 1991 gauntlet turned criminal. Over eighty women, some of them teenagers, were manhandled or sexually assaulted.

Tailhook seemed to me symptomatic of a larger ethical rot infecting the service. The Navy brass in the past had been fiscally irresponsible in chasing an unrealistic goal of a 600-ship fleet during the Cold War. An Aegis cruiser had shot down an Iranian airline jet killing all the passengers aboard. The U.S. Naval Academy was riven with cheating scandals, drug abuse, sexual assaults, even a car theft ring. Naval spies like the Walker brothers and Jonathan Jay Pollard had stolen a treasure trove of secrets. An old-boy fraternity of admirals protected senior officers from being disciplined for ethical and criminal misconduct. Then the chief of Naval operations, Admiral Mike Boorda, committed suicide rather than face public humiliation for wearing a combat decoration he had not earned during the Vietnam War.

Military reformers and even some admirals question whether giant aircraft carriers, which cost over $5 billion to build, have outlived their usefulness. The Air Force, long the rival of Naval aviation, claimed its long-range bombers can strike any target a carrier plane can, and do it cheaper. The argument sends Navy pilots up the wall. Whenever there's an international crisis, they retort, the first question out of the President's mouth is always: "Where's the nearest carrier?"

The Navy's aviators, nevertheless, seemed at first glance to be one-dimensional characters, flyboys in love with planes and themselves. Immature frat brothers with their Bermuda shorts, polo shirts, and aviator sunglasses, who hadn't yet grown up. In

the 1980s, the movie *An Officer and a Gentleman*, starring Richard Gere, portrayed green Naval flight students made men by stern drill instructors and swooned over by groupies trying to snare pilots as husbands. Tom Cruise glamorized the macho, hard-drinking, and partying world of F-14 pilots in the movie *Top Gun.**
Landing on carriers was exciting, to be sure, but Air Force officers had assured me it was little more than what their pilots did on a longer runway.

Another problem I saw in launching my project: To chronicle the training of Navy pilots I would have to fly with them. It would be the only way to appreciate fully what they went through and to convey that to the reader. But I have none of the love of flying that these aviators have. In fact, I really don't enjoy it. I am not one of these terrified air travelers. I'm comfortable in big, stable jumbo jets. But I dread small commuter prop planes. My heart is in my throat when they bump about in air pockets. I don't like heights. I despise roller coasters. And it doesn't take much to make me air-sick.

For three months, I met with Naval aviators in the Pentagon and at Naval air stations all over the country to discuss the book idea. I knew I would have to have the service's cooperation. A journalist couldn't just walk onto an air base on his own and hop into the back seat of a high-performance military jet. I had to have dozens of clearances from different commands and aviation units in the Navy. The service was still licking its wounds from Tailhook and at first wary of having an author poking around its squadrons for the two years it would take me to complete the book.

Before the Navy would give its final approval, I also had to pass a flight physical, then undergo a week of rigorous training on how to deal with cockpit emergencies as a passenger. I would be doing a lot of flying the next two years and though the service had a good air safety record the flights still could be dangerous. The Navy may have been inept in handling the Tailhook scandal, but it was smart

---

* "Top Gun" is the nickname for the Navy Fighter Weapons School (as it was then called) at the service's Fallon, Nevada, Naval Air Station. The Tom Cruise film depicted Top Guns as swaggering air aces. Actually, the school trains trainers. Each F/A-18 and F-14 squadron sent a pilot every two years to the Fighter Weapons School to spend almost ten weeks learning the latest advanced dogfighting and bombing tactics. Top Guns then return to their squadrons to instruct fellow pilots in the tactics.

enough to realize that a journalist dying in a jet would not be good PR. As they did for every other pilot, the flight surgeons would have to pronounce me fit to fly in the back seat of military aircraft and each year I would have to renew that certification with an annual flight physical. Being in my mid-forties, I was hardly a top specimen, but I passed the physical with a few waivers for the extra pounds I carried and the medicine I took for high blood pressure.

I took my week of passenger training at the Pensacola, Florida, Naval Air Station, where students began their pilot training. It wasn't fun. I had to swim seventy-five yards, then float in the water for two minutes wearing a helmet, flight boots, and about twenty pounds of flight gear. I learned how to parachute into the ocean, then quickly disengage from the chute before it filled with water and dragged me under. I was strapped to an ejection seat and shot up rails to simulate being rocketed out of a jet cockpit. It left my back sore for several days. I was placed inside a hyperbaric chamber and deprived of oxygen to simulate what I might experience at high altitudes if the air pressure regulators in the cockpit malfunctioned. It made me giddy and my ears popped for twenty-four hours afterward. I was placed in a metal cylinder and spun around to simulate spatial disorientation I might experience in flight. I almost threw up.

They saved the scariest part of the training for the end of the week. Naval aircraft operated mostly over the ocean, so pilots had to know how to escape if they crashed with them into water. Many aircraft would flip over after they entered the ocean, so the aviator would be upside down as the cockpit filled with water and sank to the bottom. For a day, I was buckled into steel cages that simulated cockpits or cabins and sent plunging into pool water, where submerged and upside down in the cages I had to unbuckle from seat harnesses and swim to the surface. For some of the dunkings, I was blindfolded to simulate escaping a submerged aircraft in black waters at night. Being strapped to a seat submerged and upside down with blackened goggles and water rushing through my nostrils was terrifying. Pilots and passengers had to go through this every four years as refresher training. I vowed to have the book research completed before that so I would never have to suffer the dunkers again.

Finally qualified to ride in back seats, I began my research. Of course, my first impressions of Naval aviators proved to be wrong. Stereotypes usually are. A 1970 Navy study claimed that the ideal carrier pilot was a firstborn male from a two-parent family with a strong-willed father. That kind of family, as well as the movie version of the Navy pilot, was becoming outdated, I discovered. The new ensigns entering flight school were far more willing to accept women in cockpits than their superiors. Navy psychological studies today found that only about twenty percent of the service's male pilots had the Tom Cruise–Top Gun personalities. (The male aviators called them "light-your-hair-on-fire guys"; the female aviators called them "Neanderthals.") More than fifty percent of the male pilots had what were termed "corporate management personalities"— aggressive but far less compulsive than the Top Gun types. The rest were as bland as accountants. The few female pilots who were breaking into the ranks often had the same personalities as the men. Some male pilots maliciously spread rumors that most were lesbians. Very few are—though they are less concerned than civilian women about their femininity. In the squadrons they worried little about makeup or their hair.

Among the more than 200 aviators I ended up interviewing, I found a far more textured and diverse and interesting lot than I had first envisioned. They all loved to talk about flying and themselves. Aviators, I soon discovered, were a very open group. What you see is what you get. They were not afraid to speak their minds, even to journalists, which probably landed a lot of them in hot water during the Tailhook scandal. During the many hours I spent interviewing them in cockpits and squadron ready rooms or sharing beers with them in officers' clubs, I rarely sensed that they were holding back from me.

Though their personalities varied, the aviators I interviewed— particularly those who fly on and off aircraft carriers—still shared a cockiness nurtured by the danger and excitement of their profession. Air Force pilots might be as skilled as dogfighters, but they don't land on a 500-foot rocking airstrip when they return home. Navy pilots barely hide their contempt for those in their own sea service, the "surface warfare" branch condemned to sailing in slow-moving ships. "Your worst day on land is as good as your best day at sea," Navy pilots like to say. Another saying among Naval aviators: "Pilots talk about flying all the time when they are on the ground because they are so passionately interested in the subject; officers who sail the seas talk about ships because they have nothing else to talk about." Still another: "No one has ever flunked out of surface warfare school and been sent back to aviation."

My goal was to fly as a passenger for every flight I described in this book. For the most part I did. The only flight I was barred from taking was catapulting off and landing on a carrier at night. These flights were too dangerous. Even seasoned pilots dreaded them. I pleaded, but the Navy firmly refused. I had to describe these flights by flying in the simulator, which I was told was surprisingly realistic, and by interviewing pilots after they landed.

But I was allowed to fly in aerial dogfights that spun me in every direction, in dive-bombing runs that left me swallowing my stomach, in formation flights where the planes seemed so close together I was sure they would bump into one another, in high-speed tactical maneuvers so low to the ground I would push up on my cockpit seat afraid that my bottom would scrape the desert floor, and in daytime carrier catapults and landings that took my breath away. I kept a brave face in front of the aviators. I tried to feign the nonchalance they seemed to have with all these "hops." But much of the time I was terrified. I felt claustrophobic strapped into the cramped cockpits with my head covered in a helmet and an oxygen mask on my face. I constantly worried about passing out when I would pull as much as six Gs during loops and turns. (One G equals the force of gravity.) I could not sleep the nights before my first dive-bombing run and my first carrier landings and catapults. The hundreds of possible accidents I had been warned about in my back seat classes kept churning in my mind. Would I remember the correct emergency levers to pull, what buckles and hoses to disconnect if something went wrong during the flight?

One problem dogged me throughout the research. Airsickness. From the beginning, I either threw up or felt miserably nauseous on every flight. The pilots always had a good laugh when I "yacked," as they called it. I tried all the concoctions they swore would settle my stomach. Pancakes and orange juice before you go up. Stuff yourself with bread or crackers with peanut butter. Try eating apples.

Apples worked the best only because they tasted about the same coming back up as they did going down. But none of the remedies solved the problem. I finally consulted the flight surgeons. I couldn't bear the thought of being bent over with my head in a bag during much of the research. The Navy doctors prescribed a combination of pills they gave to pilots with severe cases of airsickness. Before each flight I popped twenty-five milligrams of promethazine, a powerful antinausea tablet that would make me drowsy. So I wouldn't fall asleep in the cockpit, I had to take another pill with five milligrams of Dexedrine, an amphetamine. Drugged with a downer and upper, I managed to get through most of the remaining flights without yacking.

I also eventually overcame my fears and began to enjoy many of the flights. Flying upside down, soaring straight up into the sky, snapping to the left and right in hard banks, tumbling and looping and spinning gradually became exhilarating. I would find myself gazing up and to the right and left out the Plexiglas bubble that encased me in the cockpit, marveling at the spectacular view of the world that I had in a combat jet. The pilots occasionally would let me hold the stick in the back seat and fly the plane. Of course, their hands were always close to the stick to grab it in case I made a mistake. It was fun piloting the aircraft and I discovered, as the aviators had assured me, that my airsickness would disappear if I held the stick for a while and controlled the plane's movements.

Gradually, I began to feel the thrill and love of flying, which the pilots told me stirred in their souls. Returning to the ready room after a flight, I found myself talking as excitedly as they always did about what had happened during the hop. As I catapulted off and landed on the aircraft carrier in an F/A-18 Hornet, Commander Bill Gortney, who headed the squadron I was observing, radioed to me from the ship's bridge: "Welcome to the fraternity." When I climbed out of the jet after the cat shots, I strutted about with my chest puffed out. Of course, I wasn't really a part of this band of brothers and sisters. For all my flights, I was no more useful to the plane's operation than a bag of potatoes strapped in the back seat. But I could understand the pride these aviators felt in accomplishing what few had, the special bond they shared in their squadrons.

This is the inside story of how these air warriors are made. The men, and few women, who enter the training begin in Pensacola, Florida, the "Cradle of Naval Aviation," where students are first indoctrinated in the unfamiliar mores of Navy pilots and the unique hazards of flying planes over the sea. The story ends more than two years and two million dollars later—the time and cost to train each aviator—in the computerized cockpit of a sophisticated combat jet.

This book could not have been written without the help, the advice, and the patience of a number of people. My agent, Kristine Dahl, was the first to encourage me to begin this project. Dominick Anfuso, my editor at Simon & Schuster, could not have been more supportive. I am deeply indebted to my editors and colleagues at *Time* magazine who suffered my occasional absences to conduct the research for this book and who aided with advice, encouragement, and help in editing the manuscript. My thanks go to Mark Thompson and Nancy Gibbs. I am indebted as well to the pilots who read portions of the manuscript for technical accuracy and corrected many errors. They included Lieutenant Rich Jackson, Captain Dutch Rauch (Ret.), and Commander Bob Yakeley (Ret.). Of course, the errors remaining as well as the opinions expressed in this book are all mine.

I could not have begun the book without Rear Admiral Kendell Pease, the Navy's chief of information, whose early support for the project opened many doors for me among the training and aviation commands. A number of Navy public affairs officers were particularly helpful in collecting research material and arranging my visits to squadrons. They included: Commander Joe March, Master Chief Mark Malinowski, Lieutenant Mike Coleman, Lieutenant Commander Bob Ross, Lieutenant Gerald Parsons, Lieutenant Chris Sims, Lieutenant Bill Fenick, and Bert Byers.

I owe a special thanks to the squadron commanders who allowed me to roam freely in their units interviewing whomever I pleased. They were Commander David Jenkins of VT-2, Commander Bill Shewchuck of VT-21, and Commander Bill Gortney of VFA-106. In each of the squadrons I was assigned pilots to watch over me during my research. Their job was not to monitor my interviews. Rather, they helped me decipher training schedules and manuals, arranged for my flights in the aircraft, cut through bureaucratic red tape, and spent many hours with me offering advice and insights on squadron operations. They were Lieutenant Tony Chatham,

Lieutenant Felton Elders, Lieutenant Greg Schuster, Lieutenant Michael Carr, Lieutenant Joe Evans, and Lieutenant Alex Howell.

A number of senior officers also helped me at key points along the way. They included Captain Steven Counts, Captain Charles Nesby, Commander George Dom, Captain Robert E. Hain, and Commander Carroll White. I also appreciate the support provided me by the skippers of the three aircraft carriers I boarded to conduct research: Captain Mark Gemmill of the USS *Dwight D. Eisenhower*, Captain Robert C. Klosterman of the USS *John C. Stennis*, and Captain Edward J. Fahy of the USS *John F. Kennedy*. Others also helped with the project. My thanks to Tom and Ann Perkins, Lynn and Rette Ledbetter, and Matt and Susan Waller, who provided welcome refuge during my long research trips. Kelly Marra also came to the rescue with a critical piece of research.

This book could not have been completed without the forbearance and encouragement of my family. My children, Drew, Colby and David, found it amusing when Dad came home from research trips with war stories about flying. My wife, Judy, not only kept up my spirits during the long project, but, as with my other books, served as my best editor for the manuscript. Finally, this book is dedicated to the finest officer to ever serve in the United States Navy and his wife. They were my parents, Captain Thomas C. Waller (Ret.) and Barbara A. Waller.

PROLOGUE

March 22, 1995, The Adriatic Sea

TUBA wrestled into the dry suits pilots must wear when the ocean's water temperature falls below 60 degrees Fahrenheit. It would extend his survival time in the chilly Adriatic Sea by about five hours if he had to eject from his F-14 Tomcat jet fighter. The pilots called it the "poopie suit." They felt as if they were wrapped in a giant rubber band. Tuba stuck his chin under the thick collar that reached up to his neck and scrunched down to force out air bubbles that bulged inside the suit. Finished, he looked like a prune with the wrinkly suit clinging to his body.

Other F-14 Tomcat crews that had finished night flights and stripped their sweaty dry suits in the changing room began to needle Tuba. "There's a full moon out tonight," one said with a rough laugh. "Only the moon is over Norfolk, Virginia." Over the Adriatic Sea, the skies were pitch dark—"black as a bag of assholes," the pilots liked to say.

Tuba laughed too, zipping up his G suit and survival vest, then dabbing a wipette over his oxygen mask to clean it. But it was more of a nervous laugh. No matter how many times he did it, Tuba never got over the queasiness of being hurled off the aircraft carrier strapped in a thirty-ton jet into what looked like a black hole. Neither had the other pilots. Night flights were scary. The USS

23

Dwight D. Eisenhower aircraft carrier was in the last two weeks of its deployment in the Adriatic Sea, where its jets took off daily to patrol the skies over Bosnia-Herzegovina. The *Ike* was being replaced in the Adriatic by the USS *Theodore Roosevelt*, whose pilots would finally conduct massive air strikes over Bosnia five months later as part of the NATO campaign to lift the siege of Sarajevo and force the Serbs to the negotiating table. But for now, the missions over Bosnia had been boring for the *Ike*'s pilots. The Bosnian Serbs and Muslims were slaughtering one another on the ground. The only things the F-14s flying above were allowed to do was take pictures or dodge surface-to-air missile batteries the Serbs kept silent anyway. So far, the *Eisenhower* had suffered no airplane mishaps during the six-month cruise, but the air crews were now on edge that bad luck would catch up with them in its final days.

All Navy pilots were tagged with nicknames, in some cases ones they did not like. Outsiders found the practice somewhat juvenile and some of the names that once had too heavy a sexual connotation had been toned down.* But the nicknames served one purpose: the pilots found them useful when they radioed one another in combat and did not have time to use full names.

There was a practical reason John Gadzinski had been nicknamed "Tuba." That was what he played. He kept his tuba out of its case in the crowded stateroom he shared with another aviator. Gadzinski once even gave a concert for the carrier's air wing. The jet jocks were stunned that *Ain't Misbehavin'* could sound so good on the sonorous instrument.

Gadzinski was probably the only Tomcat pilot in the U.S. Navy who was a classical tubaist and a high school dropout. By age seventeen, he had already studied with the Boston Symphony and once played under Leonard Bernstein. When Boston University offered him early admission before he had graduated from high school plus a scholarship as a music major, he jumped at the chance and enrolled six months before he was to receive his high school diploma.

To this day, his high school refuses to grant him a diploma. But Tuba never regretted dropping out. Boston was exciting. He stud-

---

* A few pilots secretly kept two call signs: the politically correct one for when women were present, the incorrect one when the audience was all male.

ied chamber music with the Empire Brass Quartet. Sam Pilafian, one of the most successful tubaists in the world, taught him more about performing under pressure than he would ever learn in the Navy. He played at Tanglewood and with the Opera Company at Boston. For three years he moonlighted with the Bach Brass Quintet.

But another love began competing with music. He took flying lessons while in college and became captivated with soaring through the sky. By the time he graduated, the bookshelves in his dorm room were crammed with twice as many books on flying as on music. He finally visited an Air Force recruiting station. They were snooty, not about to put a music major in one of their cockpits. The Navy wasn't so picky.

"Do you really need to know a lot of math and science to be a Navy pilot?" Gadzinski pleaded with the Navy recruiter.

The recruiter, who also happened to be an F-14 pilot on temporary duty in the station, leaned over his desk and asked: "What's the first thing you look for when you're going to take a crap?"

"Toilet paper," Gadzinski answered.

"That's all you need to know to be a Navy pilot."

Six months later, Tuba began his pilot training at Pensacola, Florida.

He quickly discovered that music training was perfect for learning to be a Navy pilot. Music was one part talent, nine parts a head game, Tuba thought—about the same proportion for the skills needed to take off from and land on an aircraft carrier. Naval aviation also was about as nerve-racking as auditioning for a symphony or playing on cue for a radio show recording. In both flying and playing music, Gadzinski found, there came a point when you didn't think about it, you just did it. That was how he got through flying during these dark nights.

Music and aviation shared an obsession with detail as well. Gadzinski found himself using musical terms when piloting the Tomcat. Landing a sophisticated jet on an aircraft carrier involved staccato precision. The slightest misstep, positioning off by no more than a few feet, and what already amounted to a controlled crash landing onto the flight deck could result in disaster killing the aviator.

Gadzinski also found that flying jet fighters made him a better musician. Competition could be fierce for students in the music world. But Gadzinski's teachers constantly stressed that to be a good musician you not only had to play the notes well. You must make the music sing a song. For six years, Gadzinski did not pick up the tuba as he struggled through flight school, then bounced from sea duty to shore commands. But as a flight instructor in Meridian, Mississippi, in 1991, he had time on his hands. He picked up his tuba to play. He had already survived hundreds of catapults off the carrier as well as hair-raising landings. No concert performance would ever be as stressful as that.

Now he played for pure enjoyment. He found he performed better. No longer afraid to make mistakes, he concentrated only on the music and ignored the audience. With only a week to prepare, he auditioned for the Mississippi Symphony in Jackson and won a chair. Now he played music for music's sake.

Racks of flight helmets filled one bulkhead of the squadron changing room. Along the others hung survival vests, dry suits, and oxygen masks. Rosie had one corner of the changing room to herself. From the ceiling hung a flimsy sheet, behind which she could climb into the dry suit and gear, with the other, male aviators in the crowded room catching only glimpses of skin. The truth was, at the end of the flights she saw more than they. So keyed-up and exhausted were all the pilots that the males quickly stripped to their underwear as they wiggled out of dry suits, forgetting she was not behind the drape.

With her dry suit, G suit, and survival vest on, Rosie bent over as if in a formal bow to cinch up her parachute harness. The males did so as well, and for a more important reason. If the harness was loose in the crotch, male pilots could find their testicles crushed after they ejected and the billowing parachute yanked them up.

The crewmen who tended to the changing room liked to test how much kidding Rosie would tolerate. She could handle a lot. The first female aviators to break through the all-male ranks were subjected to humiliating hazing—suffering through countless lewd jokes, sexist comments, men exposing their private parts to test if the women would blanch. Most of the raunchy behavior was now outlawed, but today's female pilots still couldn't be wallflowers.

Kristin Dryfuse earned her nickname "Rosie" during a raucous port call in Israel. Her male squadron mates discovered she took ribbing well, snapping out comebacks that would make Roseanne Barr blush.

Even if she had been a male, Rosie would have had to have a thick skin to survive. Rosie was a "nugget," the pilot's term for an aviator on his or her first carrier cruise. Jokes and teasing relieved the boredom of carrier life. All pilots had to have thick skins. Nuggets had to endure pranks up to the final months of their first deployment. One junior lieutenant in the squadron had marched into sick bay because the veterans had convinced him he needed a shot for "channel fever." Rosie was the first female in VF-32, the designation for the F-14 fighter squadron. A militant feminist would not survive the rough-and-tumble of this male-dominated environment. The women who endured didn't go looking for trouble.

The *Eisenhower* was on a historic cruise. Female Naval aviators had been allowed to train in the Navy's warplanes but were barred from flying in combat with carrier squadrons. They had fiercely lobbied a receptive Congress to lift the combat exclusion rule for women. The Navy's male pilots wanted no part of mixed-gender units. Women wouldn't perform as well as men aboard carriers, they argued. Mixing the sexes for so long in what were floating ant farms would turn aircraft carriers into Love Boats. The squadron camaraderie of male jet jockeys, so critical to combat effectiveness, would be shattered with females in the cockpit. But Congress lifted the ban, and the Pentagon in 1993 ordered women aboard the Navy's combat ships. Women boarded the *Ike* in April 1994.

Though they didn't know it at the time, Tuba and Rosie were pioneers in one of the most volcanic social revolutions the U.S. Navy had ever undergone. The *Eisenhower* became the Navy's first showpiece to prove that a warship could operate as well with women aboard as with an all-male crew. Modern aircraft carriers had already become corporate war machines. Officers sprint up and down ladders with Radio Shack walkie-talkies. Cappuccino is served in the wardroom. Ship's offices are equipped with personal computers and television sets that play videos of flight deck operations or messages from the captain. The *Eisenhower*'s corporate managers spent a year preparing for women. Sanitary napkin dispensers were installed in bathrooms, doctors trained in gynecology were brought on board, the ship's barbers received classes on cutting women's hair, and mixed in with the Arnold Schwarzenegger movies were *The Piano* and *Tootsie*.

The 400 women who boarded the carrier felt as if they had stepped into a bubble with the ship's 4,600 men, as well as the media, watching them from the outside.

But the grind of sea duty finally mellowed the *Ike*. Sailors never have a day off when a carrier is deployed; it would only give them more time to grumble about how crummy life could be in the middle of the ocean. Biological clocks soon became reset. Civilians think in terms of days of the week. On an aircraft carrier, sailors live by dates of the month and are hard-pressed to remember the day of the week since every one seems like a Monday. The only thing that made Sunday different from the other days was that the mess halls served quiche at lunch. Thirty-one years old and married with two children, Tuba was feeling miserable away from his family in this floating prison. One reason he and the other pilots loved flying: it meant escape from "the boat."*

After three months of nonstop operations, the *Ike*'s crew was too worn out to be bothered by gender differences. Women, who first boarded the ship looking like scrubbed nuns, began wearing makeup again and mingling more with male shipmates. Desperate for a clip, they finally began visiting the ship's barbers. Male pilots who worried that women would be forced into cockpits before they were ready had those fears assuaged when a female EA-6B Prowler pilot was sent home because she was having trouble landing her radar-jamming jet on the carrier. The *Ike*'s male sailors loosened up and began swearing more. Even the women became earthy. Female officers called the staterooms that a half dozen ensigns would have to share "six-chick" or "six-dick" rooms. But "fuckin'" no longer became the preferred adjective for every noun. After 220 years at sea, the U.S. Navy was becoming civilized.

Tuba wasn't so sure all this civilizing was a good idea. An air warrior in the Navy must make a deeply personal commitment—more

---

* A seagoing officer or sailor wouldn't be caught dead calling a large Naval vessel like a carrier anything but a ship. Only submarines or patrol craft are called boats. But Navy pilots, to prove their irreverence, will always refer to a carrier as "the boat."

than to patriotism, which jet pilots never discussed openly, or to risking life and limb every time their planes were launched from a carrier. Drill instructors would put it bluntly for aviation recruits: "Your job is to kill people and destroy things in the name of the United States. If you don't want to do that, you don't belong here." Women, Tuba worried, were taking the *Ike*'s focus away from being a potent killing machine.

The problem wasn't the women. Tuba knew they could fly the jets and fight as well as men. Mean-spirited aviators not assigned to the ship had nicknamed the *Ike* the "Dyke," but even the chauvinists on board admitted privately that the females were competent pilots. That the *Ike*'s women had performed well shouldn't have come as a surprise. The Navy had made sure that the first female pilots aboard the ship were the cream of the crop, desperate to prove themselves and fit in on the males' terms.

No, something else bothered him. The Navy had made the carrier a debutante for its coming-out party for the twenty-first century. The ship's senior officers were under incredible pressure from the Pentagon to make the voyage a success. And in their minds, the *Ike* had succeeded. But there were deep misgivings among the junior pilots. The cruise had hardly been flawless. Fifteen women had to leave the cruise early because they were pregnant. Officers treated female subordinates with kid gloves for fear of being slapped with sexual harassment charges. The brass strongly discouraged dating during the voyage, but Tuba and the other officers knew that many crew members secretly fraternized. A half dozen couples had gone to the captain and announced that they had fallen in love. A male and female crew member had even videotaped themselves having sex in a ship's compartment.

The *Ike*'s voyage may have made its corporate managers happy. But even many male pilots, enlightened enough to realize that women performed as well in the cockpit, still did not want them dying in war. In battle, a carrier had no time for personnel problems. For pilots, combat became all-consuming. If they were not flying, they were eating, sleeping, or planning for their next flight. Could a ship adjusting to females and all these new workplace rules still win a war, Tuba wondered?

It likely could. Tuba and the other men were overreacting. The problems the *Ike* experienced were relatively minor for a crew of

5,000. The total number of disciplinary cases was less for the coed cruise than previous ones that had been all-male. The pregnancy rate aboard the carrier was far lower than the overall pregnancy rate for Navy women serving on land. The sex video stars were kicked out of the Navy. And the six pairs of lovebirds were split up—half transferred to shore jobs of their choice, half remaining on board.

But in the insular world of Navy fighter pilots, the social changes were happening too fast.

Tuba and Rosie marched up to the top carrier deck to check out their F-14 for the night flight. Tuba would pilot the jet. Rosie would sit in the back seat of the Tomcat's two-seat cockpit. Dryfuse was a Naval flight officer. Ensigns had to have 20-20 eyesight to be pilots. She did not. But aviators with less than perfect eyes could become Naval flight officers (NFOs) who sat in the back seats of planes as navigators or, in Rosie's case, as radar intercept officers who not only helped guide the Tomcat but also directed the fire of its sophisticated missile systems during dogfights. Naval flight officers had long had to fight the image of being second-class citizens to the aviators who actually piloted the planes. In F-14 squadrons, however, they ended up being co-equals, leading units or combat operations as often as pilots did. The F-14 took two people to fly; the officer in the back manipulated the radars to find targets far away. The pilot stayed busy flying the jet. When the dogfights involved four or more planes, the radar intercept officer in the back seat was key in the battle because he warned the pilot of "bogeys" coming from all directions. (Bogey is aviator slang for an enemy plane.)

All day, teams of the squadron's aviators had been flying either combat air patrols over Bosnia or training exercises over the Adriatic. VF-32's ready room just below the flight deck was never empty or quiet. Pilots would troop in to hold briefings on upcoming flights or to critique flights that had just landed. Between briefings and critiques, a compact disc player blared out rock music.

Each squadron aboard the carrier had its own ready room filled with thick padded chairs and white bulletin boards for posting daily flights. Since they were going to be cooped up in it for six months, VF-32's pilots decided to make their ready room more livable. Light blue-gray fascia walls, which looked like they belonged in the waiting room of a doctor's office, were erected to cover the dingy gray bulkheads.

On one wall was displayed two crossed silver swords. The squadron was called the "Swordsmen." Each squadron had a nickname. With the advent of women, the sexist names like the "Playboys" had been changed. Next to the swords hung framed documents commemorating Ensign Jesse Brown, the Navy's first black aviator and a member of VF-32, who died after his plane was shot down during the Korean War. (It was an ironic tribute considering the squadron now had no black fliers and in all of Naval aviation only two percent of the pilots were African-American.) In another corner hung a photo of Clint Eastwood.

Clouds had covered most of Bosnia-Herzegovina during the day so aerial reconnaissance missions had produced few pictures of Serbian armored units laying siege to the Bosnian stronghold at Tuzla. The F-14s could have flown under the clouds, but the pictures weren't worth getting shot at by trigger-happy Serbs. Coalition peacekeeping could be frustrating. Rules of engagement were so strict, fighter pilots had to be briefed by military lawyers before beginning operations over Bosnia. Tuba's air patrol had had to refuel from a Spanish tanker—a perilously close exercise because the tanker had two fuel pods strung from the tips of each wing, which two F-14 jets plugged into simultaneously and just fifteen feet apart.

The most interesting training exercise for some of the Swordsmen's pilots was supposed to have been the air combat maneuver training over the Adriatic later that day. Four F-14s squared off against four F/A-18s from another of the carrier's squadrons. Four-versus-four dogfights were nightmarishly complex for each side to control. With jets converging at 500 miles per hour from all directions, the skies suddenly got crowded, reaction times had to be compressed to seconds, and "situational awareness"—the pilot's ability to know where his plane and the others were—easily became confused. No one had left the afternoon competition happy. In the first round, the F-14s, playing the aggressors, were able to target all the F/A-18s for simulated kills, but the F/A-18s were able to bag one Tomcat in the exercise. In the second round, with the F/A-18s playing the aggressors, the Hornets suffered no simulated casualties. But they were not able to target all the Tomcats for simulated kills. Back in the ready room, the F-14 pilots argued for a half hour over who screwed up. Heated critiques, where pilots openly admitted their errors, were common when exercises went badly. The real operations were too dangerous to put up with polite talk.

It was shortly before 8:00 P.M. when Tuba and Rosie began their walk-around inspection of the F-14 on the "roof," pilot slang for the flight deck, which at this point was pitch dark except for a few dots of light from lamps and flashlights. A junior enlisted man who watched over the final preparations of the Tomcat for launch greeted them with a salute. He wore a helmet with Mickey Mouse ear cups and a brown turtleneck jersey under his life vest.

Because so many huge planes and loads of ammunition moved about quickly in cramped quarters, an aircraft carrier flight deck had to tightly control the chaos. So to make it clear what each person did amid all the noise, steam, and confusion, crewmen wore different colored jerseys to designate their jobs on the flight deck. The brown shirts were plane captains. Yellow shirts directed the movement of aircraft. Purple shirts fueled the planes. Red shirts handled weapons and ordnance. White shirts were assigned safety jobs. Blue shirts chained and chocked planes to the deck or moved them about in tractors. Green shirts hooked the aircraft to the catapults that launched them from the deck.

Tuba had already read through the Aircraft Discrepancy Book on plane number 211, looking for trouble other pilots had noted when they flew it. Each F-14 had its own personality. Pilots called cantankerous jets "Christines" after a possessed car in a Stephen King horror novel. Aircraft number 211 had had problems with its autopilot system.

Wandering around the aircraft, climbing on top of its fuselage, Tuba and Rosie now looked for dumb mistakes that had caused accidents in the past. They scanned the overall structural integrity of the plane, its outside hydraulic gauges, brake pressure, the tires. Fasteners to outside panels were inspected. Tuba pushed on the nose of the Tomcat to make sure the cone didn't flip up (it had happened before as a plane was catapulted off the deck, shattering the cockpit glass). Rosie walked to the back to check that the tailhook's hook actually pointed down. Crews had slipped up in the past and installed it with the hook up, which meant the plane could not use the arresting cables on the carrier. If the pilot could not divert to a land base he had to fly his jet into a steel and canvas barricade net that could be raised on the deck to stop the landing aircraft—a harrowing experience pilots would do anything to avoid.

Satisfied with the walk-around, Tuba and Rosie climbed into the cockpit. The pace now picked up. Rosie pulled out the nine safety pins that kept the ejection seat rockets from firing. She buckled the four fittings on her harness to the ejection seat straps, connected the oxygen tube from her face mask to the cockpit oxygen system along with the hose that would pressurize her G suit in flight. Then she reached down and hooked the leg restraints that would yank in her legs if she had to eject from the cockpit. The sudden extreme force of the ejection could cause a pilot's legs to flail out and then be fractured by the cockpit rim as he blasted out.

Rosie was lucky her legs even reached the hooks. The twenty-four-year-old radar intercept officer was only five foot four and one half inches tall. Navy jet cockpits had been built in the 1960s and 1970s for young males five foot six to six foot three inches tall. The Air Force and the Navy had been debating how to redesign the cockpits of newer planes to accommodate smaller-sized females who now could fly them in combat. But for now, many qualified females couldn't fly the jets simply because they weren't the right size. Rosie made it by a hair's width. She had pleaded with Navy detailers (personnel bureaucrats in Washington who assigned officers to jobs in the service) to let her fly in jets. But worried that she might be too short, instructors had her first crawl into an F-14 cockpit to see if she could reach one of the most critical controls furthest away from a Tomcat aviator—the handle a pilot must yank to jettison the canopy during ejection. She could. Kristin "Rosie" Dryfuse, 135 pounds, with her bob haircut and bright blue-gray eyes, became one of the shortest aviators in the Navy.

Kristin's father, a high school math teacher in the tiny town of Vermilion, Ohio, did not have the money for college. She had read James Webb's novel *Sense of Honor,* and had become intrigued by the former Navy secretary's realistically harsh depiction of life at the Naval Academy. The tuition was free, so she enrolled, majoring in chemistry.

Rosie found academy life to be everything Webb, himself a 1968 graduate, had portrayed. The sexist hazing of women had tapered off by the time she arrived, but the institution was dreary nonetheless. She could take one day of waking up at 6:00 A.M., having every minute of every hour scheduled—morning classes and drills, afternoon sports, evening study periods. But enduring the same regimen practically every day for four years became a drudge. Even when she was on liberty, she never felt completely relaxed or freed of the oppressive routine. The happiest day of her life was graduation and receiving her commission, then packing off to flight school at Pensacola.

There she met Chad Jungbluth, another student in her flight indoctrination class who would eventually end up as a helicopter pilot. Chad had been in the Naval Academy class ahead of her. He had grown up on a dairy farm in Wisconsin. They had a lot in common, both being academy grads, both dead serious about the careers they were beginning. Chad wasn't like other student pilots who chased the bimbos in Pensacola, feeding them lines to get them in the sack. The two aviators fell in love.

The Navy was still adjusting to the modern workplace where officers' wives no longer sacrificed professional careers to serve tea at ladies auxiliaries. Wives either had civilian careers of their own or, like Dryfuse, were officers themselves. The *Eisenhower* even had female pilots who had left husbands home with the kids.

Rosie had joined the *Eisenhower* in January, midway through its coed cruise. The first thing that went through her mind when she stepped aboard the *Eisenhower:* "I'm not ready for this." It seemed to Rosie that aviation training had provided only the bare minimum that she needed to know in order to operate the F-14's complex radars and navigation systems. Now she had to learn how her newly acquired fighting skills fit not only into a combat squadron but also with the operations of a mammoth aircraft carrier. Rosie also was the first woman to join VF-32.

It was a relief for Dryfuse that the first person to greet her with a warm smile at the carrier's in-processing room was the F-14 pilot she would be paired with for most of her flights, Lieutenant Jim Sullivan. (She had been paired with Tuba just for the one flight tonight.) Sully was a laid-back farm boy from Illinois. At twenty-eight he already had graying hair. Rosie would find out later that he was a stickler for detail in the cockpit.

Sully decided to take her first to the changing room where Rosie could dump off the forty pounds of flight gear she had crammed in a duffel bag. It was there that she first met half the men in her new squadron. They were stripped to their underwear, ready to begin suiting up for a flight. Stunned, the men stared at Rosie for a half second, then burst out laughing. She had now seen almost as much of them as their wives.

Rosie had gotten mixed advice from other female officers who had already spent months aboard the ship. Watch your back, warned one. The men could be mean and undercut you. Enjoy yourself, said another. They're not ogres.

Rosie decided to trust her own instincts. The men at first treated her as if she were a member of the Feminist Jihad. They were terrified of her. But without really thinking about it, Rosie worked hard to blend in. She was respectable at acey-deucy, the favorite game in the ready room. The squadron's aviators ate together aboard ship and socialized together during port calls. Pilots who didn't could be ostracized. A week after she joined the *Ike*, the ship docked in Trieste. Rosie avoided the wardroom clusters of female officers and hung out with the guys during liberty.

The men finally relaxed after a month. Rosie became as close to being one of the boys as a girl could get. Each night, after spending twelve hours on the job, she buried her nose in flight manuals for another hour in her stateroom trying to absorb the fire hose of technical information a combat aviator must know. The men found that she was a quick study, even—to their surprise—a cut above the usual male nugget.

Rosie reached up and closed the cockpit canopy. Tuba started the F-14's two turbofan engines, then began tests to make sure the instruments and cockpit lights worked. Scores of electronic boxes in the front had to be checked. Next he manipulated the stick between his legs and the rudder pedals to make sure all the trim motors worked.

With the engines powered up, Tuba began running a stray voltage check to make sure an electrical spike wouldn't fry the sensitive radios and navigation gear when they were turned on. No spikes, so he flipped on the radios for communicating with Rosie in the cockpit and with the outside world.

Rosie shined a penlight on small cardboard pads strapped to her knees. One had written on it the half dozen communications frequencies that might be used during the flight. On the other was written her "INS game plan." INS stood for the aircraft's inertial navigation system, a collection of sophisticated radars and computers that told her where the aircraft was at all times. For tonight's flight, Rosie had also drawn a square on one knee board representing the air box overhead. The exercises they would fly were air intercepts. Tuba and Rosie's F-14, along with another Tomcat, would hunt for and intercept two F/A-18s from the carrier that would pretend to be enemy planes (the F/A-18 pilots would imitate the tactics of Middle Eastern pilots flying Russian-supplied MiG-29 jet fighters). At the four corners of the box on Rosie's knee board were marked points whose latitude and longitude Rosie had already plotted as the way points the fighters would use to know where they were in the sky.

Rosie also had drawn a "fuel ladder," a matrix showing the fuel she predicted the plane would have at different times of the flight. The fuel ladder was important. One of the many nightmares for a carrier pilot was running out of gas at sea and having to ditch the plane into the water. Rosie computed her "bingo" fuel at 4,000 pounds. That was the amount the F-14 would need at the end of the flight if Tuba for some reason could not land on the carrier and they had to divert to a land base—in this case at Casale, Italy, eighty miles away. A Tomcat jet consumed about 1,200 pounds of fuel if the pilot tried to land on the carrier at night, failed, then had to circle back to try again. Rosie allotted Tuba three passes to land, which if he had to, would consume 3,600 pounds of fuel. If he failed after three times, the carrier would likely order them to fly to the divert base in Italy, consuming the 4,000 pounds left in the tanks. That meant the F-14 had to have 7,600 pounds when Tuba made his first landing approach.

Throughout the flight, Rosie would check the fuel ladder to make sure the engines were not sucking up too much gas so they would have less than 7,600 pounds upon landing. If they were, Rosie would suggest that Tuba slow down and conserve to make it through the flight. Along with the fuel ladder, Rosie also had plotted the time line for when the two F-14s should be approaching the F/A-18s for intercept. The two Tomcats planned to intercept the

F/A-18s by first picking them up on radar, then, after sneaking past them, looping back to run up on their tails. Tuba had only Rosie's radars to tell him where the enemy was on the chess board. In the movies, jet jockeys fly by the seat of their pants. In real combat, pilots planned their missions in exquisite detail ahead of time so as little as possible was left to chance.

Rosie powered up her radar and inertial navigation system. The F-14 Tomcat, whose specialty was attacking enemy planes in the air that might threaten the carrier, would be retired in 2006. But old though it was, the F-14 still had impressive gizmos. Rosie switched on the AWG-9 weapons control system, whose radar could detect enemy planes up to 195 miles out and then track as many as twenty-four of them on her scope. Next she powered up the electronic emitter that identified whether another plane was friend or foe. The inertial navigation system took about eight minutes to align, a procedure by which it began receiving signals from the carrier that told the plane exactly where it was on the earth while still parked on the flight deck. She flipped on the data link system that enabled her computers to talk to the computers of the second F-14 flying that night in their formation. Throughout the flight the two computer systems would trade information with one another on their locations in the sky.

Once aligned, the INS knew the exact latitude and longitude coordinates for the plane's present location and would continue to update their position as the carrier steamed through the ocean. These coordinates were her first set of way points and they were important. In order to determine where he was traveling in flight the pilot first had to know the exact point from which he started. With the INS aligned, Rosie now punched in coordinates for the other way points she had plotted for the flight.

A deck sailor designated the plane captain began walking around the jet once more for a final check for leaks or loose screws. He signaled Tuba to make one last check of the tailhook mechanism before the plane began taxiing to the catapult. Ordnance crewmen dressed in red turtlenecks rushed up to the fuel tanks hanging underneath each engine to arm their explosive charges that Tuba would detonate if he had to jettison them in flight.

A yellow-shirted crewman waving two flashlight wands now directed Tuba forward to catapult number two for the launch. In the blackness with steam covering the flight deck, the F-14 looked like a gray ghost weaving through the other combat jets on the deck to reach the catapult. As it rolled forward, the exhaust from the Tomcat's huge engines blew a wave of hot kerosene-smelling air across the flight deck. Tuba concentrated solely on the yellow shirt directing him. Rosie craned her neck to both sides of the plane to make sure the Tomcat the yellow shirt was guiding didn't crunch into another aircraft. Sometimes the deck crewmen could get pretty dumb and run you into things they couldn't see, she knew.

At the catapult, a rectangular section of the deck behind the Tomcat raised up like a trap door. It was the jet blast deflector so the flames from the aircraft's exhaust did not blow over ordnance or people nearby. Tuba manipulated a hydraulic lever so the F-14's nose kneeled and a green shirt could hook its front tow bar to the catapult's shuttle. Another bar, called a hold-back bar, was attached to the back side of the landing gear.

A deck crewman rushed up to the right front side of the Tomcat and held up a lighted box, which flashed "66,000." Sixty-six thousand pounds. The crew had to know the jet's exact weight in order to calibrate the power of the steam catapult, which would hurl the Tomcat off the deck at 160 miles per hour. Rosie looked at her knee board. Sixty-six thousand pounds was correct. She circled her penlight in an O to signal to the crewman that the catapult weight was correct.

The pace picked up. Tuba's eyes darted about the cockpit gauges in front of him: to the RPMs of the engines, the exhaust-gas temperature, fuel flow, oil pressure, hydraulics. He looked for warning lights that signaled trouble. He shoved the stick between his legs to all four corners and pushed both rudder pedals with his feet to test the stabilators and rudders a final time. Rosie watched outside. They worked. The Tomcat fluttered like an eagle ready to leap off a cliff. The deck crewman signaled Tuba to power up the engines for the launch. The Tomcat gave a deafening roar that made the fluids in the deckhands' chests gurgle.

Tuba flashed on the jet's exterior lights to signal to the catapult officer that he was ready to launch. Rosie grabbed the bar over her instrument panel and tensed every muscle in her body. She knew from painful experience that on launch the Tomcat dipped down, then up like a bucking bronco, slamming her head against the back seat if she didn't brace for it.

Launch!

Tuba and Rosie felt the skin on their cheeks peel back with the four-G force. It was the best roller-coaster ride they could imagine.

One second. Two seconds. That was all the time they had to decide if the catapult and jet had performed properly and they had achieved at least 127 miles per hour to fly off the carrier. If not, in the next second they had to pull either a yellow-and-black-striped handle between their legs or another one over their heads that would eject them. Ejecting from a bad takeoff could be deadly. Some pilots would be shot up into the air a hundred feet only to land back on the carrier deck—and then slam into another plane, have their parachutes or themselves ingested by a jet engine, or simply be dragged along the flight deck by a billowing chute until someone could stop them. Other pilots landed in the ocean in front of the carrier and faced being run over by the ship and ground up by its propellers in the rear.

For the first two seconds, Tuba kept his eyes focused on the engine gauges and the jet's altimeter. There was nothing to see outside.

"Good engines," Tuba shouted into the microphone in his oxygen mask. The gauges showed the Tomcat at full power.

"Two one one airborne," Rosie radioed back to the ship with the Tomcat's tail number.

The intercept maneuvers over the Adriatic went well. The F/A-18s Tuba and Rosie were chasing tried to evade detection, but Rosie kept them locked on her radar. Now Tuba headed back to the *Eisenhower*. Rosie radioed ahead with the "sierra codes"—updates on mechanical problems they had encountered with the jet during the flight so the squadron's maintenance crew could have a head start in planning for repairs. The message was sent in code so an enemy listening in wouldn't be tipped off to weaknesses in the squadron's Tomcats.

Landing on a carrier at night, called a "night trap," is the most dangerous thing a Navy pilot does. During the Vietnam War, Navy doctors found pilots' pulses raced far more rapidly when they landed on the carrier at night than from the combat missions they had just flown over North Vietnam. An air sortie could be all fouled up, but if the pilot had a well-executed carrier landing afterward he considered it a good night.

Night traps still made Tuba nervous. He would never forget his first one as a student. He had been too busy in the cockpit and worried about failing the test to be afraid of the landing itself. But over the years and the hundreds of night landings that followed, he had had more time to think about the dangers of each one. Like many veteran pilots he became more and more uneasy about them.

The problem was the lack of visual cues. Over land, a pilot could see lights on the ground or make out the horizon for some indication of where he was in the sky in relation to where he was to land. On a black night at sea, there were no lights on the ground, no horizon. Pilots could fly too high in their approach to the carrier and miss the arresting wires (it was called a "bolter" in aviation jargon) or, much worse, fly too low and slam into the back of the ship. Imagine running in a completely dark room toward a tiny, faint dot of light at one end. That was what it was like approaching a carrier at night. As far as depth perception went, you might as well be on the moon, Tuba thought.

Tuba and Rosie had already computed the exact time they needed to reach what pilots called the "marshal stack." This was a point near the carrier—for this mission it was twenty-nine miles to the rear—where more than a dozen planes would now circle at different altitudes, stacked like pancakes, waiting for their turn to approach the carrier in order to land.

One floor below the carrier's flight deck, the men and women in the air traffic control center spoke in clipped, hushed tones. The air traffic control center guided the planes back to the carrier at night or during the day when the weather was too stormy and the pilots couldn't see the ship. If it had to, the control center could land the plane itself. The *Ike* was equipped with an automatic carrier landing system, a precision guidance radar that could lock on to the aircraft at eight miles out and, with one computer feeding position updates to another, could land the aircraft on the flight deck without the pilot having to touch the stick.

The lights in the air traffic control center were dimmed. A half dozen air traffic controllers sat in front of a row of glowing screens arranged in a horseshoe. On a screen to the left was an electronic map of the area in the Adriatic where the *Ike* was steaming. On the right the same map was depicted in a brightly colored computer graphic. A sailor sat before another large radar screen to the side with sloping lines that indicated the angle of descent the aircraft must take during the approach to successfully land on the carrier.

Pressing his headset microphone to his lips, he now quietly radioed direction updates to the pilots warning them when their jets were too high or low. At radar screens in the center of the horseshoe an "approach controller" guided a pilot's plane until it was eight miles from the ship. Then he handed off control of the aircraft to a "final controller," who guided the plane to about three fourths of a mile from the ship.

At still another booth, a sailor nicknamed "Mr. Hands" moved around large white plane numbers by hand over a glowing purple display to show where each aircraft was in the approach pattern. A television camera videotaped Mr. Hands so the ship's senior officers could see where the planes were on the TV screens in their command centers. On a stormy night or when a nugget pilot was having problems landing his plane, the carrier's senior officers would turn off Mr. Hands and crowd into the air operations center to watch the radar operators talk the plane down.

In front of the air traffic controllers stood glass charts on which sailors with grease pencils constantly marked updated positions, headings, and fuel loads for planes in the air. Three chief petty officers walked menacingly behind the air traffic controllers. A lieutenant who supervised the center sat further behind in front of a control booth with earphones clipped to his head. To the left of him was a row of padded lounge chairs on which representatives from each of the carrier's squadrons sat to watch the radar screens direct the landing of their unit's planes.

At 9:40 P.M. exactly, the air traffic controllers all let out a shout: "One minute!" The cheer was done to wake everyone up and focus their minds on what would be the most intense half hour of their work shift. At 9:41 P.M., the first jet started its approach to the carrier.

With the night black and the carrier constantly moving ahead, assembling in the marshal stack required split-second timing for the pilots or they would be scattered all over the skies and it would take the carrier forever to recover them. The carrier's skipper,

Captain Mark Gemmill, wanted to avoid that at all costs. To launch or recover planes, the ship had to sail into the wind at a set speed so the planes would have the benefit of maximum lift from the air. Pilots called it "sweetening the winds." But sailing into the wind might not necessarily be the direction the captain wanted the carrier headed. Traveling at a constant speed and direction also made the ship easier to target by enemy subs or missiles. For that reason, the *Ike*'s goal was to have a plane land every sixty seconds so the ship spent as little time as possible on a vulnerable recovery course (although the pilots grumbled that the squadrons had rarely been able to meet the sixty-second interval).

Tuba and Rosie were supposed to arrive at the marshal stack at 9:44 P.M., give or take five seconds. They arrived exactly on time, began circling in a six-mile-wide oval pattern at 14,000 feet. Two minutes later, Tuba came out of the circling pattern and aimed the jet for the carrier twenty-nine miles ahead. He pulled the throttles back to idle. The jet slowed to about 260 miles per hour and sank to 1,200 feet. Flying out of the marshal stack toward the carrier, the jets were now lined up two miles apart from each other. Tuba and Rosie's jet was third in line.

Catching up with a moving carrier at night could be tricky. The ship steamed along at about twenty miles per hour. But the portion of its deck on which the jets would land was angled to the left at about 9 degrees from the vessel's centerline or keel. That meant that Tuba had to fly slightly to the right of the carrier as he caught up with it in order to land at a left, 9 degree angle on the deck. All along the twenty-nine-mile route he would have to adjust his course because the carrier was constantly moving away from him.

This move-countermove maneuver was made even more difficult by the fact that at twenty-nine miles out, Tuba could not see the carrier. He and Rosie had to depend on the navigation instruments and radar signals from the air traffic control center to keep the F-14 properly lined up.

It was 9:45 P.M. and the lieutenant in charge of the air traffic control center was cursing to himself. The first jets were little more than a mile from the carrier, less than a minute from landing, and the damn ship wasn't lined up properly into the wind. Actually the wind had shifted and the carrier's helmsman was desperately trying to steer the huge vessel in the right direction before the first jet landed. A sudden wind shift across the carrier could prove fatal to a pilot, blowing his aircraft off course as he landed. All eyes shifted to the radar screen showing the ship's direction.

The chaplain came on the ship's loudspeakers with the night's prayer.

"O God, you remind us every day that . . ."

A petty officer reached up and flipped the switch turning off the overhead speaker. The control center was becoming too tense to be distracted by prayers.

At 9:47 P.M. the ship was finally lined up properly for Tuba, who would land three minutes later.

Tuba was furious. The pilots were always having this problem with the regular line officers who steered the carrier. It didn't take a brain surgeon to keep a ship lined up into the wind, he now thought. But here he was two and one half miles from the carrier and it had just veered to the right. That meant he had to swing his F-14 wide to the left to get back on lineup, which when he did threw off his jet's rate of descent to the carrier.

Rosie at this point was sparing in the updates she gave Tuba on the Tomcat's fuel level and altitude. The relationship between a pilot and his radar intercept officer was delicate. Veteran radar intercept officers, who usually were paired with new pilots, would talk the rookies down; their lives were on the line just as much as the pilots'. But rookie radar intercept officers who were paired with seasoned pilots had to feel the pilots out on how much back seat driving they wanted. Rosie normally didn't fly with Tuba. Sully, the pilot she usually flew with, wanted his radar intercept officer to be "ahead of the plane," warning of problems before they happened. Tuba didn't like his RIO yapping in his ear as he landed. "Just back me up," he had told Rosie before the flight. "If something goes grossly wrong, let me know."

Tuba's angle of descent toward the carrier, or "glideslope" as the pilots called it, was fine considering he had had to chase an erratically moving carrier. At three miles out and less than a minute before slamming onto the carrier deck, the plane's altitude was 1,200 feet. They could barely make out flashing lights from the carrier deck. The landing gear and main flaps were already down. Tuba braked the plane again, descending it to 860 feet, one mile away from the carrier.

The next thirty seconds were the scariest. The three best things in life, so the one squadron ditty went, were a good shit, a good orgasm, and a good night trap. If you're lucky you could now have all three at the same time.

Three quarters of a mile out, Rosie radioed the carrier: "Two one one, Tomcat, ball, seven point six." Translated: their Tomcat, jet number 211, had 7,600 pounds of fuel left in the tanks. And Rosie could now see "the ball." On the left side of the carrier was an arrangement of green, red, and amber lights. If the green lights flashed, it meant the deck was clear and Tuba could land. If the red lights shone, something was in the way, such as another plane, and he would have to "wave off" and make another attempt at landing. The vertical set of amber lights were illuminated by computer-controlled lenses. As Tuba approached the carrier, the shifting lenses created the appearance of an amber ball of light moving up or down the vertical row.

The computer set the lens angles depending on the approach path needed for the plane to make a perfect landing. If Tuba was on the correct path to the carrier, the amber ball would appear to him to be in the middle of the vertical lenses. He would be "on the ball." If he saw the light above the middle, he was flying too high in his approach and must drop down if he wanted to catch the arresting wires on the deck. If he saw the light below the middle, he was flying too low and must gain altitude or he would slam into the back of the carrier.

At three quarters of a mile out, the carrier air traffic control center turned over Tuba's plane to Lieutenant Commander Tyler Davenport, a lanky Floridian with bushy blond hair and a mustache, who stood on a platform at the left rear side of the carrier. Davenport was the air wing's senior landing signal officer, who along with a half dozen assistants guided the planes down during their final few seconds before hitting the carrier deck. LSOs were seasoned pilots. Davenport had dodged Iraqi surface-to-air missiles and antiaircraft fire flying an F-14 during the first Persian Gulf War. Until Tuba now landed his F-14 on the carrier deck, Davenport and the other LSOs watched to make sure that his landing gear and flaps were down, that the attitude of his plane was level, that he was lined up on the flight deck's yellow-and-white center line, and that he was descending at the proper rate to catch one of the deck's arresting wires.

If Tuba was having trouble positioning his aircraft, the LSOs would talk him down with "candy calls." Otherwise, Davenport's men would say as little as possible so as not to distract the pilot. One did radio to Tuba that his plane was aligned properly on the center line, which Tuba appreciated, considering that the carrier had weaved back and forth so much in the dark he wasn't quite sure he was positioned correctly.

Landing on a carrier was a balletic exercise, as precisely choreographed as any classical dance. Tuba now had to fly a perfect three-and-one-half-degree-angle down as he crossed the back of the ship. That would give him exactly fourteen feet between his tailhook and the rear ramp when he passed over it. If he was just three feet too high, he would miss the arresting wires. The LSOs in fact would grade him on his landing. With so many landings under their belts, they could tell if Tuba was on a proper course just by looking at his plane as it approached.*

Rosie's eyes were focused on the yellow ball. It remained in the center. She kept quiet.

The Tomcat raced to the carrier at 140 miles per hour, its wings wobbling only slightly as Tuba quickly made final course corrections. His eyes darted from his angle of attack indicator, to the lineup of his jet on the center line of the carrier, to the meatball on the amber lenses that indicated his correct approach path. Pilots called this the "scan." Tuba's brain seemed to be processing the jet's airspeed, lineup, and glideslope at once, instinctively directing his hands holding the stick and throttles to make final adjustments.

Rosie was now just along for the ride, watching the meatball. The back of Tuba's helmet and headrest blocked her view of the blackish gray ship rushing toward them. She could only make out lights on the left and right side of the vessel.

Tuba's back was sticky with sweat. His mouth was dry as every one of his senses concentrated on landing the plane.

The F-14 roared across the stern of the carrier. Fourteen feet over the deck. Just where Tuba wanted it.

* For night flying, each jet had its own arrangement of outside lights, which the experienced LSOs could easily identify. An F-14 Tomcat had green lights on its right wing tip and under its right engine intake, red lights on its left wing tip and left engine intake, an amber light on its front fuselage, and a red light on its rear stabilator.

Rosie's eyes darted left. She could see the LSO platform zoom past her. Then the carrier's tower on her right.

The Tomcat slammed onto the deck. Tires screeched. The tailhook banged on the greasy metal deck sending sparks flying.

Tuba shoved the throttles forward to full power so, with its engines blazing, it could fly off the deck in case the tailhook failed to catch one of the arresting wires.

But this time the hook caught one of the four wires crossing the deck, yanking the Tomcat to a violent stop 350 feet down the deck. Inside the cockpit, Tuba and Rosie felt the bounce of the plane hitting hard steel, heard the heavy clatter of the tailhook raking across the deck, then felt the seat harnesses cutting into their shoulders and stomachs as the jet made its violent stop.

Twenty minutes later, Tuba and Rosie were in the squadron's parachute rigger shop peeling off their survival vests and dry suits. A delegation from the LSO platform walked in with the grade for Tuba's landing. Immediately after the landings were finished, the LSOs hiked around the carrier giving each pilot his grade for the landing.

The grades, of course, were subjective. The pilots tried to schmooze with the LSOs ahead of time and lobby for favorable treatment. It was useless to argue a grade afterward. The LSOs were like umpires. They never changed a call. But if an LSO unfairly passed out low grades, the pilots would start to complain in a more organized fashion and he'd be forced to raise his marks. On this cruise, the F-14 pilots had already begun low-level grumbling that the F/A-18 pilots were receiving higher scores because their jets were easier to handle than the bulky Tomcats.

The LSO grades were critical for a pilot's career. He was worthless to Naval aviation if he couldn't land on the carrier. A pilot with consistently low landing grades would be sent home. In each squadron ready room hung a "greenie board," a chart with color codes depicting the grades each pilot was awarded for every landing. A box by a pilot's name that was colored green meant he had flown an "okay" landing, no problems. Yellow meant a "fair" pass with some deviation from the correct approach the aircraft was supposed to take. Red was a "no grade," which meant the pilot made an ugly landing with a large deviation from the correct approach. Brown indicated that the pilot had to be waved off because he was too far off the correct approach for a safe touchdown or the deck was unsafe for landing. A blue stripe across the box meant a bolter, the pilot missed the wires and had to try again. The greenie board was up for all to see: comrades, squadron commanders, the ship's senior officers. Naval aviation was competitive enough. The greenie boards made it more so. No pilot wanted to be publicly humiliated with a lot of red boxes. Privately, the *Ike's* female pilots thought the greenie boards unnecessarily fueled the testosterone. When more women eventually reached the squadrons, they would push to retire this macho icon.

An LSO in the delegation flipped open a black notebook and read Tuba's grade. On the notebook was written what appeared to be hieroglyphics:

OK 3 S . LU . X (LO) CHLUIM

Translated: Tuba's landing was "okay." He hit the third of the four arresting wires strung across the deck—the ideal wire to catch during landing. The "S" meant he dipped just a little below his correct approach path to the carrier. The pilots called it "settling." "LU . X" meant he made a lineup correction at the start of his final approach to the carrier—mainly because the carrier was late in turning into the wind—and that caused him to fly a little low "(LO)." "X" signified that Tuba's wings wobbled a little. "CHLUIM" meant that midway through his final approach he still had to "chase the lineup."

Shortly after 11:00 P.M., the squadron's aviators gathered in the "dirty shirt" wardroom for midnight rations or midrats. In the ship's main wardroom, officers had to dress in khaki uniforms. In the dirty shirt, the aviators could wear their olive-drab flight suits. Midrats was a ritual after every night flight. The pilots were ravenous from the strain of the landing. The most popular item on the midrat menu was a greasy hamburger from the grill. Pilots called it a "slider." A cheeseburger was called a "slider with cheese." For those who cared nothing about cholesterol there was the ultimate: a cheeseburger with an egg on top. Pilots called it a "Barney Clark" after the famous artificial heart patient.

Tuba decided to forgo the slider. But on his tray was a heaping plate of jambalaya, noodles, and carrots, two glasses of fruit juice, and a piece of cake. He wolfed it down quickly. Rosie had two bowls of cereal.

They laughed and joked with the other squadron pilots. Every move and turn of the flight was dissected and debated. They pantomimed each jet's maneuvers with their hands. The aviators always talked in euphemisms about the landings—"Dark night out there" . . . "The boat was rocking." It masked the excitement and terror of each trap. The camaraderie, the pranks, the poking fun at one another were all safety valves to relieve the stress.

It was well past midnight before Tuba, Rosie, and the other Tomcat aviators had calmed down enough to fall asleep. They awoke the next morning at eight o'clock to begin work all over again.

1

INDOCTRINATION

I

INDOCTRINATION

CHAPTER ONE

The Chamber

MCKINNEY inhaled deeply the smell of nylon canvas and metal and plastic from the flight equipment room. For three weeks he had been sitting through dreary classes, about as far as it seemed to him that he could get from anything having to do with flying. Now he was finally being fitted for a helmet and oxygen mask. The helmet wasn't too small or large like the ones he had been told to wear in the past. This time it fit perfectly. A sailor measured the distance from his nose to chin for the size of the oxygen mask that would clamp snugly to his face. For the first time in his training, Charles G. McKinney II, ensign United States Navy, felt like a pilot.

McKinney walked to the room next door. Its whitewashed, cinder block walls reached two stories high. In the middle sat the chamber, a white steel box the size of a railroad car with green-tinted glass windows around it and a control panel at one end with dials and switches that would activate giant pumps to suck out the air from its insides. Before walking in, McKinney twisted off the college class ring from his finger and laid it on a red wood tray. He didn't want the ring's stone popping out when technicians lowered the air pressure inside.

Once in a while, a flight student would freak out before entering the chamber, overcome by the thought of being sealed shut in a sterile coffin gasping for air. McKinney found it almost exciting. The movie *The Right Stuff*, where astronauts become guinea pigs in tests just like this, passed through his mind. Finally he was doing something, not just reading about what it was like to fly.

It could take more than two years of training to become a fully qualified Navy pilot. But before ever setting foot near a plane, a student had to endure six weeks of what was called Aviation Pre-Flight Indoctrination. Forget the enemy for the moment. Flying itself was an unnatural act for humans. At high altitudes the world in which a combat pilot worked was alien and unforgiving. Over an ocean—the principal domain for Navy pilots—it was even more hostile. Only the most physically and mentally fit survived in this deadly environment. Preflight indoctrination taught the student the basics of the aircraft, its engine, the winds that buffeted it from the outside, the complex rules of the road that governed aircraft flight. The six weeks also began to teach the student the dangers of flight, how he must react when things went wrong in the sky and over the water—or, as McKinney was soon to discover, when there was little oxygen around him.

The indoctrination began at Pensacola, Florida, the military and cultural home of Naval aviation. In 1862, a mosquito-infested, sandy flat overgrown with pine trees and magnolias and crawling with alligators had been set aside by Congress along Florida's panhandle for the naval base. When a handful of oddball officers experimenting with contraptions called "flying machines" later convinced the Navy that it needed an Aeronautic Service, hidebound admirals in 1914 plunked the new branch at the dilapidated Pensacola yard—about as far away from Washington as they thought they could send the silly notion.

But the idea of flying planes off ships slowly caught on. By the end of World War I, the Pensacola Naval Air Station was one of the largest in the world. By World War II, the station mass-produced 22,000 pilots a year. The Naval aviation mafia became as powerful as the fiefdoms that sailed the service's ships and submarines. By the mid-1990s, the number of aviation students who were graduated was far smaller; only about 1,100 underwent the highly coveted training each year and the instruction is conducted at naval air stations scattered around the United States. But the indoctrination into flying—and a way of life like no other in the civilian world— still began at the Pensacola Naval Air Station with its lush green lawns, drooping palm trees, the gentle Pensacola Bay off the coast, steamy hot summers, and training aircraft constantly buzzing overhead.

The Navy wanted McKinney about as much as he coveted the training slot. He was black. After decades of first racism, then racial indifference, the Navy was now desperate to recruit more African-Americans into its embarrassingly white aviation ranks. Up until World War II, African-Americans were barred from flight training; the service thought they could not see as well at night as whites. Even after the military was integrated, African-Americans steered clear of Naval aviation, which was considered an elitist preserve of white males.

If he made it through the training, McKinney would become not only one of the Navy's few black pilots, but also one of its youngest. Born in Newport News, Virginia, he was something of a child prodigy. By the time her son reached three years old, Sylvia McKinney knew he was unusually bright. If he had been born in the ghetto the gift would likely have been ignored or wasted. But Charles was lucky. Sylvia was a reading teacher with a master's degree. His father, Charles, was a Naval officer in the civil engineering corps.

Sylvia took it upon herself to educate her son and to do so at a faster pace than the public school system's. Shortly after his third birthday, he was enrolled in preschool. Every afternoon McKinney came home to color-coded word cards his mother flashed before him. By kindergarten, he was reading. Sylvia had access to countless standardized tests from her school and began carefully measuring his intelligence. In kindergarten, he scored 165 on the IQ test. By age five, McKinney was in second grade. Beginning in seventh grade, he took the SAT tests every year thereafter. In his classes, he excelled in science and math, suffered through reading and writing. By age fifteen, he was a senior in high school.

Being a child prodigy could be fun. He felt proud in the classes where the students always turned to him for the right answers. Some teenage girls found it cute that he was so much younger.

But by high school, McKinney discovered that pain also came with being different. The constant moves every military family must make from one duty station to another could be difficult for children. McKinney dreaded each time he had to transfer to a new school—it always seemed to be in the middle of the academic year—and the teacher would make him stand in front of the class to introduce himself. On top of being the new kid—always with the wrong style of clothes—there was the added pressure of being younger and smarter than his classmates. He felt like a puppet in an amusement park. Some students called him Doogie Howser.

So McKinney withdrew into his own shell. He developed many acquaintances but no close friends. He liked it that way—always the outsider looking in, detached from the people around him. By his senior year, he tried to hide his age from people he met for the first time and pretended to be older. That was easy, he discovered. McKinney did look older and he acted more mature than his classmates.

Something else made him feel different. He had never experienced racism until the family moved to the Marine Corps base at Camp Lejeune when he was nine years old. The six years before, his father had been assigned to Guam, where being black or white meant little to an Asian culture and children of service families grew up oblivious to the prejudices that would later divide them when they returned to the United States. Lejeune was located near Jacksonville, North Carolina—redneck country, where a black kid, no matter how gifted he was, could still be called "nigger." At first, McKinney found it perplexing, almost impossible to comprehend. Then the slurs began to hurt.

He attended a high school in the middle of a cornfield and faced racism from all sides. Blacks, who shunned him as much as whites, considered him too well bred to be one of their brothers. The constant moves had homogenized any accent in his speech. The educational advantages his parents had afforded him now made him an Uncle Tom in the eyes of black students. He talked and acted like a white boy.

"You're not from the 'hood," his black classmates would taunt. "You don't know us."

"You're right, I don't," McKinney would answer back. What he didn't say was that he also didn't care. People automatically assumed he shared a common heritage and background with African-Americans because of the color of his skin. But he had never felt that bond. He came from a yuppie black family whose mother and father were well educated. The niche blacks expected him to occupy was alien to him. He never felt as comfortable with the African-Americans as he did with other cultures. It had rules of behavior, standards of conduct expected of him—hip, jive, slow walk, Ebonics—with which he had no experience and didn't care to have. He shared no sense of history with his black brothers. If he ever took the time to investigate, he thought he'd probably discover three or four generations of race mixing among his ancestors.

McKinney enrolled in Georgia Tech University at age sixteen and majored in civil engineering. The school was his second choice. A congressional appointment for the Naval Academy would have been no problem if he had been older, but the academy didn't accept sixteen-year-olds. He was even too young to qualify for a Naval Reserve Officer Training Corps scholarship at Georgia Tech. But for as long as he could remember, McKinney had wanted to be an astronaut. He would go crazy if he had a job where he had to sit behind a desk all his life. He wanted to fly. And in the beginning, not just any plane. He had his future carefully plotted. He would fly F/A-18 Hornets, the Navy's premier attack fighter for the future. Fast but not too fast. Easy to handle, seasoned pilots had told him, the best ticket for being accepted into the NASA program. Space shuttle pilots had to have experience in tactical jets.

McKinney thought his chances of becoming an astronaut were good—simply because he was willing to do anything to achieve the goal. The only roadblock in his way was a phrase that haunted every student pilot—"the needs of the Navy." Practically all his classmates at Pensacola wanted to fly the glamour carrier jets, the F/A-18s or F-14s. But the Navy had more than a dozen different types of aircraft—jets, propeller-driven planes, helicopters—some of which never landed on an aircraft carrier. Students were assigned to them based on their performance and even more importantly on what the Navy needed at any given time to fill cockpits. What a student wanted to fly was a secondary consideration. McKinney prayed that the Navy wouldn't suddenly need a lot of helicopter pilots when he finished flight school. He had heard of an astronaut who had begun his Naval career as a helicopter pilot, then had managed to transfer to jets in order to qualify for the shuttle program. But such moves were rare. Navy pilots were usually stuck for the rest of their flying careers with the planes they were assigned from flight school. Anything less than tactical jets would derail his dream of being an astronaut.

Men and women became officers in the Navy usually in one of three ways: through the Naval Academy, Naval ROTC in college, or Officer Candidate School. Since McKinney had been too young for the academy or ROTC, he entered OCS as soon as he graduated from Georgia Tech, which took four and a half years in order to complete its more demanding engineering degree. OCS, which for flight candidates was held at Pensacola, was designed to be a culture shock for youngsters whose only discipline in the past came from mom and dad. Males' heads were shaved. Females kept their hair cropped short. OCS candidates wore baggy green fatigue pants, T-shirts, Nike running shoes, silver helmets that rattled on their heads, and a thousand-mile stare from little sleep and constant movement.

McKinney was the type of person who always blew things up in his own mind, always a worrier. By the time he reported Sunday morning for the first day of OCS, he was petrified of what was about to happen to him.

The first week—"Poopie Week" was its nickname—fulfilled his worst nightmare. It began with officers just graduated from OCS and awaiting assignment who were supposed to indoctrinate the rookies into the regimentation of military life. The young ensigns did so with a vengeance—payback for the hazing they had just endured. Never had McKinney been yelled at so much and for so long. When he wasn't at attention, he was pushing up the ground or running or swimming. The hazing from the ensigns, it turned out, was just the warm-up for Thursday morning at five o'clock, when Marine Corps sergeants barged into his barracks banging garbage can lids to wake up the recruits.

By then, however, McKinney realized he wouldn't die from the training. The Marine drill instructors were surly and intense, but fair. McKinney's sergeant had a rich heritage. His father and grand-father had been DIs and he considered the job a calling.

McKinney had always thought of himself as a disciplined fellow. Now he marveled at how slovenly he had been as a civilian. Beds had to be made square. Uniforms had to be immaculate. Marches and salutes crisp. The DIs were maniacal about detail and preci-sion. There was a reason. A pilot had to be a nitpicker. Sloppiness in the cockpit could kill him. McKinney began to enjoy the gruel-ing training. By the time he graduated from OCS, he couldn't wear civilian clothes without making sure there were no loose strings from his shirt. OCS had taught him true mental and physical disci-pline. It had taught him to think under pressure, to assimilate large amounts of information quickly, to keep his wits when surrounded by chaos.

The air inside the chamber smelled stale. Its walls were painted white and from its ceiling hung bright fluorescent bulbs and oxy-gen tubes for instructors who would monitor the students. There was something very antiseptic to it—and quiet. Neither sound nor air could escape its hermetically sealed steel walls. McKinney felt no claustrophobia, as some students did when they walked into the chamber and the large door hatch sealed shut behind them. He liked small spaces. As a child he would always pick a corner of his bed to curl up as if in a cocoon.

Its full name was the hyperbaric chamber. Its purpose was to ac-quaint students with one of the most insidiously dangerous con-ditions a pilot faced at high altitudes: hypoxia, when not enough oxygen reached the tissues of the body. Hypoxia occurred for dif-ferent reasons. When a pilot pulled Gs in a plane (gravitational pressure from sudden acceleration), the blood in his body sank to his bottom or legs instead of where it was needed in his brain, and he blacked out. A G suit the pilot wore on the lower half of his body inflated like a balloon, squeezing his hips and legs to force the blood back to his brain. Pilots also performed a "hook" maneuver, tightening their stomach muscles and quickly expelling air with a "hook" sound, like a weightlifter straining to pick up a heavy load. It also squeezed blood to their brains.

Another form of hypoxia occurred at high altitudes, where the atmospheric pressure was not great enough to keep the blood saturated with oxygen. A human could maintain a healthy blood-oxygen saturation level (of at least eighty-seven percent) up to about 10,000 feet. Any higher, and his cabin must be pressurized or he must have an oxygen mask strapped to his face. If not, less oxygen got to the blood, resulting in hypoxia.

Before becoming unconscious, a pilot might experience any one of a number of symptoms at the onset of hypoxia: a hunger for air, an apprehensive feeling, fatigue, nausea, headaches, dizziness, giddiness, hot and cold flashes, blurred vision, tunnel vision, numbness, tingling. He might hyperventilate, feel euphoric, become confused, uncoordinated, or belligerent. His lips and fingertips might turn blue.

The problem was that no two pilots experienced the same symptoms. What's more, hypoxia would quickly sneak up on a pilot. He could easily black out before he had a chance to treat himself for the condition. At 25,000 feet, for example, a pilot could lose consciousness in three to five minutes in the thin atmosphere.

A slow leak in an aircraft cockpit or cabin might reduce air pressure, causing an aviator to become hypoxic. Cocky pilots often thought they were supermen and would not admit it. If they were lucky an air traffic controller would notice their slurred speech and order them to strap on their masks. One of the biggest causes of hypoxia was hotdogging in the cockpit. It happened, despite constant lectures and warnings from superiors. In 1989, two F-14 aviators flying at a high altitude over Arizona lost consciousness and plummeted to their deaths. They had removed their helmets and oxygen masks and donned cloth garrison caps so buddies in a nearby plane could photograph them saluting.

Students were run through the hyperbaric chamber to learn the particular symptoms each would develop at the onset of hypoxia so they could correct the problem before blacking out. The instructors also hoped it would throw a healthy scare into the would-be aviators not to play around in their aircraft.

McKinney plopped into one of the plastic bucket seats in the chamber. It held seventeen students at a time. To his left was a black control panel with colored knobs and switches that regulated the air flow into his oxygen mask and enabled him to communicate with technicians outside the chamber. The students faced each other in two rows. At one end inside the chamber sat an instructor wearing a flight helmet with its oxygen mask attached to a gray hose dangling from the ceiling. Two other instructors also wearing helmets and oxygen masks paced up and down the aisle between the students watching for problems.

The exercise was carefully regulated to prevent accidents, but there were risks. Pregnant women or students with ear or teeth problems were not allowed in the chamber. The sudden exposure to lower pressure could be hazardous to sinuses and cavities. Pilots also were not allowed to be blood donors and discouraged from smoking; both compounded the hypoxic condition at high altitudes. Students were closely monitored for an hour after visiting the chamber for signs of any complications, such as the bends, which resulted from nitrogen bubbles forming in joints or bones when the chamber decompressed. Often, they also woke up the next morning unable to hear their alarm clocks because of "post-flight ear block," caused by oxygen saturating the inner ear.

The instructors, with oxygen masks already strapped to their faces, began pantomiming orders. McKinney plugged in the jack from his communications cord to the console so he could receive directions from technicians outside. A microphone was implanted in his oxygen mask along with tiny receivers in the earpieces of his helmet. Next he attached the gray oxygen tube from his mask to the air outlet, then cinched the mask straps to the sides of his helmet so the mask fit snugly, creating a tight seal over his cheeks. He flipped the air regulator switch into the positive position. The airflow indicator blinked red, then white as he began to breathe.

The mask felt constricting. Every time he took a breath it sucked his cheeks like a plunger. He'd have to get used to this mechanical act of breathing. He flipped the switch so 100 percent oxygen flowed into his mask. It tasted cool and rich. His lungs felt invigorated.

He hit the test button that force-fed the oxygen from the tube to simulate what he might experience at high-altitude flight. His mask ballooned. It was a rush, as if someone had stuck a blow dryer into his mouth, ramming air into his lungs. He couldn't inhale it fast enough. Every time he tried to exhale it seemed that more oxygen was forced down his throat. Pilots learned a pause-breathing technique—inhale for three to five seconds, exhale for three to five seconds, pause three to five seconds—in order to avoid hyperventilating from taking in pressurized oxygen.

The pleasant voice of Susan Redding, a hospital corpsman operating the chamber's controls from the outside, piped into his earpiece. She sounded like an airline stewardess, reciting instructions to passengers.

"Okay, everybody, take a deep breath," Redding said cheerily over the radio. "No matter what you've heard, we're not going to hurt you today. Just sit back, relax, and enjoy the flight."

Redding began pumping air out of the chamber. On shelves above the students, black boxes with red digital readouts began clicking off the altitude being simulated inside. The students looked up at the boxes as if they were time bombs: 5,000 feet . . . 5,500 feet . . . 6,000 . . . 6,500 . . . 7,000. McKinney's ears began popping like crazy.

"Remember, chew and swallow on the ascent," Redding said over the radio. "If you have any problems give us a level-off sign with your hand and we'll stop."

No one raised a hand.

Hanging limply above the instructor seated at the rear of the chamber were two white rubber hospital gloves tied tight at their openings. One was filled with water, the other with air.

"Okay, we've just passed 10,000 feet," Redding said, as if she were announcing the floors in an elevator.

Both rubber gloves began to balloon slightly.

The students' stomachs began to feel the effects of Boyle's law: the volume of a gas is inversely proportional to the pressure placed upon it. The trapped gases inside the body expanded the higher a pilot flew in his plane. McKinney began to feel the two burritos and ham-and-cheese sandwich he had for lunch rumble in his belly.

"Remember, in order to vent your gases you've got two God-given ports," Redding said in a singsong. "No aiming. We don't want any fistfights in the chamber."

The students began burping and farting. Before flights, pilots avoided foods that would make them gassy, such as beans, cabbage, or carbonated drinks.

The digital boxes over the students registered 25,000 feet. Redding stopped pumping air out of the chamber and ordered the students to remove their oxygen masks.

McKinney unhooked the left sleeve on his mask so that it dangled from his helmet. He was surprised. He thought he would be gasping at this altitude. But he could still breathe. The air tasted thinner. It also felt dry in his mouth, almost bubbly on his tongue. It seemed that he couldn't produce enough saliva to keep his throat from becoming parched. The chamber also smelled like the inside of a toilet.

"Now turn, face your partner, and begin the Pensacola pattycake," Redding ordered.

McKinney turned left to Scott Pierce, a young Marine second lieutenant from Stone Mountain, Georgia, who had also attended Georgia Tech.* Like children, they began clapping their hands, patting them against each other, then touching the tops of their helmets. The instructors had the students play patty-cake to burn off energy and hasten the onset of hypoxia. It was also an easy way for the students to detect when they became uncoordinated and confused from lack of oxygen.

On the other row, Eric Turner, another Marine second lieutenant, from Oakland, California, was handed a yellow sheet of paper and pencil. Redding told him to start with the number 1,000 and begin subtracting six each time. Turner wrote 994, then 898, then 892, then 886 . . .

One minute off oxygen. Some students had silly grins on their faces, partly because hypoxia had begun to set in, but also because they felt ridiculous playing patty-cake. McKinney remained serious. He rarely laughed or joked in class. Friends would kid him about being so expressionless. That was because he was more focused than his classmates, he thought. This wasn't a game to him. It was serious business. He had learned to become a good listener in college. He found that professors noticed the students who paid attention and were more apt to help them in class. Always being serious was a character trait McKinney learned from his father, who was a strict disciplinarian. McKinney was not a hell-raiser or partier. He always ended up the designated driver because he never drank. He preferred to sit back at social gatherings, watch others make fools

---

* The Marine Corps, which is part of the Department of the Navy, and the U.S. Coast Guard, which the Department of Homeland Security oversees, send their flight students through the Navy program to become aviators for their respective services.

of themselves, learn from their weaknesses. He always wanted to be in control. He just wanted to climb into his plane and fly. An F/A-18 Hornet. By himself. Never having to rely on others.

Still, he expected to laugh or giggle as less oxygen entered his body. But he didn't. Pierce was already grinning from ear to ear. McKinney concentrated on the patty-cake as if his life depended on every slap.

"Okay, now that you've got the patty-cake down, let's speed it up," Redding said.

McKinney and Pierce began patting their hands quicker. Pierce was now blinking his eyes and laughing. They began missing beats and becoming confused with the rhythm. Pierce began patting softer. A slight smile crept across McKinney's face. His shoulders began to feel tired, as if he had been carrying a heavy load.

Two and a half minutes off oxygen.

"Okay, everyone," Redding interrupted. "What I'd like you to do is change your routine and hit your helmet twice."

Pierce could barely remember to touch his helmet once. The two now began to miss each other's hands. The other students became hopelessly mixed up with the routine. Some gave up.

An instructor tapped Pierce on his helmet and motioned him to stretch out his hands, palms down. He couldn't stop his hands from twitching.

"That's muscle tetany," Redding explained to the other students. "That's what happens when there's a lack of oxygen."

Three minutes.

"Number six, how are you doing?" Redding asked, referring to Pierce, who was sitting in seat six. "Do you feel hypoxic at all?"

Pierce did not answer. He stared out at his hands with the same grin on his face.

"Number six, are you hypoxic?"

Pierce kept staring at his hands. Afterward, he would remember nothing about his last minute off oxygen.

"Number six!" Redding shouted into her microphone. "Number six, are you hypoxic?" An instructor inside the chamber moved closer to Pierce's seat.

Pierce finally shook his head.

"No?" Redding asked, incredulous.

The instructor in front of Pierce laughed and shoved the oxygen mask up to his face. He was having what pilots called a "helmet fire." His brain seemed to be turned off.

844, 838, 832, 8 . . . Turner became more frustrated with the subtraction after every iteration.

"Number fourteen, are you good at math?" Redding broke in.

Turner wasn't so sure at this point.

"I've got a problem for you," she said. "If eggs cost twelve cents a dozen, how many eggs will you buy for a dollar?"

It was a trick question. Turner agonized over his answer.

"A little over eight dozen," he guessed.

"We'll talk about it in the classroom," Redding said, chuckling.

Four minutes.

"Now I want all of you to treat yourself for hypoxia," Redding finally said.

The students fumbled with their masks.

McKinney felt the thick, cool oxygen filling his lungs again. Air was now pumped back into the chamber. The numbers flashing on the black altimeter boxes above began decreasing rapidly. McKinney's ears clogged once more as the chamber's air pressure now increased. To equalize the lower pressure of trapped gases in his middle ear with the higher pressure outside, he tilted his head back 10 degrees, pinched his nose closed, then blew into his nose. It was called the Valsalva technique, which pilots used constantly when their jets dove to lower altitudes.

The air pressure in the chamber returned to what it was outside. The hatch door opened and McKinney stepped out of the steel box. He was worried. The other students had experienced a variety of hypoxic symptoms being in the thin atmosphere. But he could detect hardly any in himself, save for the sore shoulders and feeling worn out. His mind was lucid and disciplined after four minutes off oxygen, or so it seemed to him.

That could be dangerous, he realized. Hypoxia might sneak up without him even knowing. He would be alone in that F/A-18. That's what he wanted.

But there would be no partner to detect this condition and warn him. He would have to watch out for himself, as he always had.

CHAPTER TWO

Spin and Puke

MARY MARGARET pulled the door shut to the cylindrical-shaped metal box and put on the communications headset. The seat in the capsule—capsule number 8—felt uncomfortable. Tucked under her lap on the left side was a clear plastic bag in case she vomited.

She leaned over to her right, dialed in 1111 on a thumbwheel that displayed numbers on the right side of the control panel in front of her. She dialed the same 1111 into another thumbwheel on the left side of the panel. Then she fastened her seat belt, took a deep breath, and leaned back against the padded headrest, which had an electrical pressure sensor inside it to monitor and record when her head pressed against the wall of the cylinder. Mary Margaret waited in the dark.

Suddenly a hatch covering the glass window at the top of her capsule slid down, showing stars twinkling against a black night outside.

"Spatial disorientation! The final frontier!" The tape-recorded voice piped into her headset sounded like an announcer for a ride in a theme park. "These are the voyages of the starship. Our ongoing mission: To explore strange new sensations. To seek out dangerous visual and motion illusions. To boldly go where no man has gone before. What you see now are your instruments."

The control panel in front of her lit up two large screens, each with a lighted circle and a red dot traveling around its circumference. Above the screens sat a metal box shaped like an upsidedown T. At the top a light flashed a red dot. At the base were arranged four sets of digitally displayed numbers.

"I will again show you these instruments one at a time," the recorded voice continued.

The control panel blacked out.

Numbers in the upside-down T box illuminated.

"This is your numerical display," the voice explained. "You will be told to locate a three-digit number in one of the four digit groups."

The numbers in the upside-down T box darkened and the two screens below began glowing. The left screen displayed a reference heading to show whether the cylinder was moving in a clockwise or counterclockwise direction. The right screen showed the speed of the cylinder when it traveled in either direction.

"These are your attitude indicator lights," the radio voice continued. "Attitude" referred to how a plane was oriented in flight, whether its nose was pointed up or down, whether it was tilted to the right or left. "Now move the joystick forward, aft, and side to side."

Mary Margaret grabbed the joystick on the laptop that jutted out from the front screens. It was about the size of a penlight. She rotated it in every direction.

"Note the movement of the light," said the voice. A white dot weaved around the left screen when she moved the stick. "Use the joystick to simulate changes you perceive in pitch or roll." "Pitch" was the aviation term used to describe the nose of a plane rising or dipping. "Roll" described a plane rolling over as one wing dipped and the other rose. Mary Margaret would use the joystick to indicate the attitude she felt in her simulated plane ride. If it seemed as if the capsule was simulating the plane's nose rising she would pull the stick back. If it felt like the left wing was dipping down and the plane was banking left she would move the joystick left. Right wing dipping down, move the joystick right.

A large circle of lights on the left screen began blinking.

"These outside lights are your reference attitude indicators and they will show you your true cockpit attitude," the voice explained.

The right screen's circle of lights flashed.

"These are your rate and direction of turn indicators."

In the center of the laptop below the screen was a numerical display board. Below that board were three turn response buttons Mary Margaret could press to indicate whether the cylinder was turning left, right, or remaining stationary.

Finally, the red light at the top of the upside-down T box flashed.

"This is the point light source," the recorded voice said.

All the control panel lights now illuminated. The window shutter closed. Mary Margaret felt the capsule turning to the right.

A tape of the British pop singer Seal singing "Crazy" came on the radio in her headset.

"Now sit back, relax, and enjoy the music as you prepare for your journey through the world of spatial disorientation," the voice said soothingly.

"We're never going to survive unless we get a little crazy," Seal sang into her ears.

Marine First Lieutenant Mary Margaret Kenyon felt anything but relaxed.

Flying could play tricks on a pilot's senses. Things could be different from the way they seemed or felt. Disorientation in the sky was common as a pilot banked, spun, looped, or accelerated his aircraft in high-speed combat maneuvers. Disorientation could be deadly.

Eighty percent of a pilot's orientation in the sky came from what he saw through his eyes. He relied on two types of vision: his focal vision, which enabled him to see things ahead in fine detail, while peripheral vision helped him to sense things to his left or right. But over land at night, ground lights could easily be confused with starlight. In darkness over the sea, the sky and ocean a pilot saw often blended together creating a false horizon. Helicopter pilots often experienced a waterfall illusion hovering over the ocean, when water and mist swept up by the rotors created a false sensation of gaining altitude. If the pilot dove to correct the problem he crashed into the sea.

On lengthy night flights, an aviator might experience empty field myopia, when the eyes relaxed and became nearsighted. If he stared for a long time at the stationary light from a plane in front of him, the light might begin to move about—an illusion called autokinesis. Narrow, short runways could deceive a pilot into thinking that he was too high in his approach; when he dropped in altitude to correct, he landed short, crashing into trees. Landing at night aboard an aircraft carrier, with only the meatball and faintly glowing lights from the carrier deck for reference, could create a black hole effect for the pilot, also causing him to fly too low in his approach.

The other twenty percent of a pilot's orientation in the sky came from his somatosensory system (the scientific term for his seat-of-the-pants feeling) and from his ears. A pilot's seat-of-the-pants feeling came from the proprioceptors in his body tissues that detected gravitational pull and pressure points. Because the Gs a pilot pulled in high-speed maneuvers could fool the proprioceptors or dull their senses, he couldn't orient himself in the sky by only the gravitational pull he felt.

The ear could play even more tricks. Its semicircular canal detected a spinning or tumbling motion in the body. The canal had tiny protrusions called cupula, over which a fluid flowed. When a pilot began to spin around fast in his plane, the fluid in his semicircular canal at first lagged behind, creating a drag effect that bent the cupula. A good analogy would be speeding off in a car with a cup of water in your hand. Half the water would end up in your lap because the liquid hasn't caught up with the movement of the car. In the ear's case, when the fluid bent the cupula, hairlike cells underneath the protrusions began to move. That hair movement was then interpreted by the brain as a rotation of the body.

But sloshing ear fluid created a problem for the pilot. As the rate and speed of a turn he made in his plane remained constant, the fluid in his semicircular canal would eventually catch up with the speed his head—and thus the canal and cupula inside it—were moving. Back to the analogy of the water that spilled into your lap when the car took off. Once the car reached a cruising speed, the water would stay in the cup because it was now moving as fast as the vehicle. In the ear's case, its fluid is turning at the same rate and speed as the body. The brain would detect no movement from the ear because there was no drag being placed on the cupula and the hairs underneath weren't being tickled. At this point, even though the pilot might be spinning around, his ear would not detect any motion.

Disorienting this ear sense might not be serious as long as the pilot could still see where he was with his eyes. But it could become a big problem if the sky was pitch dark and all the pilot saw was black outside his cockpit.

The trouble could be made worse if the plane suddenly decelerated. Again, back to the speeding car analogy. Slam on the brakes and the water left in the cup splashes forward onto the dashboard. The same happens with the fluid in the ear. As a pilot halts a fast spin, the fluid in the semicircular canal continues to move in that direction at the same speed. That in turn bends the cupula and stimulates the hair cells in the direction opposite from the way the body is moving. The pilot would think he was still spinning when he was not.

Another part of the ear can play tricks as well. Its otolith organ will sense the change in gravitational pull a body feels when it is accelerated forward in a straight line. The otolith organ has a gelatinous layer with tiny crystals called otoliths. The layer changes position as the head accelerates. If the pilot tilts his head back, the gelatinous layer slides to the rear and gives him the sensation that the nose of his plane has pitched up. When he leans forward, the gelatinous layer also slides forward, giving him the sensation that his plane's nose is dipping. The problem: violent acceleration from a carrier launch can cause the otolith organ to overreact and give the pilot the impression that the nose of his plane is pitching up too high. If the pilot overcorrected, he would end up sending his plane into the ocean. A head cold, clogged sinuses, even drinking tonic water before flying could impair these senses in the inner ear and cause a pilot to crash into the ocean during a night launch from the carrier.

To give the Pensacola students a taste of the sensory illusions they could experience in flight, American Airlines and the Walt

Disney Company built for the Navy the Multi-Station Spatial Disorientation Demonstrator. The instructors nicknamed it the "spin and puke." In the center of a huge circular room sat a merry-go-round with ten white capsules the size of large oil drums. White metal bars and wires connected the capsules to one another and to the center of the merry-go-round like spokes on a wheel. Each capsule could rotate on its own in either direction as the merry-go-round also spun around. A projector in the center of the room would flash an image on the room's white circular wall. The image could have stars at night moving as fast as the merry-go-round spun so when each capsule's window hatch slid down it would appear to the student inside as if he wasn't moving. Or, the stars could move one way while the merry-go-round moved the other, giving the student the feeling that he was spinning faster than he really was. In a control booth a floor above the spinning device sat computers and monitors that manipulated the various rotations and recorded the students' reactions to the twists and turns. By combining the spinning merry-go-round with the rotating capsules and the movie projection on the outside wall, the spin and puke machine could create practically any illusion in flight that the instructors wanted.

Few students actually threw up during the ride. But the instructors, hoping to psyche them out, had made such a big deal about aiming for the plastic barf bag and not messing up the inside of the capsule, Mary Margaret was sure she would lose her lunch.

Her capsule halted its rightward rotation so she was facing inward and to the center of the merry-go-round. The merry-go-round now began spinning at 120 degrees per second, or about twenty miles an hour. Inside the blacked-out capsule, with her back facing out, Mary Margaret felt plastered to her seat. This part of the ride simulated her being launched off an aircraft carrier from a catapult, pulling almost one and a half Gs.

"Move your joystick to indicate your nose-up attitude," the recorded voice instructed.

Mary Margaret felt as if she was being shot out of the roof. She yanked back on the joystick.

"Note the difference between your perceived and actual cockpit attitude," the voice said.

Mary Margaret stared at her console screen wide-eyed. There was a huge difference. Her otolith organ had played the first trick.

Mary Margaret Kenyon still had to pinch herself to make sure she was really here, sitting in this cramped, black capsule feeling like she was rolling over Niagara Falls in a barrel. She was a thirty-two-year-old divorcée whose only companion was a mongrel golden retriever that was part German shepherd. A lot of women curled up and emotionally died after a marriage crumbled. Raising children alone. A dead-end job. Back to an apartment instead of the dream house. No more falling in love. No prince in shining armor. Life could end with a divorce.

But Mary Margaret was being given a second chance in life, a chance to start all over, to live a dream she never thought possible. Sometimes fear swept through her. What if the dream did not come true? What if she failed? She was nervous all the time. She wanted to be an aviator more than anything else in the world. This was her new life. Her second chance. Stay focused, she told herself over and over again.

Mary Margaret had begun without focus. Born in Whitney Point, New York, a small suburb north of Binghamton, she had graduated from a high school whose senior class numbered less than 200. She wasn't mature enough yet for college. She enrolled in the State University of New York at New Paltz in the Catskill Mountains in 1981, transferred to SUNY Binghamton a year later, then dropped out and entered secretarial school.

In 1985, she married a Marine and began following him from one duty station to another. They ended up at Camp Lejeune. Jacksonville, North Carolina, however, had few jobs for service wives. Mary Margaret had grown to like the military life of her husband, by now a corporal. She did something she never thought she would do in her life. She enlisted in the Marines herself. To her it was the chance for a better future, as well as a steady paycheck no matter where they were posted.

By 1992, after briefly separating from the service and earning an art history degree from Memphis State University, Mary Margaret decided to follow her husband again. He had applied to become an officer. She entered Officer Candidate School as well. Marine Corps boot camp had been a blur of screaming drill instructors, push-ups, running, and lugging around an M-16. It taught her to be a ramrod-straight Marine. Ten weeks of OCS taught her to be a leader.

It also gave her a thick hide. If the Navy was at least a generation behind in its thinking about women in the workplace, the Marines were in the Dark Ages. There were no cooks or clerks in the Marine Corps. Every soldier was trained as an infantryman. Every soldier was considered a warrior. Marine brass could barely stomach young soldiers getting married. Women in the combat ranks was an even more bitter pill for diehard leathernecks to swallow. Mary Margaret quickly learned to laugh off the slights, the cruel jokes, the insensitive remarks. Roll with the punches. *Semper Fidelis* was the Marine motto. The women in its ranks had their own private one: *Semper Gumby*, always be flexible. This was a man's organization. Accept it. Let Marines be Marines. Don't try to change them. If you fight, they will defeat you. But don't quit.

The Marines had just begun mixed-gender training units at the Basic School, which OCS officers had to take to become qualified infantry officers. Officially, the physical fitness standards the Corps set for women were less strenuous than for men. The loads on their backpacks were not as heavy. Unofficially, the male student officers and their instructors expected the females to keep up with the men's pace. And they deeply resented lighter loads for the women.

Some of the women were good athletes, strong as bulls, and could keep up. Mary Margaret was not. She could run miles and miles, but not as fast as the men. She never dropped out of the running formations. But she met the physical standards set for women. She was in terrific shape. But it still frustrated her that she was considered below average by the male students in her class. These young bucks were still learning to drive their dad's Chevy when she was already in uniform. And here they were challenging her with smart-ass questions like: "What are you doing in the Marines?"

"You wanted something more," she'd say calmly, holding her temper, trying to explain it for the hundredth time. "You wanted a challenge. So did I. Just because I'm a woman doesn't mean I don't have the same ideals, the same hopes!"

She finally thought she had the point drilled into their thick heads when the Tailhook scandal broke publicly. Mary Margaret was still in the Basic School. Tailhook dropped an atomic bomb on her world. Scores of Marine Corps aviators were among the officers implicated in the sexual assaults. Like the Navy, the Corps was forced to confront the pathological sexism that infected its ranks.

And like the Navy, the Marines reacted by withdrawing from the feminist force they perceived as contaminating their insular world.

Overnight, Mary Margaret and the other women in the infantry school became outcasts. Instructors were afraid to talk to them. Male students shunned them. It took several months before the men realized she was not a secret agent out to destroy their careers and began talking to her as a comrade.

By 1994, Mary Margaret was stationed at Camp Pendleton in California, learning to be an intelligence officer. It was then that her marriage fell apart. Picking up the pieces, she made the most momentous decision of her life. She had been jogging with a Marine captain, a helicopter pilot stationed at Pendleton. He was one of the more enlightened in the Corps.

"Do you have good eyes?" he blurted out between puffs.

"Yes," she answered.

"You should apply for flight school," he said.

"I'm too old," Mary Margaret said, puffing.

She had never considered flying as a career. It would be fun, she thought, but in the past never an option for women in the Marines.

But the combat exclusion rules had just been changed. The Corps, which had its own fleet of jets, transport planes, and helicopters to provide air cover for Marines fighting on the ground or to haul supplies, was just beginning to accept women in its cockpits.

Several days later, the captain placed a phone call on Mary Margaret's behalf to Marine headquarters in Washington. Yes, she was a bit old to begin flight training, headquarters told him. But have her apply anyway. Exceptions might be made.

Mary Margaret began filling out the paperwork. She took a battery of aptitude tests to qualify for flight school and passed. The Marines usually didn't want officers older than twenty-nine starting pilot training. But Mary Margaret, by then thirty-one, was still a second lieutenant—junior enough in rank so the Corps would still get a lot of time out of her flying in a squadron—and age waivers were routinely granted for officers who had served in the enlisted ranks.

Her waiver was approved. Headquarters ordered Mary Margaret to report to Pensacola.

She was ecstatic. Life *could* start over at thirtysomething. The Marine helicopter pilots she hung around with at Camp Pendleton also were cheering her on. But the men took her aside and delivered a private warning. Mary Margaret would be the second female officer to begin training as a combat pilot in the Marines. She would be a pioneer, and women who became pioneers in the military were treated roughly. The rest of the Corps was not as open-minded as they were. Pensacola could still be mean and unforgiving. The Corps's helicopter pilots were more accepting of women in their ranks, but the macho jet fighter community still looked on them as the enemy.

Mary Margaret wasn't worried. She was almost a decade older than the male Marines who would be her classmates at Pensacola. But she didn't think she would stand out. She had smooth tanned skin, piercing gray eyes, light brown hair that fell to her shoulders when she didn't have it tied in a bun during duty hours in her uniform. She looked like she was in her mid-twenties. Yet she felt far more seasoned than these youngsters fresh out of college or the Naval Academy. She knew her way around the Corps. She had long stopped trembling every time a major walked into the room.

Pensacola so far had not been socially intimidating. The male Marines in her flight class (they numbered about a dozen) seemed pleasant and accepting. None had confronted her with knuckle-dragging insults. They didn't appear as afraid of females or threatened by them as the older jarheads. Maybe they were the 90210 generation, where boys and girls mingled without all the sexual hang-ups of their parents, and she liked that.

But Mary Margaret didn't know what the men said behind her back. Some were part of the new generation of thinking. But not all. None of the male Marines in her class had told Mary Margaret to her face that she didn't belong here. In the new politically correct Marines, they knew a remark like that could be career suicide. If women could cut it they should be allowed to fly, many of the Marine students conceded.

But others were not pleased to have Mary Margaret in their flight class. She represented what they perceived as privileged treatment women had received back in infantry school—the lighter loads, slower running times—and they resented that. Never mind that how much a woman could carry on her back had little to do with how good a pilot she would be. Never mind that women could physically handle the service's most demanding combat jets. They still resented Mary Margaret's being in their class.

If she had known their true feeling, she wouldn't have cared. Mary Margaret was beginning life over. Nothing would stop her.

Nothing, except the wall.

Along the Pensacola Bay side of the Naval Air Station on a sandy plot of land sat the obstacle course. To pass the preflight indoctrination phase, males had to complete the obstacle course in no more than three minutes and forty-eight seconds. The qualifying time for females was four minutes and twenty-one seconds.

As obstacle courses went in the military, Pensacola's didn't stand out as particularly grueling. But it was rigorous enough. The course stretched 660 yards over loose sand that bogged down runners. It began with two rows of tires students had to hop through, then a waist-high hurdle to climb over. The men must scale a twelve-foot wall, pulling their way up on a thick rope. The women ran around that wall. The men next faced an eight-foot wall to jump and scale over with no help from a rope. The women's wall was six feet high.

After the walls, more waist-high hurdles. Then they crawled under metal pipes arranged in rows a foot high from the sand, then climbed Jacob's Ladder, which was twelve feet high. More waist-high hurdles. Next a row of monkey bars. More waist-high hurdles. A twenty-five-foot-long balance beam the width of a telephone pole to walk across. Pull yourself through a rat's maze of metal bars set up like stalls. Then sprint to the finish line.

The Navy eventually planned to replace the obstacle course with machines that would measure strength and endurance. That would cut down on broken bones from the course and failures by students who had stamina and agility (what a pilot really needed in the plane) but weren't coordinated enough to make it through the course in the required time. But for now, the O course was the standard. Flunk it and you flunked out of flight school.

Mary Margaret couldn't get past the six-foot wall. The instructor at the starting line shouted go. She hopped through the tires with no trouble, skipped over the first hurdles, scooted around the twelve-foot wall for the men. Then she slammed into the six-foot wall. It might as well have been 600 feet high. There was a technique to scaling it quickly. Begin your jump a foot from the wall. Get your armpit over the top of the wall like a chicken wing, then swing like a pendulum to hoist your trunk and legs over.

Mary Margaret did it all wrong the first time and fell back on her haunches. She tried again. No luck. She turned around, dusted the sand off her legs, and walked back to the starting line. The instructor handed her a pink slip. Failure.

She felt humiliated. Marines don't fail physical training (PT) tests. She knew the men were looking at her in disgust. She became angry with herself. She had overcome so many hurdles to get to Pensacola. Now she was stopped by that damn wall.

The instructors put her in a remedial class for two weeks. She would be given one more chance. Fail it again, and she would be before an academic board that would consider her case and probably politely tell her to find another job in the Marines.

She jogged and did pull-ups until she thought her arms would fall off. Three evenings a week she made practice runs through the obstacle course.

"You're not going to let a wall stop you!" the instructors would scream at her as she slammed into the wood. The inside of her arms turned black-and-blue.

After two weeks, she finally mastered the technique. It wasn't the prettiest jump. But she could at least get over the wall.

On her second test, she passed. Four minutes flat. She was so excited, she went out and treated herself to an expensive lunch. That weekend Mary Margaret relaxed and enjoyed herself for the first time since coming to Pensacola.

The spin and puke merry-go-round slowed its rotation a bit. Mary Margaret felt the capsule make a quarter turn to the left so that her left ear faced the outside. That slight turn of the capsule in combination with the spinning merry-go-round was designed to give her the sensation that the left wing of her plane had dipped down and she was banking to the left.

It certainly had that effect on Mary Margaret. She felt as if she was going to slide off her seat because of a hard left bank.

"Now move the joystick to the left to indicate your left-wing down attitude," the taped voice said.

Mary Margaret did as she was instructed. She knew she was just spinning around in a simulator, but the combination of a rotating capsule, that capsule being on a rotating merry-go-round, plus flashing symbols on the console screen in front of her made it seem as if this tin can she was stuck in had taken off and flown out of the building with Seal singing in her ear.

No time to think, however. The capsule made another quarter turn to the left, so Mary Margaret's nose now faced outside.

"As we change the heading, it will seem as though you have pushed the nose over and are now in a dive," the voice explained.

Mary Margaret needed no explanation. It felt like a kamikaze dive. Her body lurched forward from the centrifugal force of the merry-go-round spinning 60 degrees per second. She grabbed the console panel to brace herself. The seat belt tightened on her stomach.

"Move your joystick to show your nose-down attitude," the voice continued.

Mary Margaret did so, by now totally into the illusion.

The capsule rotated to the right another quarter turn while the merry-go-round accelerated its spin slightly to 100 degrees per second. Mary Margaret's right shoulder now faced out. The maneuver simulated a plane's right wing dipping for a right bank.

"Move your joystick to follow the right-wing roll," the voice said slowly.

Again the capsule felt as if it had tumbled sideways.

"During these illusions your cockpit has actually remained straight and level," the voice reminded her. "Only the heading has been changed to simulate these different cockpit attitudes."

Mary Margaret chuckled. She was excited. This was the best thing she had done so far in preflight indoctrination.

Now came the tricks.

"Stare at the point light source," the voice instructed. On the upside-down T box, the red dot illuminated as the merry-go-round began to decelerate.

Mary Margaret stared at it intently. Then she blinked her eyes. The light seemed to dance around. She was experiencing nystagmus, a condition in which acceleration and deceleration causes a pilot's eyes to twitch back and forth.

"Slowing or stopping in a high-speed turn can produce reflexive eye movements which can momentarily cause tilting or blurring of your instruments," the voice explained.

The red light turned off. The merry-go-round decelerated even more while the capsule rotated to place Mary Margaret facing out. To distract her so she wouldn't notice she was moving slower, the four sets of numbers at the base of the upside-down T box illuminated.

"Prepare to find numbers," the voice ordered. "Find five nine six."

Mary Margaret hunted among the three-digit groups and found the combination.

"Find eight four five."

She found them.

The merry-go-round had almost come to a complete stop, but Mary Margaret was preoccupied with the numerical display board.

"Find three four five."

She did.

"Find eight seven nine."

On the circular wall outside the merry-go-round, the movie camera began projecting stars at night. The shutter over her capsule window came down so Mary Margaret could see outside. The merry-go-round now had completely stopped, but the stars projected on the screen were already moving slowly to the right, in the opposite direction that her merry-go-round had been traveling.

Mary Margaret pushed the left-turn button on the laptop in front of her control panel to indicate that the merry-go-round was still spinning to the left.

"Look at your instruments," the voice commanded. Mary Margaret did. The dots on her screens' two circles were standing still. "You are not moving. You have just experienced circular vection." Her eyes had fooled her.

The lights on the outside screen cut off. The merry-go-round started to spin slowly to the left.

Mary Margaret again was distracted with orders to find more numbers on the digital displays. In an actual cockpit, she would be just as preoccupied with control panels full of radar screens, flashing lights, and readings from navigation equipment all demanding her attention practically at the same time.

"Find five nine two."

"Find six seven zero."

The starlight on the outside screen turned on again. The star field also was moving to the left.

The merry-go-round began accelerating in its leftward spin. But the rate of increase in the acceleration was slow, no more than an increase of about 3 degrees per second. Meanwhile, the star field on the screen outside accelerated its movement to the left at the same slow rate.

The voice kept ordering Mary Margaret to look for more numbers on her digital display.

"Find six two zero."

She glanced up for a brief second at the window giving her a view of the outside. The star field and the merry-go-round were moving at the same speed. From her vantage point, the stars weren't moving.

The capsule seemed to her to be at a dead stop.

"Which way are you turning now?" the voice asked.

Mary Margaret pushed the center button on her laptop, the one that indicated her capsule was remaining stationary.

The merry-go-round and star field continued to accelerate at the same rate.

"Find three nine five," the voice continued asking, adding quickly, "remember, push in your turn buttons."

Mary Margaret kept her finger on the center. Why does he keep asking me which direction we're headed? she thought. We're going nowhere. But there was no time to dwell on it. The voice was spitting out instructions a mile a minute.

"Find four zero nine."

The merry-go-round was now spinning quickly at 30 degrees per second. But the star field was spinning at the same rate.

"Find seven nine zero."

"Find eight nine zero. Keep your turn button pushed in."

Mary Margaret frantically hunted for the numbers on the digital display. She took quick glances out the window to see if she was moving. Nothing. Dead still. She kept the center button pressed.

"Find three five one."

Suddenly a hot flash of panic came over her. She had glanced down from the digital display to the two screens in front of her. The two red dots of light were racing counterclockwise around the circle.

She quickly released the center button and jammed in the left.

"I've got the left button on," she shouted into her mike, realizing she'd been fooled. "But I don't feel like I'm turning."

A sailor monitoring her radio traffic in the control booth above the merry-go-round chuckled. "Yeah, I know," he said to himself.

"Look at your instruments," the voice in her earpiece finally interrupted. "You have been turning for a full minute. This is the most dangerous illusion and it's called the graveyard spiral."

Mary Margaret was dumbfounded.

Indeed, it was a deadly illusion. When a pilot had his plane in a constant spin or a prolonged turn, the fluid in his semicircular canal would eventually catch up with the movement of his body. The fluid would no longer bend the cupula. The hairs underneath would no longer be tickled. The canal no longer sensed motion. The brain was being signaled that the body was standing still— when in fact the pilot might be spinning violently. If there were no visual cues outside to tell him otherwise—in Mary Margaret's case, the moving star field offered no help—the graveyard spiral illusion could spin a disoriented pilot and his plane right into the ocean.

Mary Margaret barely had time to take it all in. Suddenly the star field started accelerating in its leftward spin while the merry-go-round began decelerating.

"Which way are you turning now?" the voice asked. "Push in the button."

Mary Margaret ignored the screens in front of her. It clearly felt as if her capsule had lurched to the right. She pressed the right turn button on her laptop.

It was a mistake. The merry-go-round and the star field had been moving left all along—only the star field was moving left at a faster rate than the merry-go-round. Even though Mary Margaret's body slowed with the capsule, the fluid in her semicircular canals still kept its same momentum as before. That bent the cupula and the hairs underneath in the opposite direction of her movement.

In an aerial dogfight, where the pilot was distracted by dozens of things other than flying his plane, this sensory disorientation could also be deadly. When the pilot suddenly leveled off after a spin or high-speed turn, he might feel as if he was spinning or turning in the opposite direction. If he overcorrected for the false perception, he might end up in another deadly spin or turn, only this one in the opposite direction.

"You probably thought you were turning one way," the voice explained to Mary Margaret. "But actually you were turning in the opposite direction. You have just experienced the helicopter illusion."

The voice quickly proceeded to the next illusion as the merry-go-round picked up speed moving counterclockwise.

"Now move your head slowly to the left, locate the left thumbwheel switch, and enter in today's number. When you've completed this task, put your head back on the headrest and look at the point light source. Again, move your head slowly to the left."

Mary Margaret ignored the advice and leaned over quickly to the left. She found the thumbwheel. She could feel the capsule picking up speed as she did so. Hurriedly she rotated the dials on the wheel so it showed the number the instructor had given her before entering the device: 3195. She bolted back up and snapped her head back to the headrest.

Whoaa! she thought to herself.

The point light source in front of her—the tiny red light at the top of the upside-down T box—looked like a bouncing ball following lines of music. Her head felt like it was swelling. Her body became clammy. She leaned back on the headrest for a moment, closed her eyes, took a deep breath. When she opened them she felt dizzy. Then nauseous. Now she knew why they gave each student a plastic bag.

"Move your head to the right," the voice continued. "Locate the right thumbwheel switch. Enter today's number subtracting 101. When you've completed that task, put your head back on the headrest and look at the point light source."

This time Mary Margaret leaned over slowly. Her head was still swirling as she calculated the new number—3094—and dialed it on the thumbwheel. She leaned back on the headrest, hoping the nausea would go away. It didn't. Her face felt flush. The red dot was still dancing in front of her. If this ride wasn't over soon, she knew she would become airsick.

"As you move your head, you may have the sensation of spinning or tumbling," the voice explained. "If so, you have just experienced the Coriolis illusion."

The illusion set in when the semicircular canals became mixed up. As a jet accelerated in a turn, spin, or loop, the fluid in one canal remained steady, moving along that plane of motion. But if the pilot jerked his head up, down, or sideways, the fluid in the other ear's canal began to move. The brain received mixed messages as a result. One canal was signaling acceleration while the other signaled deceleration. For the pilot it created the sensation of tumbling in space and motion sickness. In aerial maneuvers the effect could almost incapacitate a pilot. He quickly learned not to move his head on high-speed turns or when pulling Gs.

The merry-go-round finally stopped and the lights in the room turned on. Mary Margaret stepped out of the capsule a bit wobbly-legged. She was relieved that at least the feeling of nausea had stopped. She no longer felt clammy.

But the unease remained. The lesson had been chilling. Jet fighter cockpits were crammed with navigation and directional instruments that told a pilot where he was in the sky, where he was headed. Trust them, the instructors had already begun to tell her. Trust them over your own senses.

She would. The tricks that had been played on her in the spin and puke machine were harmlessly amusing. In the capsule, the environment was safe, the illusion controlled. A pilot could make mistakes and learn from them. Tricks or mistakes in the plane could prove deadly.

CHAPTER THREE

Gouge

THE seven flight students had gathered Wednesday night at Stacie Fain's apartment in north Pensacola for the final study session before the flight rules and regulations exam the next morning. They wore cutoff jeans, T-shirts, and sandals. Several brought liter-sized soda bottles loaded with caffeine. It would be a long night, like the cram sessions they had had in college.

Mary Margaret, Brian Hamling, and Bill Perkins were sprawled on black leather couches in the small living room flipping through workbooks. Dan Smellick and Stacie sat cross-legged on the carpet around a carved mahogany coffee table—her prized possession. MTV hummed at low volume on a wide-screen television set. No one found the rock music distracting.

Overhead, a ceiling fan whirred. On her walls hung a framed Van Gogh poster and another that celebrated the fiftieth anniversary of Pan American Airlines' China Clipper. Stacked on top of the television cabinet were books on military aircraft, a Pan Am flight schedule book, a travel guide for New York City. In the adjoining dining room, also tiny, other students sat around a wrought iron table with manuals and study notes spread over its glass top.

Flight Rules and Regulations was one of five courses the students had to master in preflight indoctrination. The hyperbaric chamber, the spin and puke, obstacle course, all the other survival exercises, were the exciting parts of indoctrination. They were fun. They were hands-on, a chance to put on flight gear and test yourself physically. But the far more grueling part of preflight indoctrination was in the classroom with the "academics," as the trainers called it. Most of a student's overall grade in preflight came from the classroom work. The academics were the challenge. It was the academics where the competition began in earnest.

In the first week at Pensacola, Stacie and the other students were loaded down with a stack of manuals and study guides a foot high. There was Introduction to Basic Aerodynamics, which covered the physical and mathematical principles of flight and the effect of the airflow on a plane. In Air Navigation, students plotted routes on charts and maps, mastered the "whiz wheel" that manually converted speed, time, and distance in different measures, and calculated fuel consumption for plane trips. Aircraft Engines and Systems taught them the mechanics of prop planes and jets. Meteorology classes covered air temperature and pressure, frontal systems, as well as weather problems a pilot faced—turbulence, thunderstorms, fog, clouds, icing on the wings. The fifth week of indoctrination was nicknamed "Hell Week" because of final exams in three of the courses.

But first there was the Flight Rules and Regulations final. The students were worried. FRR was nothing but memorization. If they had had several weeks to learn it, the course would be easy. But they had only several days. A flood of picayune rules had to be soaked up quickly. After classes ended each afternoon at four o'clock, students usually drove home, ate dinner, then spent the rest of the evening studying. Partying on weeknights was out, unless a student wanted to wash out quickly. Much of the weekend as well had to be spent with manuals under their noses.

Pilots didn't have to know just altitude, but three types of altitude: absolute, pressure, and true. There wasn't just air traffic control that guided a pilot's flight but four subdivisions of air traffic control: tower control, approach and departure control, air route traffic control centers, and flight service stations. The skies were divided into controlled and uncontrolled airspaces. Within controlled airspaces, which meant an air traffic controller had radio contacts with a plane, there were five categories of control depending on the altitude at which the pilot flew and how crowded the skies around him were with other planes. On top of that, there were six more categories of airspace reserved for special uses, such as military training or practice bombing runs. There were flight rules for flying the plane using a pilot's vision. There were flight rules for flying using only navigation instruments. There were rules for taking off from a runway and rules for landing. There were rules for how much fuel a plane must have in flight, rules for the visibility of the skies a pilot could fly in, rules for cloud ceilings. There were even rules for the number of dog tags a pilot wore on the chain around his neck: two.

Stacie became the interrogator. "What three sky coverage phenomena constitute a ceiling?" she asked.

Mary Margaret laid down her manual and closed her eyes, searching for the answer in her brain.

"Broken, overcast . . ." She couldn't find the third phenomenon. She pounded her forehead.

"Obscured," Stacie prompted.

Mary Margaret slapped the manual, frustrated.

What were the fuel requirements for flight under visual flight rules or a flight under instrument flight rules where an alternate landing site was not required? Stacie continued.

Hamling flipped through his three-by-five-inch note cards hunting for a clue. To memorize the minutiae, most of the students had converted the manual's information onto a stack of question-and-answer flash cards several inches thick.

"You need enough fuel to last from takeoff to destination, plus ten percent extra, or twenty minutes' worth at the plane's maximum endurance," Hamling said haltingly, realizing he didn't know the answer without prompting from his flash cards.

Stacie already knew the answer. In fact, she already knew practically all the answers. The only Coast Guard officer in the class, she had become its unofficial den mother. That was because compared to the others, she was a veteran pilot. Stacie Fain not only knew how to fly, she had taught it. A native of Fairfax, Virginia, Stacie had attended Virginia Polytechnic Institute and State University, where she had majored in sports management.

She had always been a talented athlete: soccer, springboard diving, swimming, running, volleyball. But when she graduated, she wanted to see the world, so she worked first for Eastern Air Lines and then Pan Am as a flight attendant. After several years on the job, Stacie realized her brain was going to mush. Saying thank you for the one millionth time for picking up someone's trash. Stacie knew she was smarter than most of the passengers she was feeding peanuts.

In 1989, she had taken a demonstration ride in a tiny prop plane at an airstrip in Manassas, Virginia. The instructor let her hold the controls. She felt so at home in the sky with the stick in her hand. Maybe it was in the blood. Her grandmother had desperately wanted to be an airline pilot but women had been banned from commercial cockpits. When the plane landed, Stacie decided she wanted to be an aviator.

She quit the airlines. Over the next three years, she earned her pilot's license, her commercial license, then qualified as a flight instructor. She worked odd jobs during the day and logged 350 hours in the air.

But money was running out for more lessons and Stacie couldn't find a full-time job as a pilot. An old roommate now in the Navy sent her a letter from the Gulf War. Even the Coast Guard was deployed in the Gulf and they were short of pilots for the fleet of prop planes and helicopters the service operated, he wrote her. On a whim, Stacie paid a visit to the local Coast Guard recruiter and signed up for Coast Guard Officer Candidate School.

OCS, which was held at Yorktown, Virginia, was a culture shock for her. The Coast Guard, which worked closely with the Pentagon, was light-years ahead of the regular military in accepting women into its ranks. Still, for a twenty-eight-year-old former flight attendant whose only stress up to that point was serving passengers coffee or tea quickly, military life with instructors screaming at her was overwhelming. After the first twenty-four hours, Stacie thought she had made the biggest mistake of her life. It wasn't until her fifth week in OCS that she felt adjusted to the intense regimen.

She also discovered quickly that the military could be cruel with a person's dreams. Her Navy friend and the recruiter had assured her that the Coast Guard wanted pilots. But when she graduated from OCS, the Coast Guard wanted her in the Ohio Valley pushing paperwork for six river tenders the service operated.

Stacie was devastated. But she believed in God. She believed deeply. Maybe God had decided it wasn't her time yet. She decided to make the best of the dreary assignment at the shore station. Work hard, she decided. Impress your superiors. Fly on the weekends. And the first chance she got, she applied for flight school and won one of the slots the Coast Guard had reserved for its officers going through Pensacola training.

Stacie read out loud another problem from the questions in the back of the FRR workbook: "You're flying on instruments only to the naval air station at New Orleans. The flight takes about four hours and the closest available alternate runway is thirty minutes from the New Orleans airport. Your plane consumes 230 pounds of fuel per hour. The best fuel efficiency the plane gets is 165 pounds per hour when it's flown at 10,000 feet above mean sea level. The weather to New Orleans is forecast to be 1,000 feet overcast with three miles visibility one hour before you get to New Orleans, then the weather improves to broken clouds at 4,000 feet and five miles visibility as you land. What's the minimum fuel you're required to take off with?"

Hamling sat by himself at the corner of the long black couch, feeling overwhelmed by the blizzard of rules and numbers. He blocked out the chatter, the music from MTV, and concentrated only on the flash cards he had brought with him to the study session. He would master the flash cards first, then deal with byzantine problems like how to get to New Orleans on a cloudy day and not run out of fuel.

Brian was soft-spoken and shy. He wanted to fly the Navy's P-3 Orion, a propeller-driven aircraft crammed with radars and acoustic sensors whose twelve-man crew hunted for Russian submarines during the Cold War. By the mid-1990s, the P-3s tracked Iranian and Chinese subs as well or chased drug traffickers in cigarette boats and gunrunners in sea freighters trying to break United Nations embargos. P-3 aviators were the Nintendo warriors of the Navy, patiently staying aloft in their slow-moving planes for hours and hours, hunched over radar screens or reading yards of computer printouts from acoustical sensors searching for the enemy at sea.

The students who wanted to be jet pilots thought the P-3 enthusiasts were nerds. Hamling had just as low an opinion of the jet jockeys. All that macho playacting. They could be loud-mouthed and obnoxious. There was nothing glamorous about being cooped up in an aircraft carrier for half a year snapping towels with the guys. He was a newlywed. A P-3 assignment meant you landed at a Naval air station when the mission ended and saw your wife and kids at night.

Brian could be as intense and aggressive as any fighter pilot. He just didn't have to act like a knuckle-dragging screamer to prove it. Besides, he knew that P-3 pilots were more marketable to the airlines. They wanted stable aviators comfortable with the corporate flying a jumbo jet demanded—not some lone wolf from an F/A-18 cockpit.

Brian finally put down his flash cards and began fielding questions from Stacie. The flight students acted as if their lives were riding on this test. The competition to become a pilot had begun even before they arrived at Pensacola. Almost ten times as many college students applied for flight school as the Navy had positions available. It meant only students with the best grades in college got to fly. The students knew the odds. Some switched majors in college to find courses that gave out better grades even if they weren't interested in the new subjects.

But at least in college you could have a bad day, do poorly on a test, and it would hardly affect the grade point average you accumulated over four years. At Pensacola there were far fewer tests and even less time to recover from a poor grade. Every mistake on an exam hurt. Throughout flight training a student might be allowed three "downs"—failure on a particular test or flight that the student would be given a second chance to pass. After three downs a student could be washed out of the program. The number of downs allowed sometimes varied from year to year. But whatever the number was, a student didn't want to waste a down during preflight indoctrination. The training became far more demanding later, when a student would inevitably have a bad day in the plane and receive a down.

The pressure was a rude awakening for Bill Perkins. He sat

across from Mary Margaret, who peppered him with FRR questions. High school in South Windsor, Connecticut, had always been so easy for him. He rarely cracked a book at home. By his sophomore year he had scored 1490 on his SAT test. He entered the Marine Maritime Academy in Maine mostly because its classes were not academically demanding.

The academy was basically a trade school for kids from blue-collar families who wanted a better life, he realized quickly. But it allowed him to pursue his own interests. Being a Merchant Marine would mean plenty of time at sea for reading and thinking. Perkins was captivated by Latin. Dissecting foreign languages fascinated him. His senior year in high school he translated the twelve books of Virgil's *Aeneid.* Science fiction was his other love. He owned more than 300 sci-fi books, some of which he had read five times over. Perkins found that if he paid close attention in his nautical science classes (his major), he had no trouble instantly memorizing what the instructors said. He needed no more than fifteen minutes of study the night before each exam. The rest of his time he could spend reading science fiction. His favorites were Stephen Donaldson's two trilogies, *The Chronicles of Thomas Covenant.* His biggest thrill was once meeting novelist Stephen King.

Bill Perkins had been educated on his own terms, not based on what some school had insisted on teaching him. When he graduated from the Maritime Academy he decided he wanted to be not a seaman but a Navy pilot. Flying, he decided, would be a totally new experience, fascinating like the sci-fi books he couldn't put down. But to become a pilot, he would have to learn on the Navy's terms. Anyone could eventually become a Naval aviator. But the Navy didn't have the budget to waste time on training, so the instruction was compressed. What Perkins spent a semester at the Maritime Academy learning now had to be learned in a week.

Perkins was more nervous about this test than he had been for any he had taken at the academy. He studied for hours and hours. He had to. The numbers worked against him if he didn't.

It was the numbers that made the competition so intense. Practically every student in his class wanted to fly jets. But they all had seen the numbers. The Navy and Marines needed almost twice as many helicopter pilots as jet pilots this year. That meant most students would be disappointed. Only the ones with the best grades got their first choice of plane to fly.

The Navy had a complicated grading system. Students were scored not on how well they did, but rather on how well they did compared to other students. The numerical scores a student made on each test were totaled up, divided by 1,000, then plotted on a bell curve against the scores of the last 300 students who took the tests in order to arrive at what was called a "Navy Standard Score."

In effect, it was a floating SAT score. A student might score high on his tests, but if the 300 students before him had scored higher, his Navy Standard Score would be below average and he wouldn't fly the plane he wanted.

It meant that a student not only worried about his grades, he watched everyone else's as well. Each standard score was computed down to a thousandth of a point. Some students had even faked illnesses in order to delay their entry into flight training because they knew the 300 immediately before them had aced their tests.

That was why "gouge" became so important. It was a slang word that originated at the Naval Academy and permeated the service. Gouge was the inside scoop. Gouge was tips on what to expect next in the training, or a study aid to help a student get through the next test. Gouge could be anything from what instructors to avoid to what questions to expect on the test.

At Pensacola gouge had become a cottage industry. Students passed around old flight school tests instructors had given in the past. Or for $24.34, they could buy a two-inch-thick binder of past flight school tests from the Firehouse Print Shoppe in downtown Pensacola. The Marines and "Coasties" (the nickname for Coast Guard students) were reputed to have good gouge on each class's test. Rumor had it that the Marines had inside information on the questions for the next day's FRR test, but weren't sharing it with the Navy students. No one knew if it was true, and the Marines weren't talking. Mary Margaret and Stacie had questions from four old FRR tests that they now used to quiz the others in their group. They were compiled from questions past students remembered from the FRR tests they took. Italian students who attended flight training on a foreign exchange program were legendary for the most organized system of collecting gouge. Each memorized different sets of questions from the test, wrote them down afterward, combined the questions, then passed them on to the next group of Italians going through the school.

Gouge had its own rules of warfare. Because the competition

among them was so intense, the Navy students often didn't share their best gouge. Why help everyone else to score higher on the test, lowering the value of your grade? The Marine students were closer-knit, sharing gouge with fellow leathernecks but rarely with outsiders. So few in number and not competing with the Navy or Marine students, the Coasties were more magnanimous with their gouge.

There was a saying students also learned quickly: "Live by the gouge, die by the gouge." There were so many bootlegged copies of tests floating around with hundreds of sample questions for each subject, a student did just as well memorizing the flight manuals as the test questions. Questions in some of the bootlegged tests were outdated and the regulations had changed; memorize them and you might be learning the wrong answers. The instructors also tried to stay a step ahead of students. Each exam's test questions were changed irregularly to throw them off. And for each exam, three tests were prepared, with one of them picked randomly to be given to the students that day.

Gouge had an important function. Modern combat jets were simply too complex for a pilot to know every part in the aircraft or master every aspect of their operation. There was too much to learn in flight school. Compartmentalization was a critical psychological skill for a pilot. Gouge offered shortcuts. Gouge helped an aviator find out what he really needed to know for the task at hand. It helped him weed out the extraneous and focus on the truly important—what would help him win in combat and stay alive.

But the line between good gouge and cheating could be fuzzy. It was crossed in 1992, when 133 Naval Academy students were expelled or disciplined for passing around a bootlegged electrical engineering exam that turned out to be the exam they would take the next morning. After the "double E scandal," the academy tightened up regulations on gouge that was acceptable.

Dan Smellick had crumpled in the pocket of his shorts what he thought was gouge from heaven. Smellick's father had been a Naval flight officer in a P-3 plane. Many of the flight students were Navy brats. Smellick had wanted to fly for as long as he could remember. His room at home had been filled with model planes and books on aviation. He'd never forget the aviators his father would have over to the house. They were such an easygoing group, so close-knit, always joking and kidding one another. An air crew was like a brotherhood and there was nothing he wanted more than to be a part of it.

Smellick leaned over Stacie's coffee table and began scribbling on a sheet of notebook paper. One of the hardest things the students had to memorize in the FRR class was a one-page chart from the manual that set out the traveling rules for the flight visibility and cloud clearances of different classes of airspace. For example, in Class C airspace (the controlled airspace surrounding airports that have control towers) a pilot flying under visual flight rules must be able to see out to three miles and the cloud ceiling must be no lower than 500 feet. For Class E airspace (the skies around an airport with no control tower) the visibility must be five miles with a minimum cloud ceiling of 1,000 feet. All told, the chart had more than fifty items Smellick and the other students had to memorize.

Several hours earlier, Smellick had been sitting with a Coastie at happy hour in the Pensacola officers' club. Over free pizza and twenty-five-cent-a-mug beer, she had drawn on a cocktail napkin an easy way to memorize the chart: plot the information on a triangle with another upside-down triangle inside it. Within the four triangles subdivided from the original, have letters for each airspace class written in along with the number designations for visibility and cloud ceilings, then draw arrows running every direction to signify the relationships.

It looked like something archaeologists would find carved on the wall of an Egyptian pharaoh's tomb. Mary Margaret thought it was no easier to memorize than the chart. But the Coastie had assured Smellick: burn these triangles into your brain and you'll never forget an altitude, cloud ceiling, and airspace regulation again. That's what Smellick needed. Once he had the picture in his mind the numbers would stay there.

Brian Hamling stared at the triangles for half an hour. It worked! He had the chart memorized. Hamling smiled and relaxed for the first time that night.

Michael Ott sat munching a Big Mac and french fries in the cluttered office he shared with four other aviation preflight instructors. Ott taught the Flight Rules and Regulations course and, for the past half hour, one student after another had been interrupting his lunch with last-minute questions about the noon exam, pumping him for gouge.

He wasn't parting with much. They were acting like college freshmen before their first final, Ott thought. The FRR exam actually was the easiest of the five major tests the students took. It was rote learning. The aerodynamics test was far more challenging. The students were working themselves into a lather over FRR simply because it was their first exam and they didn't know what to expect with the test's fifty multiple-choice questions.

A Marine captain, Ott didn't take the Corps as seriously as many jarheads. He liked to begin his course with "Murphy's Laws of Warfare." He had thirty-one of them. "Never share a foxhole with anyone braver than you." "Never forget that your equipment was made by the lowest bidder." "The only thing more accurate than incoming enemy fire is incoming friendly fire." Ott had an unlikely background for a warrior. The son of Quaker parents, he had been an English major at the Naval Academy. His father still called the Marine training base at Quantico, Virginia, "the trained killer school." Ott was hardly the bloodthirsty type. He flew the CH-46E Sea Knight helicopter, a transport chopper that ferried Marines and their equipment to the battlefield.

All the preflight instructors at Pensacola were either helicopter or P-3 pilots. Jet jocks considered it beneath their skills to teach students at this stage of their training. Ott and the other instructors chuckled over this obsession the students had with jet fighters. Some were now even wearing patches for F-14 Tomcat squadrons on their flight suits. Kids. They had no concept of what being a pilot was really like. They thought it was all *Top Gun*. The fact was, more than half of them would fly helicopters. And even if it was their second choice, most of them would end up loving the aircraft they flew.

Doug Spencer walked sheepishly into the instructors' office. The Navy ensign couldn't come up with the same answer the manual gave for a study question on minimum fuel requirements for low-visibility flights a pilot made, where he had to rely only on his navigation instruments.

Spencer never left anything to chance. He was five foot six and one half inches tall. He had to be at least five foot six to meet the

Navy requirement for flying in the cockpit. Just to begin training, a student had to pass a rigorous flight physical given by the Naval Aerospace and Operational Medical Institute, whose acronym was NAMI. The slightest ailment or bodily imperfection could disqualify a candidate: allergies, migraine headaches, poor dental work, height. Officers called a disqualification the "NAMI whammy."

Even a candidate who was tall enough might be rejected if his body didn't have the right proportions. For example, the distance from his buttocks to his knees could not be too long or his legs would be chopped off by the front rim of a jet cockpit if he had to eject. Spencer had fretted that he might lose the critical half inch through a measuring error by NAMI's hospital corpsmen. He tried to stretch himself the week before the exam. A doctor had told him that people were taller in the morning than at night after the bones had settled, so he arranged for an early morning physical. He passed with his half inch to spare.

Ott worked through the problem with Spencer. The ensign thanked him and hurried back to an empty class for some last-minute cramming.

Other students wandered in to grill Ott about how questions might be phrased on the exam. Semantics could be critical. The test was written so that at least one or two answers among the four under each multiple-choice question were close to being correct. That weeded out the guessers.

Five minutes later, Spencer was back in Ott's office, more nervous than the first time. He couldn't understand the minimum requirements for landing a plane if you were a "special instrument rated pilot."

"It depends on your type of approach," Ott said calmly.

There was another practical reason for doing well on the test. The next phase of instruction was primary flight training when students began flying the T-34C prop plane. That training was conducted at Whiting Field (a short drive from the Pensacola Naval Air Station) or at the Corpus Christi Naval Air Station in Texas. Once settled in Pensacola, most students didn't want to move again so soon to Texas. The gouge also was that a student had a better chance of being assigned to jet fighters if he took his primary flight training at Whiting. It wasn't true, but the Navy rarely was able to fill the class rolls at Corpus Christi with volunteers. So it ordered the students with the lowest grades in preflight to relocate there.

In the grand scheme of things, doing poorly on the FRR test wasn't going to sink an aviation career. But the students had convinced themselves that this was a rite of passage. Ott was convinced they knew the regulations. They were just feeling intimidated by the test.

Spencer walked in for the third time. "Will there be extra pencils in the class when we take the test?" he asked.

"Yes," Ott said with a smile.

Three P.M. The students paced nervously near a bulletin board where the FRR grades would be posted. They all had grim looks on their faces. They realized that everyone had done well on the test. But that was the problem. Everyone had done well, so the average score would be high. Anything less than a 96—or two questions missed on the fifty-question exam—might be below average. Stacie worried she had missed more than two questions. She was angry with herself. She knew all the regulations. She'd taught them. The test must have psyched her out, she thought.

The grades were finally tacked to the bulletin board. No names were listed—only the students' Social Security numbers beside the grade so no one would be embarrassed.

It was not good news. The average score was calculated at 95. Three questions missed and you were below average. Stacie and Spencer had scored 92s. They were angry with themselves. Smellick was ecstatic. He had aced it with a 100. That triangle diagram had worked. One student in the class had scored 80, another 86. A notation was made beside each of the two grades: "see instructor." A gentleman's C, even a B, meant failure.

CHAPTER FOUR

Bimbos

REED DUNNE always approached this class as if he were dismantling a sea mine. For the past year, the Navy lieutenant had been the Pensacola training command's human resources officer. With the dry title came the job of giving the student pilots the sexual harassment lecture.

In the years since the Tailhook scandal, the Navy had undergone profound cultural changes. The service in the past had been the most impervious of the three to civilian influence. The Air Force airmen spend most of their professional careers posted at military bases in the United States, where they mingle freely with the civilian world and spend most nights at home with families. A sailor could spend one third of his life at sea, away from civilizing influences. Until the 1980s, squadrons on port calls would routinely rent hotel rooms for boozing and prostitutes. Adultery in the past was endemic. "What goes on at sea, stays at sea" was still the unofficial motto of Naval officers and sailors. In a few squadron ready rooms hung large plaques engraved with the names of pilots and dates. Outsiders were told they commemorated daring missions. They actually marked when pilots acted wild or scored sexually in a foreign port.

The Vietnam War had exacerbated the hedonism among Navy pilots. The aviator's motto then: "I don't want to die horny and sober." The ensigns and lieutenants of Vietnam became the commanders and captains of the 1980s and early 1990s, who allowed a social pathology to fester in the service. The entire military had long been guilty of sexism, but the Navy was its worst offender. A 1991 survey found that more than half the women in the Navy who were polled complained of some type of sexual harassment.

But in the last five years the service had taken what for it was giant leaps in attacking institutional sexism. For the first time, the Navy's senior leaders decreed that even de facto sexism would no longer be tolerated. Enlisted men and officers received stern lectures on sexual harassment. There were still abuses. By 1996, the number of Navy women reporting sexual harassment in surveys was still unacceptably high, but the number had fallen dramatically from the nightmare years of the past.

Sex now had the Navy wound up tighter than a steel coil. Dozens of new regulations had been issued governing practically every form of interaction males and females in the service might have. Every officer and enlisted man was under a spotlight. Classes on sexual harassment were mandatory and each time a sailor moved to a new duty station he was required to take a refresher course. Some measures were comic. Color-coded brochures were passed out with instructions on how men should act around women in the workplace. A polite compliment was in the "green zone" and acceptable. "Red zone" behavior included promising promotions for sex. The slightest infraction brought a convoy of admirals down on your back, Dunne knew. Sailors watched their language. Officers dutifully repeated the new party line that women were accepted as equals. If they had other opinions, they kept it to themselves.

Yet prejudice and bitter resentment toward women in the service still simmered below the surface. When Dunne gave his refresher class to senior officers and enlisted men, the session would inevitably erupt into bitter debates about the new regulations.

Happily he found many of the young student pilots far more relaxed about the subject. They were from a different generation. They were raised in a sexually integrated environment. Most of their mothers had careers. Girls were treated equally in high school and college classes. Few universities were segregated sexually. There was less emphasis on dating and getting laid. Male and female students hung out together in groups rather than always being paired in dates or going steady. Girls and boys often as not were best friends rather than sex partners. The male students were not as hung-up about women flying in combat planes as the old men in the Navy. They were freshmen in college when the Tailhook scandal broke. They couldn't understand all the fuss the Navy was making about sexual harassment. It was like a parent overreacting.

Still Dunne knew he had to be careful about what he said in this class. Not all the male students were open-minded. Many of the Marine students resented women in their ranks. Some of the young men were from the South, where attitudes about women in combat were still hard-line in opposition. If asked, they would say that women should be able to fly combat jets if they were qualified. But privately they thought that no woman was qualified to fly jets.

They had all heard the rumors spread by male pilots in the force that women were being given breaks to get them through the training. It was true in one respect. What the males never mentioned was that everyone who went through the training got breaks. Few pilots would make it through the rigorous instruction without them. It might be gouge, or extra help in a qualifying flight, or more downs before a student was washed out. How many special favors a pilot—male or female—received depended on how much the Navy needed aviators at any moment and on how much extra time and effort the service was willing to spend to train them.

Some of the men in the class already were on edge that week about the subject of women in the military. The ABC television network had aired a docudrama on Lieutenant Paula Coughlin's sexual assault at the Tailhook convention. Coughlin, a tough-talking helicopter pilot, had been the first female officer to go public with her attack. Some eighty male officers, including thirty admirals, were eventually disciplined in the scandal. Many officers at Pensacola had watched the movie, but few talked about it the next day. It reopened painful memories they preferred to keep buried. Coughlin had been driven out of the service by hostile aviators who blamed her for the scandal. An underground newspaper, *The Gauntlet*, was even circulated among aviators accusing her of being a temptress and malcontent.

Several male students had gathered outside a classroom the day after the show ran. A few began making jokes about Tailhook. Others complained that Coughlin was "a bitch" who cashed in on the scandal with a lawsuit. They quickly halted the conversation, realizing that even in a private group of males-only they were treading too far into a subject that could get them in trouble.

Dunne wanted the lecture to make the students think, but not be so provocative that he'd have them squabbling. He passed out a sheet that antiseptically summarized the Navy's regulations on sexual harassment. Punishment now could be quick and severe for flagrant infractions, such as grabbing or fondling, or offering rewards for sexual favors. The Navy had declared "zero tolerance for sexual harassment," which meant offenders could be out of the service within a week in clear-cut cases.

Comical though they were, the service's color-coded zones of behavior had already become ingrained. Sailors aboard ships could now be found suddenly shouting "yellow zone" to interrupt a conversation moving in a sensitive direction. Yellow stood for questionable behavior depending on the situation. Male and female shipmates now quickly established where the yellow zone lay in workplace relationships.

Still, the yellow zone was fraught with uncertainty and peril. The Navy was feeling its way with men and women serving together as equals. Dunne always began his class with a preamble: "As we talk about the regulations, some of you may have a visceral reaction. Maybe you don't agree with them or would be reluctant to say anything for fear of offending me. Hey, I didn't write them. I'm here to tell you the way it is. If you want to take exception to anything that's going on, by all means do so." But air it in the classroom where the debate could be painless, he urged, "so you don't go out there in the real world and make a blunder."

Dunne began with the first danger area in the yellow zone: "sexually explicit or sexually offensive language. Who decides if it's sexually offensive?"

The students debated whether it was the person being offended or some higher authority. Or simply common sense.

"Let me pose a hypothetical to you," Dunne continued. "Let's say you're all doctors. These three rows specialize in proctology. These three rows specialize in urology. And these two rows specialize in gynecology."

The students all laughed.

A Christian fundamentalist walks into the class several minutes late and is shocked by what he hears being spoken in mixed company. He complains to you because you're the senior officer in the group.

"What are you going to do?" Dunne asked.

The students couldn't agree on an answer. Talk to the person who complained, explain the context of the conversation and the fact that these were professionals, said one student.

What if the person who made the complaint stands fast, Dunne prodded. "This is what my Bible says. This is how I'm going to interpret it. Therefore I'm staying true to the maker rather than true to any of your arguments. And I'm not going to yield on this issue. How are you going to handle that?"

"He had the opportunity to leave the room, so you ask him to leave," answered Alwin Wessner, an ensign sitting in the back of the classroom.

"What's the natural fallout of that going to be?" Dunne asked back. "I came in here and voiced an objection and get thrown out? I'll probably start rattling some cages. It may go to the skipper. I may get on the 1-800 Dan Rather hotline."

Wessner didn't have an answer.

"The thing you need to be aware of," Dunne explained, "is if somebody goes so far out of his way to voice an objection—which is a pretty bold move—they've probably got pretty strong feelings on this. If they do, the prudent course of action is to acknowledge them. Say I'm sorry. No harm intended." Then refer the case to the commanding officer, who will be the one who takes the heat anyway if the fundamentalist "goes out and calls Dan Rather."

So handle any sexual complaint gingerly, Dunne warned. "It's a bomb waiting to go off in your face."

Next case: fraternization. The Navy tries to stay out of the dating game. Male and female sailors can ask one another out. Only when an ardent suitor won't take no for an answer and the other party complains, will the service intervene, said Dunne. Socializing between officers and enlisted sailors was strictly forbidden. But it happened, sometimes inadvertently, and preventing it in a mixed-gender service can prove tricky, he conceded.

"I'm at a club and some guy in civilian clothes with a short haircut comes up and asks me to dance," Mary Margaret Kenyon explained. Is he an enlisted man or officer?

"Ask to see his ID," Dunne said jokingly.

The students laughed.

"What about when we go ask a female to dance?" Ross Niswanger, another Navy ensign, interrupted. Enlisted females "don't have to cut their hair."

But with ten times more men than women in the Navy, "the chances of you meeting an enlisted female are a lot smaller than my chances of meeting an enlisted male," Mary Margaret retorted.

The mixed-gender service had created other anomalies. An officer couldn't marry an enlisted person. But an officer could be married to an enlisted person. Both a husband and wife might begin as enlisted sailors, but one later is commissioned as an officer. To guard against nepotism, the Navy would make sure they worked in different jobs. And when they were in uniform, the Navy required that the enlisted spouse salute the spouse who is an officer.

Another tricky subject: "displaying posters or calendars of nude or partially clad individuals." *Playboy* centerfolds, *Sports Illustrated* calendar girls were now clearly out. So were the Chippendales.

What if all the females on a ship want to put up Chippendale posters and all the males want to put up *Sports Illustrated* posters, the students asked? Can they go up by mutual consent?

"It's still inappropriate," Dunne answered, even in the berthing spaces. A visitor might still walk in who would be offended. "How much rent are you paying for your office space in the Navy? Zero. What right of ownership do you have to that space? Nothing." It's the Navy's property and the service has now ruled that sexy posters can't hang on its walls.

What if someone complains about a picture sitting on your office desk of your wife in a bikini? Dunne asked.

The students laughed, then argued about wives' pictures.

"Just put the picture in your desk until he leaves, then put it back out," Mary Margaret said, sounding a bit frustrated. She had come to hate these sessions. Just when you had succeeded in blending in with a unit, you had to go through a sex harassment class that made you stand out again.

"Okay, that's one option," Dunne said noncommittally.

Somebody could find something offensive with everything, other students complained. This could get out of hand.

"You have to go with the letter of the law," insisted Wessner, who now seemed to have become the class lawyer. "If you can justify that that's my wife in a bikini, then somebody will say, 'Well, why can't I have this girl on the calendar.' You're opening up a can of worms if you allow one husband to keep the picture in the open."

"Or better yet, what if you're married to the girl on the calendar," Dunne added.

The students howled with laughter.

But Wessner had a point.

"To give you a real-world example," Dunne continued. "I had a guy in my last squadron whose wife had won bikini contests in Hawaii. He had a lot of pictures of her on his desk winning the competitions. This wasn't an otherwise charismatic officer who everybody wanted to be around and talk to. But I guarantee you, twelve hours a day at work there were a half dozen junior officers around his desk. It did become quite a distraction for the others who had to work with him in the office. Eventually a couple people went to him and said they were tired of having all these moron lieutenants hanging around the office. Hide the pictures. We bet they'll leave. Sure enough, he hid the pictures and he was not the most popular guy in the squadron anymore."

The point is "it doesn't matter if the picture is of your wife or not. The objection is with the content and not who the person is."

Niswanger held up his hand. "I know this is the Navy policy," he said, irritated. "But is this going to continue? Does it look like things are going to relax any time in the next five years? This PC has just gotten out of hand."

"I'm not here to argue where it's going to go," Dunne answered. "I'm just here to tell you where it is."

Bret Hines and Danny Johnson nursed two plastic cups full of beer in the bar at the Pensacola officers' club. On Friday afternoons the club served free food with the beer so the two always hung out there until the buffet table closed at six o'clock. The large wood-paneled lounge had television sets on perches in its four corners, all playing videos of aircraft landing on carriers. Rock music roared from a jukebox. Stickers for the squadron patches that pilots wore on their uniforms were plastered on the mirrors behind the bar. Above, hanging on wooden pegs, were rows of white mugs with squadron logos engraved on them. Squadron plaques also hung from the walls. A machine in the corner belched out popcorn for the drinkers, some of whom played dice games.

A retired aviator well on his way to getting drunk for the night sat with Hines and Johnson, keeping their cups filled with the pitcher of beer he cradled. He made several slurred attempts to flirt with the female student aviators who walked by, but they all ignored him.

Bret Hines and Danny Johnson had been the best of friends since their first year at the University of Virginia. They were both Virginians—Johnson from a small town off the Chesapeake Bay called Parksley, Hines from Richmond. Johnson was like his father, a former Marine sergeant. Full of nervous energy, he could never sit in front of a television set for longer than an hour. Hines still had the cultured Southern accent Richmonders were famous for. He also had freckles and bright red hair. The female students at preflight indoctrination thought he was cute. They had both joined the Marine Corps after college and had trained in the same platoon during the Basic School for officers. They had kept up a friendly competition throughout college and their military training.

Sometimes the competition scared Johnson. He and Hines had gotten this far because so many things had fallen in place. They had both been standouts in high school, both had been awarded full ROTC scholarships in college, both made As and Bs. Now they both had scored high enough to win aviation slots at Pensacola.

But it could all fall apart with just one slipup. At the Marine Basic School the competition was so extreme during the six months of training that just one bad day could kill a recruit's class rank. Hines had done poorly on the obstacle course one week and Johnson had scored low on a combat orders test; they both fell out of the top ten percent of their class as a result. The point spread between the trainee who was number one in the class and the trainee who was number fifty was just two percentage points. Now you're just waiting for the day when something else goes wrong, Johnson worried, some misstep that ruins your chances of flying.

By 5:30, the bar was crowded with aviators, instructors, and flight students. Then the first of the groupies began to trickle in. They looked like beauty pageant contestants with heavy makeup, bleached-blond hair, brown roots, and vapid stares. Johnson marveled at how they always seemed to come in packs of three. The female student pilots detested them. They called them the "Pensacola bimbos." Their purpose in life—or so it seemed to the female student pilots—was to bed down male aviators. When they walked in, the female pilots would begin giggling and fiddling with their hair to imitate them.

But groupies have hung around Navy pilots for decades. On Friday nights they filled up the officers' clubs at Naval air stations. Until the scandal, the Tailhook convention had become a magnet for bimbos from all over the country, even from abroad, looking for their Tom Cruise. College coeds, party girls, prostitutes, they would drive pickup trucks into the parking lot of the Las Vegas Hilton during the convention, strip off their clothes, wiggle into hot pants and leotards, then march up to the hotel's third floor, where the raunchy squadron parties were in full swing.

They would lean out of hotel windows. "Show us your tits, show us your tits," the aviators would chant from below. The women would oblige or drop their pants and moon them. Many willingly strutted down the Third Deck's gauntlet, the same one where other females had been sexually assaulted, so they could be fondled by the aviators on each side of the hallway or have "zaps" (stickers with squadron logos) plastered on their naked bottoms. It was called "zapping."

The bimbos had their own games, like "package check." After a plane took off from a carrier it would sortie near the aerial refueling tanker to make sure the tanker's hose would pass fuel properly when the jet needed to fill up later. The procedure was called a package check. At Tailhook, a pilot would walk down the gauntlet with a zap on his crotch. A bimbo would yell "Package check!" and grab it.

The bimbos inflated the pilots' egos. They fed the chauvinism that almost brought down the aviation force after Tailhook. But the bimbos remained. In Pensacola, they swarmed around the young men learning to fly.

Stacie Fain found them disgusting. They came in three varieties: single party girls with low-skill jobs, divorcées (always, it seemed, with two kids), or students from the nearby University of West Florida with dreams of finding a husband among the officers. Whatever, their brains had long been fried by peroxide and the sun. The running joke among the female flight students: Pensacola sold enough hair dye to float a battleship. At parties, Stacie always made a point of first letting strangers know she wasn't one of the locals.

But for some of the male students it was like being a kid in a candy store. Mary Margaret had seen the bimbos at the local clubs sometimes walk up to the freshly scrubbed ensigns and plant a French kiss on them or grab their crotches. Snaring a pilot was a blood sport. Some of the male students tried to spend practically every night with a different bimbo. They'd slow down when their grades nose-dived. The Naval Academy graduates often were the worst, Stacie found. After four years of regimentation, curfews, and few women, they stormed Pensacola eager to make up for lost time.

The bimbo chasers among the student pilots were fewer than in years past, however. The air station was undergoing a generational shift. The Tailhook era of older pilots who drove Corvettes, busted up bars, and screwed any female in sight was fading. One lieutenant commander on base still made it his life's calling to try to talk every female flight student into the sack. But he was more an icon to a lifestyle the Navy now frowned upon and many of the young ensigns found outdated. The most popular "ensignmobile" now was a Jeep convertible. Some of the male students had already settled down with wives. Others had girlfriends or fiancées back home. They would talk a good game about the Pensacola nightlife, but most remained faithful. A few were deeply religious. The bimbo chasers in their class might be sexual heroes when they later joined macho squadrons, but for now they were called "hosebags" by the female students. Even the men called them "the male sluts."

That kind of lifestyle could now be dangerous as well. The gouge being circulated in the class claimed on good authority that Pensacola had one of the highest rates of AIDS cases in the country.

The rumor was somewhat off the mark. Florida did rank third in the country in adult cases, but Pensacola's rate wasn't much higher than the national average. Still the gouge had chastened the chasers.

The older pilots considered Pensacola their spiritual womb, the cradle of Naval aviation where you were supposed to think you'd died and gone to heaven. Hines and most of the young students thought it was a cultural landfill. Pensacola actually was a city with colliding cultures. It consisted of laid-back Gulf Coast Floridians, born-again Christian fundamentalists, Deep South rednecks spilling over from Alabama and Mississippi, retirees who migrated from the North, out-of-town tourists who visited year-round, and the Navy families who came and went every two years. The students had nicknamed it "Gumpville."

Hines and Johnson began their cruising for the night at Señor Frog's, a seedy cantina in downtown Pensacola. They had long given up hope of meeting girls like the ones they dated at the University of Virginia. The UVA women partied but they were smart. You could talk to them. There were no rocket scientists in Pensacola. You weren't going to meet the girl of your dreams in the bars, Hines quickly realized. Certainly not the girl you'd bring home to mother.

Hines and Johnson had decided to be creative and had begun running in long-distance races on Saturday morning, hoping to meet nice girls who were into physical fitness. Maybe from the University of West Florida. So far no luck. Even if he found one, what the hell would he do, Hines wondered. No girl in her right mind should get serious with a student pilot. He would be moving from base to base for the next two years, preoccupied with flying. What kind of meaningful relationship could anybody have under those conditions?

The drinks were cheap at Señor Frog's. The waitresses were sassy and some of the bimbos who wandered through the tables sported bikini tops and tight-fitting jeans shorts. Hines and Johnson decided to stay for an hour and just enjoy the view. Johnson nursed only one beer. He had agreed to be the night's designated driver. The Navy would now expel a student pilot if the police caught him driving drunk, so every time they went out one of them stayed sober. "We're the post-Tailhook generation," Johnson would say with a laugh. He couldn't believe the scandal ever happened, that Navy pilots could act that way. The service was so puritanical now. Get into any kind of trouble off base—a drunken brawl, the cops raid your party—and you could be kicked out of the program. It almost made him scared to be a twenty-two-year-old. They thought they would have a young and reckless life just out of college. Instead, they felt like seminary students.

The one Pensacola nightclub that had become almost a historic landmark for Naval aviators was Trader Jon's. It looked like a combination saloon and someone's cluttered attic. The walls were covered with framed photos of Navy air aces, admirals, generals, politicians, movie stars, European royalty, all of whom had sipped beer at Trader's. Low-hanging ceiling fans pushed around air that smelled of stale alcohol and snuffed-out cigarettes. From the high-raftered ceiling hung squadron flags, Naval Academy class banners, and more than a hundred model planes. One half of the saloon housed threadbare pool tables and a dance area with a country-western band. Young pilots often rented it out for "wingings"— parties to celebrate their officially becoming aviators and being allowed to wear the gold wings on their uniform. The other half of the establishment had a horseshoe-shaped bar where tanned young women in halter tops poured beer. Beside the bar stood a transplanted signpost for the corner of Billy Mitchell Boulevard and Supersonic Avenue, along with a beat-up barber's chair. Before the Tailhook scandal, the bar also displayed the original papier-mâché statue of the rhinoceros, built by a Marine squadron, which could dispense drinks from its penis.

Trader Jon's has been called the Smithsonian of Naval aviation. There was even a Trader Jon, who owned the place. His real name was Martin Weissman, a New Yorker who never served a day in his life in the Navy. But everyone called him Trader. He had been passing through Pensacola in 1937 when he fell for a girl and decided to stay. He'd long forgotten how he got into the bar business. Trader now had snow white hair—he wouldn't say how old he was but he had to be well past seventy—and he spent most nights stooped over the cash register sporting a pink Panama hat and ringing up receipts.*

---

* Weissman died in 2000 and Trader Jon's closed in 2003, its memorabilia displayed in a museum thirteen years later.

Hines and Johnson had never been to Trader's. Most of the student pilots considered it a museum for the older aviation crowd—too dark, too cluttered, too dirty, too many bimbos over forty with leathery tans. They liked the beach bars better. The most infamous was the Flora-Bama Lounge, which stood squarely on the Florida-Alabama line at the Gulf Coast shore of Perdido Key. It looked like it was built by a drunken architect. A mismatched collection of shacks, grass huts, picnic tables, and dingy white canvas tents that covered about an acre of the beach. Under one tent, shaped like a Quonset hut, a rock band blared. A bluegrass group played in another gray shack that housed a dark, fetid bar. Steps led up to a second-level bar, which extended out to a pier. Several kiosks hawked beer and tropical drinks and temporary tattoos. Flora-Bama, which didn't start rocking until midnight, packed them in by the thousands: Pensacola's pilots, Alabama good old boys, students from nearby Auburn University, bikers, underage drinkers with fake IDs, rough-talking women, college girls, along with a sprinkling of middle-aged tourists who felt out of place. Each year Flora-Bama hosted a mullet toss on the beach to see how far contestants could throw dead fish from Florida to Alabama. Parked in front of the lounge every night—on their respective sides of the state line—were Florida and Alabama cops to round up the rowdy drunks.

Hines and Johnson decided to leave Señor Frog's at 10:30 and end their night at Seville Quarter, a labyrinth of bars, discos, restaurants, and courtyards linked together by hallways that took up almost a city block. Seville Quarter was Pensacola's largest meat market, more upscale than Trader Jon's or Flora-Bama. In one courtyard surrounded by brick walls, a folk singer belted out poor imitations of the originals. Another disco looked transplanted from *Saturday Night Fever* with a lighted dance floor surrounded by balconies full of bar tables. A sports bar had pool tables and television sets, the channels turned tonight to baseball games. Bartenders with black bow ties and waitresses in khaki shorts and polo shirts hustled drinks. Aviators, student pilots, and sailors from the base—most in jeans or cutoffs—packed each bar. Sprinkled among them were OCS candidates, heads shaved and wearing their dress white uniforms to impress hometown girlfriends here for the weekend. And by the hundreds, bimbos in spandex or tight-fitting minis with breasts bulging out roamed the hallways and bars like packs of wolves. Many had already found pilots they were clinging to.

Hines and Johnson would never forget their first visit to Seville Quarter. A drunk woman walked up to Johnson and grabbed him by the arm.

"I just wanted to feel your muscles," she said breathlessly. They struck up what barely passed for a conversation. Within two minutes, she asked: "Do you want to see my tattoo."

"Okay," Johnson said warily.

She unbuckled her belt, pulled down her shorts, then her thong panties, bent over and proudly displayed a Harley-Davidson motorcycle logo painted on one of her cheeks.

"I have a friend who has two tattoos," she said, wobbly now. "Wanna see 'em?"

What the hell, Johnson thought. "Okay."

She yelled, and her friend, just as drunk, walked up and bent over. She had a moon painted on one cheek, a dolphin on the other. Johnson and Hines stared at them dumbfounded. Welcome to Pensacola.

Tonight, they decided to sit in one of the open-air bars that had a reggae band.

Shortly before 1:00 A.M., two bimbos wandered up to their table and sat down. Both were in regulation uniform: short dresses that seemed painted on, hair frizzed, high heels.

One claimed she was a preschool teacher. Johnson thought it more likely that she hadn't gotten past preschool.

"What do you do?" Hines asked the other one.

"I just hang around," she said vacantly.

"No, I mean what do you do for a living?" Hines asked again. He sometimes found it hilarious talking to these women. The other night they were with Dan Smellick at Flora-Bama when he said to a bimbo in the course of the conversation, "Oh, you're being facetious."

"What does that mean?" she said, totally perplexed. Smellick, Hines, and Johnson sat silent for a second, then burst out laughing.

Hines asked for it. Now he got this woman's life story, which turned out to be pretty short. Well, it was like this, she began, shouting over the reggae music. She used to work in a clothing store. But like the manager was really creepy, you know. So like now she's unemployed and waiting for the sexual harassment money to come in. You see, she sued the manager for sexual harassment. She's also suing the bank, something to do with her last month's check that the store owner never gave her. Hines couldn't figure what the connection was between the bank and the store owner. And he didn't care.

He gulped down his beer and got up from the table. Hines and Johnson quickly excused themselves. They couldn't remember if they'd given the women their names. God, they hoped they hadn't.

CHAPTER FIVE

The Dunkers

BILL MALLORY eased himself into the dripping wet metal box. Two Navy divers standing on platforms on each side of the box bent over and quickly began buckling Mallory's parachute harnesses to the cockpit seat he now occupied. He was in full flight gear with helmet. As metal clanged against metal and gates were shut around him, the divers rattled off last-minute instructions: use your right hand to hold the metal control stick between your legs for controlling the aircraft as it crashes, left hand grabs the yellow-striped lever that would eject the canopy, push the pedals in the cockpit forward, head back, enjoy the ride.

Enjoy the ride. Right.

Mallory was strapped into the device with the most fearsome reputation in preflight indoctrination: the Dilbert Dunker. The ocean could terrorize a Navy pilot. Military aircraft that had to be ditched at sea usually sank like a stone, trapping aviators who didn't escape their cockpits in watery graves. Even if he ejected, a pilot parachuting into the ocean could face a horrifying death if he didn't disconnect quickly from his parachute before it filled with water and dragged him to the bottom.

Navy trainers had built a variety of contraptions at one of the Pensacola air station's large indoor pools to simulate the water hazards an aviator faced. A student would be hooked to a parachute suspended from the high ceiling, then dropped with the chute billowing into the pool so he could practice untangling from its lines and unbuckling from his harness before the nylon sheet filled with water and dragged him down. Also strung over the length of the pool was a steel clothesline with a parachute harness on two large lines hanging from it. A student was hooked to the harness, then a motorized pulley on the clothesline dragged him quickly across the surface of the pool to simulate the wind catching his parachute after he had landed and dragging him along the ocean.

Then there was the Dilbert Dunker. Named after the cartoon character, it simulated a plane crashing into the sea and quickly flipping end over end. It consisted of two forty-foot rails anchored at an angle onto heavy metal scaffolding with the bottom end of the rails extending about ten feet into the pool water. Near the top of the rails sat the metal box, painted red, with no top and the side walls made of a thick screen. It simulated an aircraft cockpit with the canopy jettisoned that would quickly fill with water when it crashed into the ocean. Inside the box was a cockpit seat, into which the student was strapped, along with the navigation stick, floor pedals, and canopy ejection lever he would pull back before his plane dove into the sea. Arched over the box was a thick metal crash bar padded with foam rubber and wrapped with white plastic tape. Looking like the crash bar on automobile convertibles, it represented the front rim of the cockpit canopy that remained with the plane after the rest had been jettisoned.

When a diver on the platform yanked a disconnect lever, the metal box went hurtling down the rails and splashed into the pool. Almost in the same instant that the box hit the water it flipped end over end so that at the bottom of the rails the cockpit, with the student pilot strapped inside, was upside down and underwater. The student then had to quickly unhitch himself from the seat belts, crawl out of the cockpit, and swim to the surface.

Mallory had seen the movie *An Officer and a Gentleman.* All the students had made a point of watching it before taking their dunker training. In the 1982 film, which depicts Navy flight school, the Dilbert Dunker is portrayed as a potential death trap. One of the trainees in the movie, Daniels, panics underneath the water and can't disconnect himself from the cockpit's seat belts. Seconds pass. Then the movie's hard-bitten drill instructor, played by Louis Gossett Jr., dives into the pool and rescues the trainee, who has almost drowned.

The Dilbert Dunker was far less dangerous in real life. A Navy diver was underwater at all times ready to quickly unhook any student who panicked and pull him out of the box up to the surface. Also the dunker trained the students for an emergency they likely wouldn't face. The only plane in the Navy inventory that would flip end over end after it crashed into the water was the T-34C trainer, in which the students first learned to fly. The prop plane's engine was up front, making it nose heavy and prone to flip over after hitting into the ocean. Other Navy planes just sank, or, as was the case with fast-moving, high-performance jets, practically disintegrated when they struck water. Pilots, no matter what they flew, were better off ejecting rather than trying to stay with an aircraft and ditching it at sea.

Mallory was excited about the ride. He had made sure to be the first in line for Dilbert. He quickly snapped together the seat belt buckles, then grabbed the stick and canopy ejection lever.

This was competition. Bill Mallory versus the dunker. He thrived on testing his physical stamina. He was a natural at it. That's how he'd gotten here.

Mallory was a Navy brat. His father, who eventually retired as a lieutenant commander, had dragged the family from one base to another, sometimes changing addresses once a year. The last move had been the most difficult for Mallory. He began high school in Panama City, Florida, a blond-haired, blue-eyed boy the girls found attractive, but who was shy by nature with no friends in this strange new world full of its cliques that ignored him.

But Mallory had one attribute that soon made him popular. He was a gifted athlete. He joined practically every team the school had—soccer, wrestling, football, gymnastics, basketball. He learned each sport quickly. When he eventually excelled at one sport, he usually became bored, then moved to another. He played not so much for the game but for the physical challenge. Sports became the icebreaker. Sports became the way to win friends in high school, the ticket to popularity every teenager craved.

By his senior year of high school, Mallory had an A average and somewhat of an interest in math. He was a natural for the Naval Academy, always hungry for jocks with good grades.

At Annapolis, Mallory majored in mechanical engineering. His grades were average and he found the social life depressing. The institution was still reeling from its own sexual harassment scandal—male classmates had handcuffed a female midshipman to a urinal—so he didn't dare date any of the academy women for fear of making a faux pas that would get him kicked out. But again, he thrived with sports. The academy's gyms were as modern and fully equipped as any big university's. He boxed and competed in gymnastics. He joined the wrestling team and won two tournaments in the 190-pound weight class. His senior year, he played rugby and was named to the All-America team.

Mallory had wanted to join the Navy SEALs, the most physically fit warriors in the military. But the SEAL slots were filled by academy grads with higher grades than his. So he chose aviation. F/A-18s were his dream. He arrived in Pensacola in December 1994. He bought a pickup truck to carry his mountain bicycle and a sleek, black Kawasaki ZX-11 motorcycle, one of the fastest production cycles available on the commercial market. He once powered it up to 165 miles per hour. Its engine whined like a jet and he could barely see for the wind beating against his face.

Mallory hit a bureaucratic stall when he arrived at Pensacola. The station often had students backed up waiting to begin their aviation training. So they were placed in "stash jobs," make-work assignments while they waited. Mallory's stash job the next six months was a meaningless administrative chore during the day. His more important assignment, as far as the air station was concerned, was playing on the base rugby team in the afternoon.

Mallory also overdosed on the nightlife. He was now ruggedly handsome, not a bit of fat on his muscular body. At one point, he was dating five Pensacola women at the same time. He bar-hopped and partied practically every night. But after several months, he dreaded waking up every morning with a throbbing hangover and his wallet twenty dollars lighter. He soon tired of the bimbos. No brains. Mallory returned to athletics. Not only rugby. He joined an amateur football league for local men that suited up in full pads and played tackle ball.

Mallory took one last look at the shoulder and lap straps that belted him to the dunker's cockpit seat. They all were hooked together at his stomach by a butterfly clip that could be opened quickly with one sweep of the hand. He quickly went through the hand movements in his mind that would free him from the box when it was underwater.

No sweat, he thought. Keep calm. That was the secret.

He looked up and gave the divers on the platform a thumbs-up.

An alarm bell rang behind him, the signal that he was about to ditch the plane. With his left hand, Mallory pulled back the striped lever to simulate ejecting the canopy, then reached up and grabbed the overhead crash bar to brace for the impact.

A second later, the diver jerked up the handle that released the box.

Mallory could hear metal scraping against metal.

He managed to suck in one last breath.

The box crashed into the pool water. His head whiplashed forward, then back. His left arm braced against the overhead bar felt jammed.

The jolt shocked him. In the next half second he was upside down in the metal box, the noise of it slamming into the stops at the end of the rails ringing in his ears, bubbles in the light green water streaming all around him.

Mallory knew he could hold his breath underwater for at least a minute. But that was right side up with air bubbles trapped in his nose to keep the water out. Upside down was a different story. The bubbles in his nose blew out and the chlorine water now rushed in, burning his nostrils and for a brief moment choking him.

Upside down in the water, he felt completely disoriented at first. A slight tinge of panic swept through him. His first instinct was to get the hell out, and quickly. The same feeling he once felt while scuba diving in Florida. He had swum into a small dark cave when his air hose malfunctioned. They would have fished him out and stuffed him in a body bag if he hadn't overcome that first paralyzing flash of fear and turned on the tank's safety valve behind his neck, which finally pumped oxygen into his lungs.

Mallory now willed himself to remain calm.

He counted in his head—one second, two seconds, three—the required time he was supposed to wait for the box to settle upside down in the water until he began disengaging from the seat.

He remembered what the instructors had told him before the ride and with his right hand pulled the control stick to his crotch. He let go of the stick and moved his hand up his stomach until he felt the buckle that hooked the shoulder and lap straps.

All the while, he kept his head arched back and pressed against the headrest. The Navy had found that upside down in a plane, a pilot came out of the aircraft in the direction that his head was pointed. If Mallory emerged with his head pointed down looking at the strap buckle, he would somersault forward out of the cockpit and run into the front propeller. If he kept his head back and unlatched the buckle simply by feel, he would pull out of the cockpit at the required 45 degree angle and miss the propeller.

Mallory did as he had been instructed. With his head back, he unsnapped the buckle with his right hand, then reached down to the overhead bar. With all his might he pulled himself from the cockpit using both arms.

His hands hit the bottom of the pool. But he had escaped at the right angle. He pushed up to the surface, relieved to be free of the box. The students were underwater no more than ten seconds. It seemed to most like ten minutes.

Piece of cake, Mallory thought as his head bobbed up from the water. He looked behind him. The Navy diver who had been underwater watching him swam to the surface and gave the graders on the platform a thumbs-up. Mallory had passed. It had not been as bad as he'd feared. The instructors had told him it was easy.

It was the next device that everyone hated the most. For good reason.

More than half the students in this class would likely fly helicopters. But when a Navy copter had to be ditched at sea, there was no escape for the crew. No ejection seats. No parachutes. The airmen and passengers inside went down with the aircraft. Over water it could be a horrifying experience. The crash was violent. There was always chaos in the cockpit and rear crew compartment. Equipment would be damaged. Power systems blown. Loose wires dangling from the ceiling. Debris scattered everywhere inside. People injured.

Almost immediately water came rushing in with a thundering roar. Trying to escape then was hopeless. The flood of water jammed windows and hurled crewmen back inside the aircraft. When the cockpit and rear cabin finally filled up, something happened even more disorienting for those inside. Helicopters are top-heavy creatures. The engines, transmissions, and blades are all above the fuselage. As it sank into the ocean the aircraft quickly rolled over so it was upside down. The crewmen could still escape, but they had only seconds to collect their wits and find a way out. If not, they were trapped, buried at sea.

Few survived helicopter crashes into the ocean until the Navy built a training device called the Helo Dunker. It came close to simulating such an accident, without drowning the students in the training. In order to stay qualified to fly, even veteran Naval aviators had to suffer the Helo Dunker every four years for a refresher class. They dreaded it like Chinese water torture. But studies had shown that the dunker training doubled a pilot's chances of living through a helicopter crash at sea.

Next door to the pool with the Dilbert Dunker was a second pool where the Helo Dunker had been built. Suspended under a steel girder platform over the pool was a large metal cylinder painted blue that simulated the fuselage of a helicopter. The cylinder had three windows on each side as a chopper might. In the front was the cockpit with two seats for the pilot and co-pilot, which was partitioned by a metal wall with an entranceway that led into the rear compartment. In the rear cabin were arranged six seats, three on each side, into which passengers were strapped with seat belts.

A diver manning a control booth on the platform would power motors that dropped the cylinder into the pool. As it quickly sank, filling with water, heavy steel cables attached to both ends of the cylinder would flip it upside down, simulating a helicopter inverting after it crashed. Other beefy divers were posted underwater on each side to watch the eight students in the upside-down cylinder struggle to unsnap seat belts and crawl out of the giant can. If any student panicked and couldn't break free, the divers would grab him roughly by the collar and pull him to the surface.

Each student had to make four rides in the Helo Dunker. Ross Niswanger was petrified for the first. The inside of the cylinder was painted a dull gray and smelled damp and chlorinous. The students had been told that for this ride they were to find the nearest exit and escape. Niswanger sat in the seat beside the main door to the rear cabin. Good seat, he thought to himself. Spin around quickly and you're out of this coffin.

A diver walked through the dunker barking out final instructions. The students, all of them in full flight gear with helmets, fumbled with the buckles to the seat belts draped across their laps and over their shoulders, then cinched them tight. The diver stepped out of the cylinder. "Assume the crash position!" he ordered. Knees together. Grip your seat belt straps. "Enjoy the ride."

They always said that before trying to drown you, Niswanger realized.

He began hyperventilating. Thousands of students had gone through the Helo Dunker, the instructors told them. No one's died.

The assurances made no difference to Niswanger. He was still scared witless. He grabbed the back of the seat to his left to brace for the fall. Again he quickly rehearsed in his mind which direction he would swim to escape.

It must have been the submarines that now made him so afraid of this dunker. Niswanger was one of several students in the class who had served in the Navy's enlisted ranks. In his case, it was as a nuclear mechanic for the USS *Ohio*, a sub that fired intercontinental ballistic missiles tipped with atomic warheads.

His mother had divorced and remarried. Niswanger became as close to his stepfather as his real father. He grew up thinking it was normal to have a mom and two dads. He was still immature after graduating from high school in California. School was a hoop you had to jump through. He hadn't really been interested in learning. College was postponed. He wanted to learn how to live first, so he joined the Navy at eighteen and spent almost a year in nuclear engineering schools training for submarines. It was the thought of water filling up this cylinder to the top, over his head, that now panicked him. Flooding in a compartment was a submariner's worst nightmare. If it couldn't be stopped, there was no escape. You were trapped inside. You sank to the bottom of the ocean and waited for death.

His third year in the service Niswanger had won an ROTC scholarship. His dream all along had been not to remain cooped up in a sub, but to fly for the Navy. He enrolled in Oregon State University and this time studied hard. It was his ticket to becoming a pilot. He had come to believe fervently that if he worked hard, the Navy would always reward him. He graduated with an A average from Oregon State.

But the reward didn't come this time. His paperwork for flight school had been accidentally misrouted into the bureaucratic ozone. By the time aviation personnel officers retrieved it, all the pilot slots in training school had been filled for the year. The only openings left were for Naval flight officers, which the service was short of, so Niswanger was assigned as a back-seater.

It was a crushing blow for him. As crushing as the day he was told his father had committed suicide after discovering he had been diagnosed with terminal bone cancer. Niswanger had done everything to become a pilot. He had worked hard. His test scores on the flight qualification test had been high. Good grades in college. Perfect eyesight. Everything. Now he was stuck as a Naval flight officer.

He didn't buy any of that crap from the Navy that the NFO was just as important in the cockpit as the pilot, that he was the radar officer who fired the missiles and managed the dogfight, that he was promoted through the ranks as fast as the pilot. The pilot still flew the plane and the NFO was still the passenger. The pilot was the real commander of the aircraft, the one who landed on the aircraft carrier and became the air ace. No one remembers the flight officer in the back seat, who spends the dogfight hunched over a radar screen puking his guts out. All the students knew that NFO really stood for "no fucking option."

Niswanger vowed that the first chance he got he would apply for a transfer to become a pilot. Switching seats wasn't easy but the Navy occasionally allowed it. After he had finished training as an NFO and served in a squadron for several years he would send in his paperwork to be returned again to flight school for training as a pilot. This was just one more hurdle he had to jump over. He still believed the Navy would reward him if he worked hard.

But first he had to survive this dunker. The diver at the control booth punched the button to begin the simulated crash. There was a clang of metal hitting metal. The giant cylinder splashed into the pool. Niswanger felt the cool water creep quickly up his ankles, then to his legs, his lap, his stomach.

He gulped down one last breath to hold as the waterline reached his chest. In the next half second cables attached to both ends of the cylinder flipped it upside down as it sunk into the pool. It turned to the right. Niswanger felt himself jerked to the rear, the back of his head plunging into the water. In the next instant he was submerged, turning upside down. The water rushed into his nose, clogging his sinuses. He opened his eyes but could see only blurry images obscured by a light green haze and thousands of bubbles. He could hear the faint sounds of metal banging as the cylinder settled down into the pool upside down.

He reached for his belt buckle but willed himself not to unsnap it. There was a set procedure the students had to follow as the cylinder sank. Once underwater, count eight seconds before unbuckling. The cylinder had to be completely submerged and turned over before the students could release themselves from the seats. Otherwise, in a real crash, the inrushing water would toss them about as if they were in a blender. If a student unbuckled too quickly, the divers underwater would spot it and order him to repeat the ride. Niswanger wanted to avoid that at all costs, so he waited several more seconds until the cylinder's rotation had completely stopped before disengaging his seat belt.

Once free, he had to remember his reference points. He had plotted his escape route before the ride, but that was when everything was right side up. Now that he was upside down, every move seemed to him to be in the opposite direction he had anticipated. He was confused at first. But thankfully, he recalled the instruction drilled into him before the ride: as the helicopter sinks, quickly pick out spots in the cabin you want to grab with your hands in order to escape. Rehearse in your mind the movements you should make, then forget about direction, just perform the movements.

Another rule: no swimming and kicking inside the cabin to escape. You'll strike other passengers. Pull yourself out with your hands and arms reaching for one reference point after another as if you were climbing up a rope. Keep calm. That way everyone gets out as quickly as possible without beating one another up.

Niswanger wheeled himself around with his arms. He saw the right cabin entrance in front of him, which now was on the left side. He grabbed both sides of the entrance and pulled himself out. His lungs felt like they were about to burst. But as he swam to the surface it seemed as if he was shedding the paralyzing fear that had gripped his body.

He bobbed to the surface feeling exhilarated. The fear had been conquered. Another hurdle jumped over. He would not be afraid of the next three rides.

Mary Margaret Kenyon felt anything but exhilarated. Her nostrils burned. Her sinuses ached from the water forced in them. Rubbing her forehead didn't help. Other students were giving each other high fives after every ride. Mary Margaret felt no bravado. She just wanted to be rid of the damn dunkers, thankful she would only have to suffer through them every four years.

A complication was introduced into the second ride. In a real crash, the helicopter's windows might become jammed. There might be no way out other than through the one entrance in the rear cabin. All the passengers and crew would have to keep calm and file through that one opening.

Just her luck. Mary Margaret drew seat number eight in the capsule, one of the furthest from the rear cabin door.

The capsule sank again into the pool. This time it flipped to the left. The diver manning the control constantly changed the direction the capsule flipped so the students wouldn't become used to the rotation. Mary Margaret took a deep breath before submerging. But still it didn't seem enough. When the capsule had stopped upside down she grabbed the seat to her right to pull her way to the entrance.

If any student shoved others to get out of the entrance the divers underwater would make him repeat the ride. Mary Margaret hoped no one got in her way. All she could think of was, Get out. Get out now!

She made it to the entrance. She didn't know how. There was a little bumping among the students as they clawed their way to the opening. But not enough to disqualify anyone. Mary Margaret hoisted herself up from the capsule to the water's surface. Two down. Two to go.

Stacie Fain had swum competitively in high school. She was comfortable in the water. But this felt completely foreign—being upside down underwater, strapped to a seat in this metal can. And now blindfolded.

Helicopters flew at night. If they crashed, it would be completely black inside the submerged aircraft. The passengers and crew would see nothing. They would have to feel their way out of the chopper.

For the third ride, the students wore swimmer's goggles under their helmets whose lenses had been painted black so they couldn't see outside when wearing them. The only saving grace for the third ride was that they were allowed to escape from the nearest window or exit. Stacie was assigned the co-pilot's seat. But it was still nerve-racking with her head pointed down and her eyes covered. She had taken her last breath too late before submerging and had water in her mouth.

She raced through the eight-second count, unbuckled, and quickly had her grip on the window ledge to her left for the pull-out. She hoped the divers hadn't noticed that she had moved too quickly. But she felt as if she were choking on the water in her mouth and what had rushed up her nose. Students were submerged in the Helo Dunker no more than twenty seconds but it seemed to Stacie like an eternity.

She made a hard breaststroke up to propel herself to the surface. She was desperate to get to the top.

When her head broke the surface, Stacie ripped the blackout goggles from her eyes.

"Fuck!" she said, water sputtering from her mouth. One more ride.

But the fourth one was the most difficult. Again it would simulate a night crash. The students would have to wear the blackout goggles once more. But this time the windows would be jammed shut. They would all again have to make their way to just one entrance, but this time blind.

Twenty-six-year-old Alwin Wessner didn't care how difficult they made this last ride. They could blindfold him and tie his hands behind his back. It was nothing short of a miracle that he was even sitting in this wet can. Wessner had beaten the odds already.

He had been born Alwin Ernesto Rodriguez in Phoenix, Arizona, the son of a South American father and Spanish mother. His father left when Alwin was five years old, Alwin said. By seven, according to Alwin, his mother had a nervous breakdown and was committed to a mental institution. He was sent to live with a succession of eccentric aunts and uncles, who neither wanted nor cared for him. He was left largely on his own in front of a television set. By sixth grade he had dropped out of school. The relatives no longer bothered to send him. When truant officers dropped by, the relatives told them he no longer lived with the family. Eventually the truant officers stopped coming. Alwin had fallen through the cracks.

He was destined to become a juvenile delinquent, a runaway, a lost soul who would likely end up in prison. But by age thirteen something inside Alwin told him he didn't want to be a bad kid. He didn't enjoy being mean to others and he didn't want to feel worthless himself. Millions of troubled youths never had an inner voice telling them they could do better. Alwin did.

He began sneaking to the library whenever he could, reading books to try to teach himself. By fifteen, he was working as a dishwasher in a restaurant, hiding his money so relatives wouldn't confiscate it.

But finally someone noticed Alwin Ernesto Rodriguez. Carol Ross managed the family restaurant in which he worked. She felt sorry for this awkward, shy teenager with his curly black hair and dark brown eyes. After they closed each night, Alwin would load his bicycle into the trunk of her car and she would drive him home.

She soon discovered he didn't really have a home. Slowly Alwin began to tell Carol his story. She hated the way he was being forced to live. It was no way to raise a young boy, particularly one who had the potential she saw in Alwin. He was a nice kid who desperately needed someone to love him, to give him a family. She was deeply religious. She prayed to God for guidance. What should she do? Finally, Carol Ross, a forty-year-old divorcée with two sons and a daughter of her own, made a momentous decision. She told Alwin she wanted him to live with her.

Alwin agreed. The aunts and uncles didn't care if he left. They never so much as telephoned Carol Ross after he walked out.

Alwin assured Carol he wouldn't be a bother. Just give me a bed and a room and I'll let myself in and out like a boarder.

"No, Alwin," Carol said firmly. "You're going to be part of the family."

Family was a foreign concept to Alwin. He had trouble adjusting to being in a warm and loving home with people who cared about him. It was almost a year before Alwin could open his heart to Carol. But she eventually became the mother he never had. At eighteen he walked into the local courthouse and signed the papers to have his name changed. He wanted a complete break with his old family. His last name now would be Wessner, after the maiden name of his mother's mother. He did want to preserve some tiny link to the real mother he never knew.

He was now Alwin Ernesto Wessner. He asked friends to call him Wes.

Carol forced him to make another important decision. He should do more with his life than clean dishes in a restaurant or work in a factory. College was out of the question, Alwin thought. He'd only finished the sixth grade and he wasn't about to suffer the humiliation of returning to junior high and high school at his age. But there was the military, he decided. The service would accept him without a high school diploma if he could pass the GED test (General Equivalency Diploma), which was considered the equal to a diploma.

Wessner bought a thick manual on how to pass the GED test and began studying it. He could read well and from what he had absorbed from books and television he had a fair mastery of basic science. But math was the stumbling block. You didn't pick it up watching game shows. Algebra and geometry looked like Greek to him.

Alwin studied for days and days. He finally learned enough to pass the test and earn his GED. He marched into a Navy recruiter's office proud that he had earned a GED, but still ashamed that he never went to high school. That didn't matter to the recruiter. The GED was enough. Wessner later scored respectably on the basic entrance exam the Navy used to screen new recruits. He was assigned to be a "fire controlman"—a sailor who fixed the computers on missile systems—and packed off to the Naval base at San Diego, California, for boot camp.

He became red meat for the drill instructors. Wessner had no social skills. Growing up, he had had few friends his age, no school life for learning how to interact with others. He was a misfit, a screwup, not particularly popular with the other recruits. He was "nonreg," in Navy parlance, the kind that DIs set upon like packs of dogs. And they did with Wessner. The instructor who ran his training unit despised him, loaded him up with extra duties and push-ups, made his life hell.

But Wessner refused to wash out. He had nowhere else to go. The Navy was his only chance to make a life for himself.

Wessner finally passed basic training, even earned a small promotion to E-2, the next step up from raw recruit. He shipped out for basic electronics school, where the instructors dumped twenty manuals in his lap and told him to go learn them. The classes were mostly self-taught. Computer surveys run on each student predicted how long it would take him to master the material based on his educational background. The average was six weeks. The computers predicted it would take Wessner twice as long. The instructors thought he would never complete the course.

But Wessner had also become deeply religious. God had blessed him with a capability to understand things, he felt. He flipped through the pages of the electronics manual.

God hadn't prepared him for this. He was stumped. It was all gobbledygook.

He went to the study center on base one night and stared at chapter one of the first manual. It explained the basic concepts for electricity. He stared at it for four hours, rereading the chapter over and over.

Then something clicked in his brain. As if by magic, he suddenly understood the basic concepts of currents and voltage, the principles of electronics. Then the complex formulas finally made sense. The understanding and learning snowballed. Wessner completed basic electronics in far less time than the computer predicted. More importantly, he now for the first time believed he might succeed in the Navy. He wouldn't be just chipping paint on a ship. He had the ability to learn a valuable skill, even rise up in the ranks.

Wessner finished advanced electronics school in the top ten percent of his class. He then started thinking about becoming an officer. He decided to take the Scholastic Aptitude Test. Two classmates who were strong in math began tutoring him in advanced algebra (he still couldn't bring himself to tell them he never attended high school).

Wessner scored 1020 on the SAT, hardly spectacular, but enough to get him into the Naval Academy if he first attended the service's prep school at Newport, Rhode Island, for a year. The prep school had been set up for high school graduates with low grades, particularly athletes the academy wanted for its teams, who would be tutored for a year to bring them up to the academic level of the other plebes who were accepted. The Navy each year reserved a certain number of plebe slots at the academy for the enlisted sailors who wanted to become officers. Many of those sailors also had to spend a year at Newport brushing up on study skills.

Wessner by now was assigned aboard the USS *Missouri*. He impressed the chief petty officers and commissioned officers who supervised him. He was a hard worker. He definitely had the potential to be an officer. But don't get your hopes up, the *Missouri*'s officers warned him. The academy took only a small percentage of the applicants and time was quickly running out for his applying. He was almost twenty-one. Even if he was accepted, he would have to spend at least a year at Newport. Twenty-three was the cutoff age for being accepted to the Academy. Wessner was cutting it close, particularly if he wasn't accepted the first year he applied. He rushed to mail in all his paperwork.

Captain David W. Davis sat in his den on a Sunday afternoon in Annapolis, staring at three application packets. Davis was the Naval Academy's director of admissions. He always reviewed closely the applications of the fleet sailors who applied. They could be rough around the edges, but he knew seasoned sailors ended up being good junior officers, more mature usually than the kids who came to Annapolis straight from high school.

Davis had one slot left for the freshman class and three applications from sailors in the fleet, one of which was Wessner's. He had glowing recommendations from officers on the *Missouri*, including its skipper. But the SAT score was pretty mediocre and Davis couldn't believe what had been left blank on the application. The section where Wessner was supposed to write in his father's name and his mother's name, blank. High school name and transcripts, all blank. Did this seaman come from nowhere?

Davis had already telephoned the *Missouri*, which was in port. "You guys got to be kidding," he said.

"No, he's done well," the *Missouri*'s executive officer had assured him. "He deserves a chance."

"Wait a minute," Davis said, skeptically. "This guy hasn't gone to high school. He's never even taken chemistry—"

"You've got to believe us on this one," the executive officer interrupted. "We've just got a feeling he'll make it."

Davis now stared at the three applications for more than an hour. Finally, he called his wife, Susan, to the den and asked her to look at the three packets.

She read the three carefully.

"Pick this one," Susan finally said, pointing to Wessner's file.

"Why?" Davis asked.

"The whole purpose of the prep school is to give kids a chance," she said. "This kid has never had a chance."

Several days later, Wessner was summoned to the office of his division chief on the ship.

"You made it," he said, tossing Wessner the cable ordering him to report to Newport. "Now don't let us down."

Wessner loved prep school. Though he was three years older than most of the other students, this was the high school experience he had never had. He finished in the top third of the class. The school's commander took a liking to him. The commander had also begun as an enlisted man and had risen through the ranks. He telephoned Academy officials to look out for this kid. By now, the story of Wessner's deprived childhood had been circulating the grapevine. Every sailor had a "sea daddy," a mentor who helped him along the way. Wessner now had a boatful of sea daddies, chiefs and officers he had met in his three years in the Navy who had heard his remarkable tale. At each of his new duty stations now, a quiet call had usually preceded him from one of his sea daddies. "Look out for this kid. We want him to succeed."

But Wessner almost didn't at the Academy. And he had no one to blame but himself.

He began his plebe year the summer of 1990 and immediately got on the wrong side of the Academy's upperclassmen. It was a problem many enlisted sailors had when they entered Annapolis. Wessner had already been through Navy boot camp and endured months of taunting by drill instructors. But at least the DIs had been senior chiefs, twice his age. But here he was a twenty-one-year-old sailor, with sea duty under his belt, being screamed at by senior midshipmen his own age.

The high school graduates who were plebes almost peed in their pants, they were so intimidated by the upper-class hazing. Wessner considered it a pain in the butt. Taking crap from these martinets, who were no older than he, was tough. Wessner wasn't impressed and it began to show. He was quickly labeled a fleet sailor with an attitude, a wiseass, who didn't show respect, who never saluted fast enough. The upperclassmen heaped more hazing on him.

Several times he wanted to quit. Davis had become his faculty sponsor and Wessner quickly adopted the captain's family, spending weekends at the Davis home pouring out his troubles.

You have to go through plebe year, Davis lectured him. Don't throw everything away. Gut it out.

Wessner did. He later realized he did have an attitude problem that first year. He buckled down and snapped to. The hazing stopped.

The courses at Annapolis were demanding, but Wessner managed to graduate in the top half of his class. He also managed to win something he never dreamed possible—an aviation training slot at Pensacola. He had fantasized about being a pilot. He took his first plane ride at eighteen when he flew from Phoenix to boot camp in San Diego and found it thrilling. The summer before his junior year at the academy he had visited Pensacola and flown in the T-34C training plane. He asked the pilot to make him sick during the ride. The pilot chuckled, then flew up into a stall, spun around, and swooped the plane down. Wessner didn't throw up. It was a blast. He enjoyed every part of the roller-coaster ride.

For the fourth ride in the Helo Dunker, Wessner buckled himself into the seat just across from the rear cabin's main exit. It could have been worse. He could have been stuck in seat number seven, which was the furthest away from the main entrance in the capsule's rear corner. In seat three, Wessner figured that all he had to do was reach forward, grab the edge of seat four in front of him, pull himself to it, then reach to the left to get ahold of the side of the main entrance.

He pulled the blacked-out goggles down over his eyes and replotted the moves he would make in his mind. It would be easy. Keep calm. No sweat. After all, the dunker wasn't what he was worried about at the moment. Tomorrow, that was when the most critical test came, at least for him.

Not only had Wessner never learned to interact with others his age because he didn't attend secondary school. He missed out on athletics as well. He was uncoordinated. At the Academy, he finally won a spot on the squash team, more out of determination than physical skill. PT tests were always an uphill battle for him. At Pensacola it took him two runs before he passed the obstacle course. But the O course turned out to be not the problem. It was push-ups. Pensacola required forty-two push-ups to pass preflight indoctrination, and they had to be perfect. The first time, the instructors failed him because the push-ups were sloppy. The second time, his back was straight, but he fell two short of the number required to pass.

Wessner went before a review board, which had to decide whether to kick him out of the program for failing two physical tests. He pleaded to be given one more chance. Okay, the board ruled. You have two weeks to get in shape and pass the push-ups test. Otherwise, you're out. Wessner promised he wouldn't disappoint them.

But now he didn't know. Push-ups were a silly requirement to flunk. Did he really want to be a pilot? Was something inside him telling him that he wasn't fit to fly? If he didn't have the physical stamina to do forty-two push-ups, did he have it to land a huge jet on an aircraft carrier?

The Helo Dunker sank and flipped over for the fourth ride. Wessner followed the movements he had rehearsed in his mind. He had no problem escaping the capsule blindfolded. Tomorrow, he would take his third push-ups test. If he failed, the dream would end. He'd be washed out of the program.

2

FIRST FLIGHT

CHAPTER SIX

VT-Screw

JENNA eased back the control stick to the propeller-driven plane. The turboprop engine in the T-34C (the Navy used T to designate training aircraft) gave a low rumble like a muscle car on a drag strip and lifted her to 8,500 feet above the ground. On clear cool days, and this was one of the few of them over southern Alabama, a pilot could see across the Florida panhandle from this altitude and out to the glistening Gulf of Mexico.

Ensign Jenna Hausvik, a twenty-three-year-old Navy brat (her father had risen through the enlisted ranks to become a lieutenant commander), looked out the side of her cockpit at the farmland cut into dark green and brown patches below. She was about to send her plane spinning toward the earth—the pilot's euphemism for it was "departure from controlled flight"—and would need some reference point on the ground to which she could fix her eyes so she wouldn't become disoriented.

"What's a good reference point to use out here?" Mike Consoletti asked. Consoletti was a Navy lieutenant and her instructor. The T-34C was a two-seat aircraft that looked a lot like one of the old World War II fighter planes. The student pilot sat in the front while the instructor strapped himself into the back seat, peering around the helmeted head in front of him to make sure the aircraft wasn't going to drill a hole into the ground. The instructor had most of the controls in front of him that the student had and could grab them if the student was about to crash him—which happened occasionally.

Not that Consoletti minded the duty. At thirty-three, still with a boyish-looking face, he had spent his career in the Navy flying utility helicopters. Piloting the T-34C was a refreshing break, a chance to be back in a fixed-wing plane. The Beechcraft Turbo Mentor, the plane's official name, was no head turner. The pilots nicknamed it the "Turbo Weenie." Still, the aircraft was powerful and rugged enough to do loops and barrel rolls and wingovers. Plus instructors were flying constantly, which was every pilot's dream, even if it had its occasional hair-raising moments with students.

The constant flying—six hours in the air on busy days—could also have a payoff. Training duty often was a way-stop for many instructor pilots wanting to leave the service for the airlines. Here they could build up flight hours in a prop plane and make themselves more marketable to Delta or USAir or anyone else who was hiring on the outside. Some training squadrons even had gouge manuals in the instructors' ready room that were crammed with sample application forms for the airlines.

"A city," Hausvik answered into the helmet microphone pressed to her lips. A city was as fine a reference point for the spin as any, she thought. She saw Evergreen, a tiny Alabama town off to her left.

"That's good," Consoletti agreed. "See the black pond over there?"

Jenna turned her head to the left and spotted Little Pond, which was shaped like a gourd just below. "Yes sir," she said. The intercockpit communication system let her and Consoletti chat without the rest of the world listening.

"That's a good reference point too," Consoletti said. "The airfield south of Evergreen is also great."

The instructors were always second-guessing the students. It was their job.

Jenna now powered up the plane to about 150 miles per hour. She looked around her seat for any loose objects in the cockpit that might smack her in the face for what she was about to do, then double-checked to make sure her seat belt and shoulder harness were locked and tight.

"Autoignition . . . on," she said into her mike.

She scanned the panel in front.

"Engine instruments . . . checked. Stall checklist complete."

She looked to her left and down one more time. The Evergreen airfield would be the reference point, she decided.

Jenna began a clearing turn to her left at a 45 degree angle so she would be nearer her reference point. When she rolled out of the turn, she grabbed the power control lever on the left side of the cockpit near her lap and yanked it back to 200 foot pounds of torque. Torque was the measure of horsepower for the shaft that drove the engine's propeller. Consoletti watched his power control lever move back to the 200 reading.

Jenna checked to make sure her wings were level, then smoothly pulled back the control stick between her legs. The nose of the plane reared up. Suddenly the entire aircraft began to shudder as if it had the chills.

Jenna was stalling the plane. An aircraft flies as long as its wings create enough aerodynamic lift. But when the lift is completely lost, the plane stalls and begins falling to the ground. Stalls could occur through a combination of the plane's wings being tilted up too high—in flying lingo it was called an "excessive angle of attack"—and the aircraft's airspeed being too low. Because Navy pilots often flew at high angles of attack in aerial dogfights or flew slow at low altitudes to land on aircraft carriers, they had to know—if they wanted to stay alive—what conditions would cause their planes to stall and how to recover from it.

Jenna had already practiced how to regain control if the aircraft stalled because of lost power (push the stick slightly forward to start the plane gliding at 100 miles per hour) or while it was banking in order to land (lower the nose, level your wings, and add power). Now she was stalling the plane to put it into a spin.

When engineers design a plane, they must compromise between

stability and maneuverability. Civilian airliners were designed to be stable and difficult to spin. The passengers don't want too much maneuverability or dinner trays start flying. High-performance fighters, on the other hand, were designed to turn, roll, or loop violently in aerial combat. But the more maneuverable a plane was, the easier it could be stalled. And in some cases when that happened, the nose pitched up, the aircraft rolled over on its side, and then spiraled nose-down toward earth as if it was swirling into a drain.

During World War I, fighter pilots intentionally entered into spins as a way to descend quickly through clouds. In those days planes had no fancy flight instruments to guide aviators through clouds, where they could easily become disoriented in the fog. So pilots would spin down through them, and hope there was enough space between the cloud ceiling and the ground for them to recover.

Today, the only time military pilots intentionally performed spins were during air shows or test flights or training runs. Jenna not only had to recognize when an airplane entered a spin, she also had to know how to recover from what could be an extremely disorienting experience. If she couldn't recover by the time the T-34C plunged past 5,000 feet, she had to bail out.

Bailing out could be just as disorienting. The T-34C had no ejection seat. Jenna would have only about thirty seconds to unbuckle her seat belt and shoulder harness, yank the emergency lever to blow back the canopy, tumble out the side of the cockpit, then pray that she was not flattened by one of the sweeping wings.

As the plane shook, Jenna jammed in the left rudder pedal with her foot and yanked back the control stick. The aircraft lurched up, then flopped nose first on its left side. It felt to Jenna like twirling around on high heels, then falling over on one shoulder. The next second she was looking straight toward the ground, the front of her chest straining against the shoulder straps.

The plane twisted around in a corkscrew movement as it plunged to earth. Now Jenna felt as if she was strapped to a seat and upside down with someone swinging her around and around like a lasso. The first time a student spun could be terrifying, but Jenna had performed the maneuver ten times and now never felt out of control with the plane. In fact, now it was fun.

"Here's one," she said to herself as the aircraft made its first rotation. Her eyes darted from the instrument panel to Evergreen field, her reference point so she wouldn't become disoriented. Warning beepers blared irritatingly from the cockpit controls to signal that she was losing altitude quickly.

"Here's two," she said after the second rotation.

"Right full rudder," she continued and jammed in the right rudder pedal to begin the recovery.

"Give it a little bit more rudder," Jenna continued talking to herself. "Now stick forward."

Consoletti was impressed. Hausvik was one of the few students he had ridden with who talked her way through the procedures. Most beginning pilots were too preoccupied with reading their instruments and getting out of the spin to recite the procedures at the same time for the instructor.

Jenna always talked to herself in the plane. She considered herself a verbal person. She hated to write essays. She was always blunt, to the point. On book reports in school you had to blather on for pages. She either liked the book or didn't. No in between. Why waste words. Reciting cockpit procedures out loud helped her concentrate as she performed them. It was also a good way to keep the instructor in the back calm during the flight. Consoletti wasn't a screamer, but other instructors were and could be easily panicked if they sensed the student didn't know what he was doing. It was even better if you rattled off the procedures matter-of-factly, Jenna had found. That way you sounded like you knew what you were doing, even if maybe you didn't. It was a trick she had learned at the Naval Academy. Don't ever let them see you lose your cool. Especially at the Naval Academy. Be confident. Jenna had a handshake like a vise grip. Never show weakness.

The turboprop responded perfectly to the corrections Jenna had entered in with the rudder and stick. The nose began to rear back up. Jenna kept her stick centered between her legs. She reduced the power, then leveled the wings so they returned parallel to the horizon. Next she began the pull-out. As the speed increased she could feel the Gs tighten her stomach and shoulders.

"Oil pressure okay, oil temperature is right," she said, going through the checks she now had to make after recovering from the spin. "Gyro and RMI okay." The attitude gyroscope in the upper center of the control panel showed Jenna how her plane was tilted in the sky. RMI stood for the radio magnetic indicator, which gave the direction the plane was heading.

"That's good," Consoletti said over the microphone. "You had a touch too much forestick, but nothing really to complain about. Otherwise, real nice."

No student would recover perfectly from a spin at this point. Not at primary flight training.

It seemed almost preordained that Jenna Hausvik would join the Navy. Not only was her father a career serviceman, her mother was in the Naval Reserve and her brother had joined the Navy as well. Even before high school, Jenna knew she wanted to be in the military. It offered security, a steady job. A corporation could be cold and heartless. Employed one day, laid off the next. The military looked out for you, Jenna thought. It protected you. She had seen the life her father and mother had lived in the Navy and could think of nothing else for herself.

The Naval Academy was eager to accept Jenna. Not only was she smart, Jenna had also been a star athlete in high school. At five feet ten inches tall, she had played basketball and run track. She had even set two track records in high school, in the high jump and heptathalon. She was the kind of well-rounded kid Annapolis was always hungry for.

Sports soon became her refuge at the Academy. She had been assigned to the company from hell. Its upperclassmen were sadistic. They made a sport of harassing plebes to see if they could run them out of the unit. Jenna wasn't going to be scared away. But she joined the Academy's basketball and track teams for women as a refuge. They became her oasis. In her dorm and classes a rigid and stifling caste system prevailed with the upperclassmen acting like feudal barons. On the basketball court or track field, everyone was equal. No one pulled rank. People looked out for one another on the team. Athletic ability, not the number of stripes on your shoulder boards, was what mattered. Just walking to the gym every day, Jenna would undergo a mental transformation. She would clear her head of the snap-to orders and discipline for the pure joy of physical competition. It was almost like moving in and out of enemy territory every day. Athletics became her reality check from the artificial regimentation of the company from hell.

Jenna wasn't one of those kids who knew she wanted to fly before she could walk. When she was commissioned, she had intended to apply for the Navy's supply corps, her father's job in the service. But ironically, the Academy slots for that branch filled up quickly her senior year with students not physically qualified for other duties such as flying. Jenna was overqualified in that respect. She was strong as an ox. So a week before she had to decide her branch of service, Jenna chose aviation, practically on a lark. She had never been interested in planes. She was even afraid of heights. (She tried to conquer that fear at the Academy by taking parachuting lessons.)

Preflight indoctrination at Pensacola had been easy for her. She found the classes far less demanding than the Academy's. And again she excelled in sports. She was the first woman ever to win the school's athletic award for the obstacle course. Her time bested many of the men's. Jenna, in fact, was disgusted that the women were given more time to complete the O course and didn't have to scale the twelve-foot wall that the men did. She could hop over it with ease and found it humiliating that she was forced to jump over the six-foot baby wall. It only reinforced the prejudice among the men that women were being given special treatment in the flight program.

The physical fitness standards were the biggest sore point in the program. They caused the most friction between the men and women. Few of her female colleagues would agree with her, but Jenna thought the standards should be the same for women and men. If the women couldn't keep up with the men they should leave. She could keep up. In fact the men were eating her dust.

All she wanted was to be just one of the guys. She could be as earthy as any sailor. She had dark brown hair parted down the middle and cut even around her neck, and wide, dark gray eyes that seemed childlike. She was thin and muscular and walked with a lope. She wore no makeup while on duty, but didn't have to; her features were striking.

The next step after preflight indoctrination was learning how to fly. For Jenna, that meant driving thirty miles northeast of Pensacola to Whiting Field. Whiting was stuck in the middle of cotton and soybean fields just north of Milton, Florida, a sleepy panhandle village at the mouth of the Blackwater River. During the 1800s, Milton had been a bustling sawmill town where logs floating down the snaking Blackwater from Alabama were cut, then stacked aboard ships that sailed up from the Gulf of Mexico. The Depression and deforestation wiped out the lumber industry. The Opera House at the corner of Willing and Caroline streets, where country singer Roy Acuff had once performed, had been converted into a museum to preserve the town's few remaining relics. Now one of Milton's largest employers was the air station.

Whiting had served as a POW camp for German prisoners during World War II. Now it had one of the busiest airports in the United States. More than a thousand students were trained there each year. Its control tower juggled about 440 training flights per day. In the mid-1990s, every Navy pilot—whether he ended up in a prop plane, jet, or helicopter—began his flight training in the T-34C (which has since been replaced by the T-6B Texan II). Congress had long questioned this. Army helicopter pilots began their helicopter training in a helicopter. Air Force jet pilots started in a jet trainer. Money could also be saved if the Air Force and Navy combined their primary training into one program, congressmen argued.

The two programs may eventually be combined. But the Navy has fiercely resisted giving up the turboprop plane as the first aircraft its pilots fly. Sleek Air Force jet trainers were far easier to handle. But the Navy was like a parent who wanted a teenager to learn to drive a car with a stick shift rather than an automatic transmission. The Navy believed it produced better pilots if all of them started on the turboprop.

The Turbo Weenie taught them the basics of aviation better simply because it was so difficult to fly. The T-34C, which could reach 280 miles per hour in a dive, was inherently tough to control in the air. Its souped-up, high-torque engine always wanted to twist the aircraft in the direction that the propeller was moving, much like a balsa-wood model plane with the propeller wound up tightly by a rubber band. What's more, the high winds the propeller washed back over the plane forced its nose to go right. It was called the "prop slip-stream effect."

Students were constantly making adjustments with the stick and rudders. That could be maddening for beginners. Every adjustment they made in the controls seemed to produce an opposite reaction out of the Turbo Mentor, which the students would then have to counter with another adjustment. Add power and the plane wanted to climb and turn right; then when you corrected by easing off on the power it would dip and turn left. The aircraft responded to every twitch of the control stick. Flying became a constant battle between inputs, outputs, and corrections for inputs and outputs. If the corrections weren't made, the aircraft would wobble along like a tadpole in water. Students had to use every muscle, every sense in their bodies to keep the plane on course.

It could be overwhelming at first. Students would hold the control stick in a death grip. The instructors had a favorite lesson to correct this.

"When you go to take a leak, do you grab your dick like a baseball bat and swing it all around?" the instructor would ask.

"No sir," the student would answer meekly.

"Then don't do that with the control stick."

The instructors used a different example for the female students.

At first, the plane would fly itself better than when the students had their hands on the controls. But eventually, if they could master the Turbo Weenie, flying a jet would be easy, the Navy believed.

Jenna spent her first three weeks at Whiting sitting in a classroom learning the complicated insides of the T-34C or in a cockpit trainer memorizing its maze of switches and gauges. Students were blindfolded and required to point out the location of every knob and button in the cockpit. After ground school, she received her assignment to a training squadron.

Jenna cringed. Of the five training squadrons in primary, she'd been stuck at VT-2. Navy squadrons were often identified by letters and numbers. The T stood for training. The V was a holdover from the 1920s when the service used V to designate its heavier-than-air aircraft and Z for its fleet of blimps. VT-2, whose insignia was a black crow for the "Doerbirds" (its instructors "do more" with students, so the legend went), was the oldest of the primary training squadrons. It was also the meanest, according to the gouge Jenna had heard at Pensacola.

There were all kinds of horror stories about VT-2: The instructors were hard-asses and screamers in the cockpit. Students were treated like plebes rather than the officers they now were. The washout rate was high. "Attrition is our mission," was the squadron's unofficial motto. If you wanted to fly jets, stay away. Instructors were so stingy with grades, the best you'd end up with were helicopters. The students' nickname for the squadron was "VT-Screw."

Of course the gouge didn't quite fit reality. VT-2's washout rate was no higher and the percentage of its students getting jets was no lower than the other squadrons'. The unit had only a few screamers. In fact, its instructors were a fairly relaxed group, surprisingly cohesive considering that they came from all types of aircraft and three different services. The Navy, Marines, and Coast Guard all were required to supply the squadron with teachers.

The VT-2 instructors, however, liked to keep the bad-guy myth alive. It was good to have the students a little scared when they checked into the squadron. The instructors were told by the front office not to cut their pupils any slack. Don't be afraid to fail them on flights. These kids could kill themselves in a Turbo Weenie just as easily as in a high-performance F/A-18 jet. The squadron took pride in the fact that it was demanding. VT-2 had the highest percentage of graduates eventually earning their wings of any primary squadron in the program.

Jenna began her flying with what were called familiarization flights, or "fams" for short. Her first flight in the T-34C, called "fam one," was fun. It was the "ooh and ah!" flight. The instructor behind her actually flew the plane while she experimented with the controls and switches in the front. It surprised her how responsive the stick was to her touch, like a joystick in an arcade game. The plane twitched at its every movement. She would shift the stick to the right and instantly the plane would bank right. But the nose would dip at the same time and she would have to remember to pull back on the stick slightly as she also turned. She leaned her head back and gazed out the plastic cockpit bubble surrounding her. She could see the sky and earth rushing by, all around her. It was like being in a three-dimensional film at Disneyland.

The fun quickly stopped. Familiarization sounded more benign than it actually was. Fams became overwhelming. *She* had to start flying the plane on fam two, her second flight. That was when the instructors picked you up and threw you into the pool, she realized.

Just taxiing to the runway was difficult. You drove with your feet. The rudder pedals, not a steering wheel, turned the aircraft right or left. Jenna snaked down the taxiway the first time. At the runway, the pilot had to keep the brakes on while he powered up the plane for one final check of the engines. The brake was activated by pressing lightly both rudder pedals at the top with the balls of the feet. Jenna was so nervous she jammed the brakes with all her might. Her knees began twitching from fatigue. The plane looked like it was doing the mambo on the runway.

Her takeoff was passable. She managed to accelerate to the correct speed and keep the plane pointed somewhat straight ahead down the runway. As soon as she was high enough, she banked right to clear Whiting Field's airport. She was delighted. The first time at the controls and at least she hadn't crashed. But in the air she had to learn how to move forward, right and left, up and down, all at the same time. It was driver's ed in three dimensions.

Then came the "trim monster." Keeping the plane in balanced, level flight could be difficult if it wasn't properly trimmed. An aircraft rolled right or left and tilted up or down because of movable flaps on its wing surfaces. The main wings in front had ailerons near the rear wing tips. Moving the control stick right or left caused the ailerons to flip up or down, which in turn caused the wings to roll up or down and the plane to bank right or left. The vertical wing in the back, called a vertical stabilizer, also had a flap attached to its rear called the rudder. When Jenna pumped the right or left rudder pedals with her feet, the rudder on the vertical stabilizer would swivel right or left, causing the plane to yaw right or left like a car turning on a level road. The final flaps were the elevators attached to the horizontal stabilizer wing in the rear. When Jenna pushed the control stick forward or back, the elevators flipped up or down, causing in turn the plane's nose to pitch down or up.

Simple enough, except that every time an aileron, rudder, or elevator broke the smooth surface of its wing when it flipped up or down—or, in the case of the rudder, right or left—that produced a countervailing resistance from the wind. Think of driving a car down the highway on a hot summer day with the windows down. You flip your flat palm up and down outside and the wind buffets it up and down making it hard to keep your arm steady. The same kind of turbulence occurs when the smooth surface of a wing is broken in flight by a flap going up or down. The pilot feels that buffeting on his control stick and rudder pedals.

That was why each aileron, rudder, and elevator had a trim tab attached to it. The trim tab was nothing more than a second flap attached to the first one. When, for example, the aileron flipped up, the trim tab flipped slightly down. The tab, in effect, acted as a countervailing force to the turbulence created when the aileron was moved. It moderated the effect of the wind hitting the aileron when it moved up.

In high-tech jet fighters, much of the trimming was done by computers. Pilots called it "fly by wire." In the T-34C the trimming was performed manually by turning three clunky wheels beside the pilot's left thigh that manipulated the aileron, rudder, and elevator trim tabs. The students joked that the wheels were borrowed from antique blimps. But the instructors believed that manually manipulating the wheels helped instill one of the most basic skills of learning to trim an aircraft.

Students had it drilled into their heads that after every change they made in the plane's direction they had to follow it by trimming. Three words were repeated like a mantra: power, attitude, trim. Whenever you changed the power of the plane, you had to adjust the attitude because the plane's nose probably dipped or tilted one way or the other because of the power change. Then you had to trim out the pressures you felt on the stick and rudder because the aileron, rudder, and elevator had moved. Trim the rudder first, then the elevator, then the aileron. If he didn't trim, the pilot was constantly fighting the plane. Keeping the aircraft on a precise course became difficult. The stick and rudder pedals would begin to feel like lead weights on a tight spring, wearing out the pilot.

Jenna was first introduced to the concept of trimming during preflight indoctrination classes at Pensacola. Reading it in a manual really didn't register with her. Now it overwhelmed her in the cockpit. For every turn, every dip, every speed change, every maneuver she made in the plane, she had to remember: power, attitude, trim. Rudder trim, elevator trim, then aileron trim. Jenna could manipulate the stick and the rudder pedals just fine. But rotating the three trim wheels at the same time became a distraction. It was like patting your head with one hand, rubbing your stomach in a circle with the other, and hopping on one foot. The more complicated the aerial maneuver, the more distracted she became with the stick, rudder, and trim wheels. And the more distracted she became, the less attention she paid to where she was flying. The instructors called it "getting behind the aircraft" and Jenna felt as if she was barely hanging on by its tail.

She soon discovered another problem with flying: airsickness. At the end of her second flight, she reached down to the zippered pocket on the leg of her flight suit, pulled out a white paper bag and vomited into it. She had eaten little before the flight because other students had warned her that the T-34C gave a rough ride. She felt like her stomach was being scraped out and she was throwing up bile.

For the third flight she ate a hearty breakfast. This time she filled up most of the barf bag. The fourth ride she tried a concoction another student recommended to keep her stomach settled: peanut butter and ginger. She didn't throw up but she still felt nauseous. The fifth flight, fam five, she vomited again. And again, during the sixth flight.

Jenna became racked with doubt. She could take the puking every flight. Well, almost take it. But it didn't seem to her like she was improving her airmanship. She was always behind in the cockpit procedures. And there was something else, something she was too ashamed to admit to anyone else.

She wasn't enjoying flying. She was studying air manuals every waking minute between flights. The primary training seemed like graduate school crammed into six months. Students had to fly almost constantly or the skills they had barely learned each day became perishable. It was total immersion in aviation. When they weren't in the planes, or in class, or studying by themselves, students talked about flying or quizzed each other on procedures as they ate or drank beer on the weekend. You couldn't even take a crap or a leak without studying. Flight regulations in glass frames were hung over urinals and in the stalls of the johns in the squadron's bathrooms. For Jenna, the flights themselves were not fun. It was all too much, too fast. No time to catch your breath. The minute you thought you had one maneuver mastered they threw a new one at you.

She began to pray for bad weather so her training flights would be canceled. Then she felt guilty when they were scrubbed. Was this what she really wanted to do? After all, she hadn't been like most of the other students. She hadn't wanted to fly since she was in diapers. She'd only chosen aviation at the last minute. Was she constantly vomiting because she really didn't want to be a pilot? Because she wasn't cut out to be a pilot?

Jenna would learn later that most of her fellow students hated their first flights as well—even the ones whose rooms back home had been filled with model planes. It was the fams. They weren't meant to be enjoyed, the instructors warned. In the beginning, flying wasn't fun. A lot of students threw up during fams. It even took the instructors, particularly those who came from smooth flying planes like the P-3s, at least a half dozen flights in the bumpy Turbo Weenies before their stomachs stopped feeling queasy.

Jenna would eventually get over the barfing, the trainers assured her. She went to the squadron's flight surgeon, who gave her airsickness pills for the next three flights. They helped, but the surgeon would give her no more, not wanting her to become dependent on the pills for every plane ride. She threw up again when she wasn't under medication.

But she soon began to live with the vomiting. After the seventh fam flight, the students were given a break and sent to basic instrument school where for two weeks they learned to fly the plane watching only its instruments inside the cockpit. The Navy flight training program emphasized instrument flying. Its pilots would have to operate at night and in all kinds of miserable weather so they had to be able to fly when there was nothing to see for a reference point outside. A civilian pilot could divert to another airfield if the one he planned to land at was socked in by bad weather. In the middle of the ocean, there was no other place to land than on the aircraft carrier. In instrument training, Jenna spent much of her time in the cockpit trainers, which were run by crotchety retired aviators, who were nonetheless easy graders. She also flew in the real cockpit with a cloth bag draped over her, so she could see nothing outside and had to depend solely on her instruments. But the instrument training hadn't seemed as stressful to her as the familiarization flights.

After ten instrument flights, Jenna returned to the fams. She felt far more confident during her next three familiarization rides. Not as many new procedures were thrown at her during each flight. She now was given more time to perfect the skills she had learned. The puking and flying became a routine she thought she could live with if that was how her years as an aviator would have to be spent.

As they sped south back to Whiting Field's airport, Consoletti had an uneasy feeling about Jenna's flight. It was her twelfth one in the fam curriculum. The next flight was her "check flight," when another instructor would test her to see if she could handle the plane safely enough to fly solo on her fourteenth flight. Consoletti knew he could take any kid off the street and eventually teach him to fly. Moving the stick, pumping the rudder pedals, flipping the cockpit switches were all "monkey skills," as the instructors called them, which anybody could learn given enough time. But the Navy didn't have a lot of time or money to spend on teaching the monkey skills.

Consoletti knew that civilian flight instructors could be gentle. If the student arrived for a class unprepared or flew poorly, he just came back the next day and tried again. The civilian instructor wouldn't rant or rave. Why should he? He'd earn more, spending more days teaching the student. But Consoletti wasn't paid by the hour. He had a set amount of time to teach a student how to fly and no days could be wasted. Nothing drove him and the other instructors up the wall more than when a student came to a flight unprepared. The trainers considered it a personal insult. A student who hadn't studied enough ahead of time was wasting their time. Instructors would even refuse to climb into the cockpit with a student if in the student's briefing before takeoff he hadn't demonstrated that he knew the procedures he was supposed to memorize for that flight.

The Navy instructors were also far more picky about procedures and flight maneuvers in the plane than their civilian counterparts. Students had to learn more about stalls and spins, as well as aerobatic maneuvers, which would be needed for aerial combat. Landing patterns had to be flown more tightly. A civilian pilot could make a wide sweep around the airport and take his time lining up to land on a long runway. Jenna had to fly a smaller circle around Whiting Field with little time to mull over her flight path and even less margin for error, then hit the runway at a precise point. She would need those skills for landing on the floating postage stamp called a carrier.

Mike Consoletti had come to the training squadron six months ago, after spending two years flying transport helicopters aboard a carrier. It was a welcome break. He had a wife and a one-year-old son. Now he had to figure out some way to stay on land in his future Navy jobs. He never again wanted to spend six months away from his family on sea duty.

Instructing students was still a new experience for him. He tried to be more a teacher than an intimidator in the cockpit. But that would soon change, the other instructors had warned him. The longer you stayed in VT-2, the more impatient you tended to become with the students. Every student seemed to make the same mistakes. Seasoned instructors almost knew when a student would screw up, even before he screwed up. There were students who were natural-born fliers, who picked up the skills quickly. But they were rare. The instructors called them "Chuck Yeagers," after the famous test pilot. The students who never got the hang of it were called "rocks." The substandard ones always behind on procedures were "plumbers." Often the rocks and plumbers applied to flight school for the wrong reasons: their father was a pilot and they felt compelled to keep up the tradition, their best friends at the Academy signed up for aviation and they didn't want to be left out, or they thought they looked cool in the leather jacket and aviator glasses. The rocks and plumbers usually dropped out.

The rest of the students were called "cones," for coneheads. The Navy brass liked to gush that their new recruits were smarter than the older generation. Indeed, most of the flight students had higher SAT scores and came from better schools than their teachers. But Consoletti and the other instructors weren't impressed. These students were a smart-assed, whiny bunch and the Navy had become too politically correct in its training, they grumbled. In the old days a student who fouled up was ordered to lean over after he'd landed and the instructor would jerk his control stick right and left, whacking the kid in front on the sides of his helmet. Do that today, the ACLU and a hundred congressmen would be on your back. The Me Generation kids got too many breaks in training. They came here with an attitude, thinking it was a given they'd be a pilot because they had already passed preflight indoctrination. The instructors quickly dispelled that silly idea; one third washed out at primary. Nevertheless, these kids constantly questioned authority, the instructors griped. They even questioned the grades they got—unheard of when the old-timers went through school. The training was losing its edge. The students didn't seem lean and hungry anymore. When the instructors were in school, practically everyone was single and wanted desperately to fly jets. Now a majority of the students were married and half didn't care what plane they flew. Some were even being forced to take jets, the ultimate sacrilege.

Whether the complaints were valid didn't matter. The instructors clung to their gouge as much as the students held on to their notions of VT-Screw.

Jenna was one of the cones, no better or worse than the average student coming through Whiting, Consoletti thought. She knew the procedures for flying the plane cold. She also didn't need to sweat studying all night the insides of the aircraft engine. Jenna was a student who came to each flight better prepared than most. Instructors liked that. She controlled her airspeed well on maneuvers. Her final approach before she landed was always excellent. And she was almost flawless on stalls and spins.

But she made some mistakes that would flunk her on her thirteenth familiarization ride, the critical check flight. Jenna's "air work," the ability of pilots to hit precise targets in landing patterns and on maneuvers, was ragged. It was not enough to know the procedures on paper. She had to be able to use the monkey skills to make it happen in the plane. She had to make the plane do what she wanted it to do. Those skills were mastered faster by some students, slower by others.

Jenna had to clean up more loose ends when she flew her landing pattern, Consoletti decided. He jotted down notes on a pad strapped to his knee as Hausvik flew the T-34C back to Whiting. Her angles of bank in and out of airfields weren't perfect and she was landing too far down the runway. Consoletti had kept his left hand resting lightly on the trim wheels in his cockpit. They moved whenever Jenna operated her trim wheels and Consoletti quickly noticed Jenna wasn't trimming the plane after turns and dives. That meant she was working too hard to make the plane hit precise way points during maneuvers.

Consoletti always came back from a flight with a certain feeling about a student's performance. The instructors could give students one of three grades on a flight: "above average" (to get jets a student needed a lot of them and VT-2 wasn't generous); "average," which most students received; "below average" (students could count on receiving a few of them as well); or a "down," which meant the student had failed the flight and had to repeat it. Consoletti flew back feeling that Jenna's was an average to below-average flight.

He mulled it over some more as Jenna landed on the Whiting runway. She was enthusiastic and certainly conscientious. She talked his ear off in the plane, rattling off every procedure in the book. Maybe she talked too much. Maybe she needed to just do it, and skip the talking. Maybe he ought to give her an incomplete on this flight. Instructors could do that if they thought the student didn't deserve a down but just needed another day in the plane to straighten out the few remaining problems.

No, Consoletti finally decided. On the grade sheet in front of him he marked average for the box on basic air work, below average on her landing patterns, and above average on procedure. That way she netted out average for the flight. Jenna was fairly average at this point in her training, Consoletti thought. Nothing spectacular, but she'd probably make it through the program. She had a sixty percent chance of passing her next check ride, Consoletti decided. It depended on whether she cleaned up her landing patterns. Consoletti hoped he had made the right choice giving her the average grade. Instructors hated to see one of the students they had just passed flunked by another instructor on the check ride. It was embarrassing. No one ever said it outright. But when a student fouled up, it always reflected badly on the instructor who had taken him up in the plane the flight before.

Jenna stood in front of a blackboard in the long study room for students in VT-2's main squadron building. She drew the landing patterns she had performed with Consoletti. The study room was as quiet as a library. At one end, couches were arranged in a square where students sat with flight manuals in their laps. Further down, several rows of comfortable seats were lined up in front of a lectern for classes the squadron held. Behind them were maps and navigation charts posted for all the areas the students flew in their training flights. In one corner sat a mock-up of the inside of a T-34C cockpit, into which students could climb to practice identifying gauges and flipping switches. The students called it "chair flying."

Jenna marked down one plus for today's flight. She hadn't thrown up. But she realized her landing patterns weren't perfect. It seemed to her that she was flying too tight in them, turning too much at some points, flying too low at others. On the other hand, Consoletti seemed calm during the flight. The instructors were there to take the controls if you were going to crash. They could do it in an eye blink. Consoletti never grabbed them from her. She had held her breath when she finally parked her plane after landing. There were all kinds of horror stories about instructors screaming at their students after bad flights. "Okay, that's it. You're done with this program. Go home!" Consoletti was calm and deliberative in his critique.

But it frustrated her. The easy part of flying, or so the other students kept telling her, was hitting the right points on the landing pattern. The hard part was the final landing. Jenna landed fine but had trouble with the pattern.

Mike Harris walked in and interrupted her thoughts. Harris was another Navy ensign. Jenna and he had been stuck in the same company at the Academy, the company from hell. They had become almost like brother and sister, commiserating with each other when the hazing became intense. Mike was a pessimist, who had few interests outside of learning to fly. Get a hobby, Jenna would tell him. She'd love to have his problems. He'd been scheduled for flight training ahead of her and had just finished primary with top grades.

Harris had been eavesdropping in the instructors' ready room in the west line shack. The line shack was just off the plane parking lot on the airfield, a small brick building where the instructors and their students were scheduled for flights and where they held their briefings before and after they flew. The line shack had two ready rooms. One had rows of straight-backed chairs and looked like a dentist office waiting room. That's where the students sat nervous and quiet most of the time, waiting for their flights. The other ready room had comfortable couches. It was where the instructors joked and laughed and swapped horror stories about students who tried to put them in a smoking hole. Students had to knock first and receive permission to enter the instructors' ready room. To walk through the entrance way unannounced—crossing the "line of death," as the instructors called it—could get a student an ass chewing. Harris, nevertheless, had been passing by the entrance when he heard the instructors talking about Jenna.

"Greco is saying your check flight is iffy," Harris told Jenna.

Greco was Captain Tony Greco, one of the Marines who taught in the squadron. The Marines had a reputation for being terrors as instructors. Tony Greco more than lived up to it. A second-generation Italian with a razor-edged crew cut, Greco began as a screamer. He had mellowed the past year. He now would sing "Born to Be Wild" from the back seat of the plane in order to relax nervous students before aerobatic maneuvers and he hadn't yelled at a kid in the past four months. But he was a perfectionist in the cockpit and had no patience for any student who saw flying as a job instead of a calling.

It seemed appropriate that Greco's other duty besides instructing students was serving as the squadron's safety officer. After he had nearly severed his thumb in a high school boating accident, doctors said he would never be able to manipulate the controls in a plane. He proved them wrong and now he would be damned if he'd do anything stupid in the cockpit or let a student fly unsafely. In the 1950s, the Navy had about 2,000 major plane accidents that killed up to 600 pilots every year. By the mid-1990s, the service had no more than two dozen big accidents annually. VT-2 had been lucky: no major accidents in the past three years. But primary training could be dangerous nonetheless. The airspace over Whiting Field was so crowded, most students had close encounters with another plane or a bird by their fourteenth flight. A VT-2 commander had died in a plane spin three years ago. Just eight months earlier an instructor died and a student was severely injured in a T-34C mishap at the Corpus Christi air station. Even Greco had once been forced to make an emergency landing. A thirty-nine-cent plastic part no bigger than a pencil stub broke in his fuel control unit and forced him to shut down the engine.

Greco had been a pussycat with Jenna. She had flown with him during an instrument training flight and had done poorly. But Greco had been patient in explaining her errors and gave her a better grade than she thought she had deserved for the flight. His prediction now didn't scare her. She'd show that jarhead.

• • •

Magnus Leslie rolled the car window down and let the cool wind of the late afternoon blow in as he drove on Highway 87 away from Whiting Field. He couldn't get the flight off his mind. He banged the steering wheel and gritted his teeth. They were stupid mistakes. Stupid, stupid, stupid, he said to himself over and again.

Ensign Magnus Leslie was a twenty-three-year-old tormented perfectionist. Maybe it was the Norwegian and Scottish blood that roiled in him when things went wrong. He had walked into the squadron the first day with a worried look and it hadn't left his face since. The flight, his eleventh one in the familiarization course, had actually not gone badly. He had received a grade of average. But Magnus's goal during primary was to have four above-average flights and so far he had only three in his folder. This one could have been above average, he kept telling himself, had he not committed so many small errors.

Oddly enough, if the flight had gone horribly, Magnus would not have been so irritated with himself now. He had had those kinds of flights before. Usually if something went wrong in the beginning, he'd lose his composure and the whole training session would unravel for him. But if Magnus bombed, he would simply study harder for the next flight and correct all the problems.

But it was when he had the basics mastered yet still messed up on the little things—extras that spelled the difference between an average and above-average grade—that made Magnus furious. On one stall as he turned to make the approach for a landing, Leslie didn't use enough rudder to recover the plane from the out-of-control maneuver. He performed the stall again and flew it perfectly the second time, but one out of two wasn't good enough for an above average. On touch-and-go's, where the plane set down on the runway and without stopping immediately took off, he had landed too far down the runway several times.

And then there was lining up properly over the runway as he circled around it. The students were required to circle the runway in a tight oval pattern to line up for the actual landing. The entire landing pattern was usually flown in about three and one half minutes, which meant that Magnus did not have much time to be confused or disorganized. The pattern had little room for variation if a student wanted a passing grade.

Planes landed against the wind for maximum lift if they needed to take off quickly after they had touched down. Against the wind, a student, for example, would fly first at 170 miles per hour along the right side of the runway at an altitude of 1,100 feet. This was called the "upwind leg." As his plane flew over the end of the runway, he banked it left, which was called the "beginning of the break." The plane then circled around the end of the runway, flying crosswise to the wind (in aviation lingo it was called the "crosswind leg") and slowing down to 100 miles per hour. As the plane came around, the student lowered its altitude to 800 feet.

Then came the tricky part. The student now had to maintain a certain distance from the left side of the runway as he flew the downwind leg. Magnus gauged this by looking out the left side of his cockpit down at his wing. He should see the edge of the left side of the runway cutting underneath his wing at exactly three quarters of the distance out to the wing tip, or at the point where the furthest aileron hinge connected to the plane.

Magnus always found it difficult to line up the runway edge at the correct wing tip distance. His view of the correct distance could be thrown off if his plane was too high. It could even be thrown off if his cockpit seat was raised too high or low.

By the time his plane was near the front of the runway on its left side, Magnus had to quickly lock his cockpit harness, lower his landing gear, make sure his parking brakes were off, take one last scan of his instruments for any signs of trouble, then check that his landing lights were on and finally that his wing flaps were down. As he banked the plane to the left for the final circle to the strip, he had to slow his airspeed to between 90 and 95 miles per hour. He dropped the altitude first to 400 feet, then to 150 feet, and finally to 100 feet as he lined up straight ahead on the runway and sloped the plane down for the landing.

All day, it seemed to Magnus that his wing tip distance was slightly off, or his altitude wasn't quite right in the pattern, or his airspeed was a bit too high. It was frustrating. These were minor problems but just enough to lower his grade. He couldn't study his way out of it. It was like playing a sport, he had decided. You couldn't read a bunch of books and be a star soccer player. These were piloting skills he would only develop with practice.

Magnus liked his new on-wing. Each student was assigned one instructor, called an "on-wing," who taught him through most of his familiarization flights. The on-wing was not only a teacher, he was also a mentor, a counselor, and an adviser for the student, a shoulder to cry on if he was having problems in the training. The on-wing's goal as well was to prepare his student for his thirteenth check flight, when another instructor certified that he flew safely enough to solo. The on-wing had been important to Leslie. Primary flight training could become a mill. With the squadrons having to meet quotas of pilots for the fleet, Magnus at times felt like a manufactured product. The on-wing was the closest thing he had to a friend in the system.

That was why Magnus was nearly devastated when after his tenth familiarization flight, his on-wing told him he was transferring out of the squadron. Magnus felt like a dog in a pound with no owner.

It was a relief when Dave Morey picked him up. Morey was a blunt-spoken Navy lieutenant from New York who flew helicopters. He had grabbed Leslie in the ready room after he learned that his first on-wing was abandoning him. Morey flipped through Magnus's flight folder quickly and looked up. "You look like a good student. I'll pick you up," he said abruptly. That was it. Magnus had an owner. A lot of instructors didn't like to be stuck with new students near the end of their familiarization training. If they were problem children, the new instructor had little time to correct the mess before the check flight. But Morey could tell from Leslie's folder that he'd have no trouble with this kid.

Morey loved to talk it up in the cockpit. If a student was doing poorly by his eleventh familiarization flight, Morey would try to build up his confidence with encouraging words. If the student was doing well, he would give him a grilling so the check flight with another instructor would seem easy. But with Leslie, Morey had sat back and stayed quiet. He didn't know Magnus so he wanted to give him a chance to perform, to see what he could do.

After the flight, Morey gave his critique with the bark off. "I saw some stuff I liked. I saw some stuff I didn't like," he told Leslie, and then ticked off the particulars, pro and con.

Morey's words, particularly the critical ones, kept churning in Magnus's mind as he drove down Highway 87. He worried constantly about his performance in flight school. Was he ahead or behind the other students? He wanted to be ahead. But was that being realistic? Was he where he should be in the program?

Magnus always put this kind of pressure on himself. He had been an A student at Macalester College, a small Minnesota school near his hometown of Minneapolis. At preflight indoctrination school in Pensacola, he had graduated second in his class. He had done everything he could to prepare for primary, even taking ten hours of civilian flight training before he arrived so he would have a leg up on the other students. But other students had the same idea. Civilian flight training helped in the beginning of primary school, but by the twelfth familiarization flight the students with no prior flight instruction had already caught up with the ones with prior training. Sometimes it was a drawback; the students had to unlearn bad habits they had picked up in civilian classes.

When Magnus transferred to Whiting, he told himself he shouldn't expect to be number one or two in the squadron. He realized he had to set more realistic goals for himself. In college, he was always a basket case before a big exam. If he was like that at Whiting, he'd have no fun flying and he probably wouldn't perform well as a result. Just do your best, he kept telling himself.

Magnus looked down at the speedometer. He was speeding. He lifted his foot off the gas pedal and slowed down. He was going to worry himself into a car wreck if he didn't watch out.

He walked into the tiny apartment he shared with a roommate. His wife, Amy, was in the living room with the same relaxed smile she always seemed to keep on her face. Since they had been married a year ago, they had been forced to live apart because Amy was finishing up a master's degree in public affairs at the University of Minnesota and Magnus had been preoccupied with flight school in Florida.

Magnus didn't want to talk about the day. He was smelly, tired, hungry, and still angry with himself. He had to wake up early for a six o'clock flight the next morning. All he wanted to do now was shower, eat, and go to bed. When he had a bad day, he would withdraw and keep his thoughts bottled up inside. The instructors warned students not to take their problems home. But he couldn't help it. He didn't want to be gloomy in front of Amy. She was only visiting for one week and it wasn't fair to her for him to be moody, he realized.

Amy could tell he was depressed and edgy. He complained that a neighbor's music was too loud. He became peeved at little things. She used to get angry with him in college for being so uptight about tests he would then ace. Now she found that when Magnus was in one of his dark quiet moods it was best to just stay out of his way. Magnus always worried, she knew. He would find the most obscure things and fret silently over them. Like a dog with a cold wet nose, he seemed to be healthy if he was worrying, she thought. On the other hand, very little bothered Amy. She had a cheery personality. She had to, being around Magnus.

They had met their sophomore year at Macalester and fell into instant compatibility. From the beginning, their friends predicted they would one day be married. They never fought, never even had the slightest disagreements. They acted like a settled married couple even when they were just starting to date. Amy found Magnus so thoughtful, always such a gentleman, always remembering anniversaries or special moments they had. He was smart and, when they were by themselves, funny. He took care of her and when she was around him she felt warm inside.

Amy was an oasis for Magnus. Freshman year had been confusing for him. Macalester was swept by liberalism, it seemed to Magnus, and all his friends were changing before his eyes. He was conservative politically and discovered he was one of the few Republicans on campus as well as one of the few students who voted for George Bush. But Amy was a Republican as well and so down-to-earth. Not fickle or protesting the Persian Gulf War as his friends were. The college was becoming too touchy-feely for him, too U.N., one world.

Amy and Magnus saw each other as rocks of stability. They would find themselves laughing together at their friends who put on liberal airs. He would never forget one night at a campus party when everyone was carrying on around them, Amy lay her head on his shoulder and smiled softly at him. It hit him like a ton of bricks that she was so different from the others. He had never been in a relationship with a girl for more than four months and in the beginning he expected this one would fall apart as the others had. But he was always happy being around Amy.

Amy decided by her junior year that she wanted to marry Magnus. But he wouldn't ask her until his senior year, when he had enough money saved for a ring. But before he did, he sat her down to explain how hard life could be living with a Navy pilot: the long separations, the constant coping with the dangers.

"If you don't want me to fly, tell me," Magnus said, looking her straight in the eye and secretly praying that she would have no objection. If she had, he was prepared to give up his dream.

It was a helluva question to be asked, Amy realized. She couldn't imagine breaking off the relationship because Magnus wanted to fly. She would have hated his giving up flying just to please her. Ten years from now, would he hate her for not letting him chase his dream?

"Do what you want to do," Amy said simply. She had a brother in the Air Force. She knew what military life would be like.

After several visits to Florida, Amy decided she wouldn't be like the other students' wives. Many had given up trying to find jobs and instead spent their days keeping tiny apartments clean and gossiping with other wives. They formed a tight clique, out of boredom, Amy thought. She wanted no part of it. She wouldn't join the ladies auxiliary. She was determined to have her own career even if it meant changing jobs when they had to move every couple of years.

Amy would also never adjust to the fear. One day she had been fiddling with the metal dog tags Magnus wore around his neck.

"Why do they have Magnus's religion stamped on the tag?" she asked out loud.

"So they know if he should be given the last rites after a crash," Magnus's roommate had said matter-of-factly.

It shocked Amy. She gasped inside.

The late afternoon winds gusted to almost twenty miles an hour over Santa Rosa Sound just north of the Gulf of Mexico, bouncing the T-34C that David Perrin tried to keep on a steady course. Perrin burped the remains of the greasy bacon cheeseburger and spicy fries he had eaten for lunch. He had never gotten sick in a plane. Spins, rolls, loops . . . nothing made his goat's stomach uneasy. He didn't even bother to carry a barf bag when he flew.

On the ground, he'd stick a pinch of snuff behind his lip, but in the air he just munched away on a piece of gum, waiting for the next "helmet fire," the term pilots used for when their thinking became scrambled with too many things to do in the cockpit.

Mike Hicks still wanted to pick up the pace today, to knock down Perrin with emergencies and see if he could recover. "There will be all kinds of good things happening to you," Hicks had told him with a laugh before they took off for Perrin's eleventh familiarization flight.

Perrin expected to be slammed. He knew it didn't take a rocket scientist to fly a plane. The hard part was dealing with the emergencies that could kill a pilot in the air. Some students froze. The instructors called it the "sandbag syndrome." They just sat in the cockpit like a lump until the instructors warned them of the danger or grabbed the controls. Much of fam training involved getting students to react quickly and instinctively to the hundreds of possible emergencies during a flight.

Perrin banked the plane right so it would circle over East Pensacola Bay. At the ripe old age of twenty-four, he thought he now had flying all figured out. Much of it was simple memorization, learning thousands of parts inside a plane and the hundreds of procedures for flying the aircraft. It took a Navy pilot almost five years before he completely memorized the operation of the aircraft he would eventually fly. Perrin kept a two-inch-thick stack of three-by-five-inch cards in his flight bag with items jotted on them that he had to commit to memory.

Then all you had to do was wait until flying just clicked for you, Perrin thought. He was convinced that it already had for him. After graduating from Virginia Tech, Perrin's stash job before Pensacola was five months at the Pax River Air Station, the Navy's test pilot school in Patuxent, Maryland. It had been a dream assignment. The student test pilots, all jaunty lieutenants with at least 1,000 hours of flight time under their belts, let him ride in the back seat of their jets. And when they weren't busy, they let him fly the aircraft. It was the best way to learn to handle the stick. No books to study. No radio calls or cockpit procedures to worry about; the test pilot in front performed all of them. Perrin just concentrated on the stick and rudder pedals. He was allowed to land jets, take off, and put them through aerobatic maneuvers. Hell, his stick skills were as good as those of the instructors teaching him, he thought. He was a natural. Perrin was eager to solo even though that was still three flights away.

Hicks wasn't ready to name Perrin the ace of the base. Perrin was a cocky sonovabitch, Hicks thought. He had all the trappings Hicks had seen with fighter jocks: black hair cropped short, the saunter, the Mustang convertible that he drove fast, and the wiseass attitude. Perrin was even learning to play golf, a favorite pastime for fighter pilots. But the jet jock mentality should come only after you learned to fly the jet, Hicks knew. Having the flair before that could get you in trouble during flight training. A student had to earn the right to be cocky. Perrin's stick skills were good enough for jets. He was relaxed in the cockpit, which meant he could think in flight and stay ahead of the plane. But he was still lazy on flight procedures and Hicks often had to nag him to keep up with his studies.

Perrin was a fast talker. But Hicks, one of only three black instructors in the squadron, had seen his share of fast talkers and he wasn't snowed. Mike Hicks had been born thirty years ago in northeast Washington, D.C. It had been a nice neighborhood, but he could easily have been sucked into the shuck-and-jive world where a black kid ended up with just a high school education and no job. Hicks's father, however, had been in the Air Force so the family escaped D.C. for duty stations all over the country. Hicks's mother had also seen to it that he went to private schools until the tenth grade. He graduated from college cum laude with a degree in computer information systems and went on to pilot electronic reconnaissance planes for the Navy. As a hobby, he liked to build computers from scratch.

Hicks would joke with his students in the cockpit. He gave them all nicknames. After an early familiarization flight, Perrin became "She-Ra," for the heroine in the *Masters of the Universe* cartoons. Hicks had been running him through trim drills and had given him the stick after spinning out of whack the aileron, elevator, and rudder trim wheels.

"Whoah!" Perrin shouted when he grabbed the control stick. It had about sixty pounds of pressure coiled up in it from the trim wheels being misaligned and practically flew out of his hand.

"You wimp." Hicks laughed and started calling him She-Ra because he'd been outmuscled by the stick.

But Hicks was a stern and uncompromising taskmaster on cockpit procedures, Perrin's weak point. He once chewed out Perrin for not being three flights ahead in his studies. As soon as a student thought he had caught up, Hicks would throw more tasks at him in the cockpit. If a student answered a question correctly, he would keep asking him follow-ups until he got one wrong. There was no letup. Knock them down, let them get back up, then knock them down again. Particularly with Perrin. That way his thirteenth check flight would be a breeze.

Over East Pensacola Bay, Hicks decided to let the fun begin.

"Your right fuel cap is streaming fuel," he announced to Perrin over the intercom. Each wing had an opening into which fuel could be pumped. Hicks was simulating a dangerous emergency where fuel was leaking from the cap that secured one of the openings.

Perrin knew what to do first. With fuel streaming out of the plane, he had to land it fairly quickly before the fuel ignited and blew him up. He peered outside the cockpit, searching for a landing field. He spotted one nearby at the town of Holley just south of the bay. Perrin flipped off nonessential power to the wing systems, then cranked the landing gear down by hand. Lowering it with the electrical motor might create a spark that could ignite the leaking fuel.

Then he rotated the dial marked 14 on his radio console for the frequency of the Holley field control tower.

"Come in Evergreen," he said and realized immediately that he had his airfields mixed up. Evergreen's field was north in Alabama. He was talking to the Holley field just underneath him in Florida.

"What the hell you doin'?" Hicks shouted into his mike, laughing.

"Aw right, aw right," Perrin answered and laughed himself. "So I'm a dumbass." Perrin was always relaxed around the instructors. The other students were terrified of them and treated them like gods. But they were only lieutenants. His father had been a Navy commander and they had once lived next door to an admiral. Lieutenants didn't scare him. He felt comfortable kidding with them. Hicks once had to shoo him out of the instructors' ready room when he plopped down in a couch like one of the guys.

Perrin started flying the plane in a bow tie pattern to kill time while he cranked down the gear.

"Well damn, I can smell fumes," Hicks said. "It's getting pretty nasty in here."

Perrin began the procedures to eliminate fumes from the cockpit, first turning up the cockpit air conditioning. Then, while he reduced his airspeed and lined up the plane for the emergency landing at Holley, he clipped his oxygen mask and its communication cord to his helmet and began breathing in pure air.

"Man, this stuff is really stinging my eyes back here," Hicks whined.

Perrin gave him an exasperated look in the rearview mirror. Hicks was really piling it on.

Perrin reached above him with his left hand and pulled back the cockpit canopy. The onrush of wind sounded like a tractor-trailer truck driving over him.

"Wow!" Perrin shouted into his mike. The view from the open cockpit was spectacular.

"Fly straight," Hicks chided.

"I'm trying," Perrin said.

Nearing his final pattern for the landing, Perrin radioed a May Day call to the tower below to warn that he had a plane in distress.

Then he made his second mistake. He flipped the switch near his left arm that electrically lowered the plane's wing flaps for landing. In a real emergency that might also have created a spark to ignite the fuel.

"What are you doing?" Hicks griped. "Remember your emergency procedures?"

Perrin instantly realized his mistake. "Aw fuck," he muttered.

With the fuel emergency drill finished, Perrin turned the plane south, then west along the Santa Rosa Sound. He leaned over to his left to see where a recent hurricane had washed sand over the sound's beach homes, then turned north to the city of Pensacola's regional airport. The Navy had a contract with the airport that allowed its planes to practice touch-and-go's on the long runway, so Hicks radioed the Pensacola tower and asked that they clear the airspace for Perrin. He always enjoyed holding up commercial airliners circling impatiently overhead so one of his wards could practice on the field.

Perrin executed a perfect landing, "just like the airline pilots do," he bragged to Hicks.

Hicks laughed. The airline pilots had practically half the panhandle in which to land.

Perrin lifted up off the regional airport runway, then banked south over the city of Pensacola, hoping he'd spot the house he rented down below. He didn't.

He turned the plane northwest for an outlying field the Navy rented at Saufley.

Hicks interrupted his calls on the radio. "Look, there's the Blues," he announced on his radio and pointed to the left.

The Blue Angels were the Navy's precision aerobatics team whose home base was the Pensacola Naval Air Station. They were flying their weekly practice over the station. Perrin made a U-turn in his plane to watch for a few minutes. The six F/A-18s, all painted bright blue with gold trim, were flying in a diamond formation, which was so in unison that from Perrin's vantage point it looked like a giant wing angling in the sky.

Perrin smiled in awe. To be a Blue Angel pilot, that was the ultimate, he thought. But that was so far down the road, he didn't need to even think about it. Not with the Chinese fire drill Hicks was giving him now.

"I smell smoke," Hicks said, interrupting his daydreaming.

Perrin began looking for another field to make an emergency landing.

CHAPTER SEVEN

Going Down

MAGNUS LESLIE banked the plane south for Whiting
Field. Blue skies had broken through the thin blankets of gray
clouds over the panhandle and the temperature had been warm-
ing up from the low 70s it had been that morning. Magnus was
dripping wet in his flight suit. He would have been sweating from
nerves anyway, but the plane's broken air-conditioner had made
matters worse. The cockpit felt like a sauna. Still, the check flight
was over. It was in the bag, Magnus felt. Morey had prepped him
well during his twelfth familiarization flight. The day before, they
had spent the session running through all the maneuvers, landings, and emergency procedures and Magnus had performed them
all with no glitches. Morey had even given him an above-average
grade.

The critical thirteenth check ride, when another instructor certified that a student flew safely enough to solo in the plane, was the most difficult flight a student made in primary. "Study hard," Morey had advised Leslie. "You should be a little uptight and stressed for this flight." For Magnus that was an understatement. "But don't let it upset your flying." Magnus would try desperately not to.

"Call the other guys to get some gouge on the instructor who will be doing your check ride," Morey said finally.

Magnus had made a few phone calls but had found out little about John Prickett, the Navy lieutenant who was now sitting in the cockpit's back seat marking notes on his grade sheet. Prickett was calm during flights, he had learned. He didn't play tricks on students during their preflight checks before takeoff. Some instructors loved to do that to make sure the students really were inspecting their instruments, Magnus knew. The students even had a gouge sheet on the favorite instructor tricks, like screwing up the gyroscope heading indicators before takeoff or fiddling with the transmitter that governed the plane's fuel flow. Prickett never got cute with the instruments. But for the past hour he had been a question machine.

Magnus didn't mind that. In fact, he liked the mental sparring with instructors. He was already wound tight as a spring, checking and rechecking his instruments to make sure little mistakes didn't trip him up. All the what-if questions kept him from dwelling too much on the flight and calmed his nerves.

Magnus set the plane at the altitude it needed to be for the flight into Whiting's airspace.

"I don't see any problems," Prickett finally said out of the blue over the cockpit communicator.

Magnus grinned. He was in the clear. Only an act of God could now screw up the rest of the ride. He had landed at Whiting so many times, there was no chance of making a silly mistake on that leg of the flight.

Prickett kept cranking out the questions.

"What are the normal and maximum ranges for the oil pressure?"

"Sixty-five pounds to eighty-one pounds," Magnus answered.

"What's the normal range for the interstage turbine temperature?"

"Four hundred degrees to 695 degrees."

"What's the maximum temperature at idle?"

"Six hundred and sixty degrees."

Prickett gave up for a while.

Magnus quickly banked the Turbo Mentor to the right. Prickett looked up from his notes surprised.

Magnus had seen a helicopter ahead of him and had turned to avoid it several seconds before the nearby Pensacola approach control station had radioed that the two aircraft were getting too close.

"Good job," Prickett conceded.

Magnus now had a slight smirk on his face. *He* had seen something before the instructor. Usually the vigilant instructors spotted other planes long before he did. Chalk up one little victory for the student, Magnus thought to himself.

Prickett, in fact, was mildly impressed with Magnus's flying skills. He knew the kid was a worrier, but Magnus actually seemed calmer to him than most students on their check flight. Prickett could usually tell from a student's preflight briefing if the ride would be uneventful or a nightmare. If the student was fuzzy on basic facts during his briefing, Prickett would begin nipping him with questions like a terrier to find out what else he didn't know. He wouldn't think twice about giving a "ready room down," canceling the flight and failing a student based just on a poor preflight briefing. But Prickett could tell immediately that Leslie had done his homework.

The other instructors called Prickett the "fiery Welshman." Prickett was of Welsh descent. He was short with curly blond hair. He had graduated from Boston University with a history degree and had traveled to St. Catherine's College at Oxford for his master's. He wrote his thesis on the Social Democratic Party of Britain. But he was a political independent now and didn't think he was a fanatic about his heritage. Of course, he was a regular at Celtic festivals—he enjoyed the music—and northwest Georgia had a large Welsh community, some of whom still spoke the mother language.

Prickett wanted to be "a little asshole" in the cockpit, as he put it. He loved to pose problems for the students, scenarios that they wouldn't find in their flight manuals, to force them to think beyond the procedures they had memorized for dealing with emergencies. Before climbing into the cockpit, Prickett threw such a problem at Leslie.

"You have streaming fuel, but let's say the manual crank doesn't work to lower the landing gear. What do you do?" Prickett asked.

The question wasn't that hypothetical. Prickett had just returned from a cross-country flight with a student. Taking off from an Air Force base near Atlanta, Prickett had noticed fuel seeping out the rim of the fuel cap on one of his wings. Other instructors later blamed it on a lazy Air Force crewman who hadn't secured the cap tightly enough. Prickett was no stranger to mishaps. Three months earlier, his radio had caught fire on a night flight and he had to land in the dark with his electrical system shut down. For this latest problem, he quickly grabbed the controls from the terrified student and ordered him to begin hand-cranking the gear down. They were about 800 feet off the ground. Prickett made a hard U-turn, then radioed an emergency message to the tower. The student pushed the emergency landing gear handle on his right with all his might but it wouldn't budge. Then they heard a cable snap. Suddenly the hand crank spun freely. There was no way now to manually lower the gear.

Prickett decided that he would have to electrically lower the gear and take his chances with a spark igniting the leaking fuel. With fire trucks lined up on the runway, he waited until he had the plane no more than 200 feet from the ground before lowering the landing gear at the last minute. That way if a fire started, they would be on the ground quickly and perhaps have enough time to jump out before the flames reached the cockpit. At least, that's what Prickett hoped.

He activated the electric mechanism. The gear rotated down. To his relief, there were no sparks. They landed safely.

The flight manuals only prescribed lowering the landing gear by hand as the emergency response to streaming fuel. Magnus took a guess at what he might do if the hand crank broke.

"Set the plane up for a landing on its belly," he answered.

Wrong. Lowering the gear electrically had less chance of creating sparks than the metal fuselage scraping along the runway. Prickett didn't expect Leslie to know the right answer. He just wanted to get him thinking about emergencies. "If something isn't covered in the book, use your best judgment," Prickett lectured.

"The big thing the instructors need to know is that your head's squared away and you can get back safely."

As Leslie prepared to land at Whiting, Prickett totaled up the grade in his head. Magnus used common sense in the cockpit. Just out of Whiting on their way north to Alabama, Prickett had asked where the best place was for an emergency landing.

Magnus thought about it for a moment. The obvious choice was to turn around and land back at Whiting. But he worried that the answer was too simple.

He decided to try it anyway. "Whiting," he said.

Good answer. Students usually picked a field ahead of them, never considering that the field they'd just left might be closer.

Magnus had performed the early emergency exercises safely enough for a passing grade. During one he had a choice of two farmer's fields split by a road. He picked the field north of the road.

"Good choice," Prickett told him. Magnus would likely have crashed into the road's power lines if on his south-to-north route he had tried to land in the lower field. Most students wouldn't have spotted the power lines until it was too late.

Magnus had actually picked the north field because it was the longer of the two. Only later did he actually see the twenty-foot-high power lines. But if Prickett thought he had seen the lines earlier than he did, Magnus wasn't about to tell him otherwise. Like an ice skater who nails his first triple jump, Magnus now felt the rest of the flight would go well since it had begun smartly.

On his touch-and-go landings, however, Magnus still tended to be too tight on the downwind leg; the left edge of the field crossed at the two-thirds point on his wing rather than the three-fourths distance, Prickett noticed. He also flew too low on his downwind leg. Magnus always landed safely but the touchdowns were long on the runway and slightly off the centerline.

Magnus became frustrated, then defensive about the inconsistencies in his landing patterns. He began making excuses in his mind. Maybe there was some quirk in the plane that made him line up wrong each time in the landing patterns. Even when Prickett took the controls for a lineup, he was slightly off, Magnus thought. But he didn't mention it to Prickett. That would appear to be quibbling, which the instructors hated. Magnus didn't want to sour the flight, especially when he knew he'd likely pass.

As he taxied to his parking space, Prickett peppered him with more questions.

"What do you do if your right brake fails now?" he asked.

"Use maximum beta reverse," Leslie began. On top of the power control lever that Magnus grabbed with his left hand was a red button marked BETA. When he pressed that button down and yanked back the lever past the idle setting to the beta setting, the pitch on the propeller blades shifted to blow air forward of the plane instead of to the rear. Pilots used the beta setting, which produced reverse thrust, to stop a prop plane when it landed on a runway.

"I'd also push in the rudder on the failed brake side to keep the plane from yawing to one side," Magnus continued.

Prickett gave Leslie an above average for his knowledge of procedures, a below average for the landing pattern work, and averages on the other categories. It was an average flight.

"Have fun on your solo," Prickett finally said.

Walking into his apartment an hour later, Magnus saw that Amy had an anxious look on her face.

"How'd you do?" she finally asked.

"I passed," he said, with the first smile she'd seen in two days.

She ran up and hugged him.

Magnus drove Amy to the Pensacola airport for her flight back to Minnesota. It always took him several days to recover from the gloom he felt after she flew back. When he returned to his apartment, he spotted a note she had left in the bedroom. "Whatever happens, I love you," it read. "I hope your flights go well." The note was leaning against a stuffed teddy bear in an aviator's suit.

Rob Hoehl got his first inkling that he was in for a rough day as the plane sped toward the Alabama state line. Ensign Hausvik, who sat in the front cockpit seat, had taken off fine from Whiting Field. Her briefing before her thirteenth check flight was excellent. Crisp, to the point, not a single checklist item missed. After Jenna had leveled off the Turbo Mentor at 5,500 feet, Hoehl thumbed down the switch on his power control lever that activated the cockpit communications radio.

"Set up for a turn pattern to the right," he ordered. Clearing turns to the left and right were the simplest maneuvers a student learned.

"Turn pattern to the right, yes sir," Jenna said smartly. "Clear right, turning right."

She was a talker, Hoehl already realized. The turn, however, wasn't smooth. Jenna seemed to be bobbing up and down on the horizon.

"Sir, it looks like my gyro is off," Jenna complained.

Hoehl checked it. The gyroscope worked fine. Jenna hadn't adjusted it correctly. He suspected she was paying too much attention to the instrument and not enough to the horizon outside, which would just as easily show her that the plane wasn't flying level. He also noticed that the trim wheels in his cockpit moved little after each turn. Jenna was neglecting her trimming, which could account some for the plane's erratic flight path.

At twenty-nine, Rob Hoehl was tall and thin with a receding hairline. He had played guard on the University of North Carolina's junior varsity basketball team but hadn't been good enough for Dean Smith's varsity team. Jenna hoped the fact that they both had played college ball might have some bonding value for the check ride. It didn't. Lieutenant Hoehl was probably the toughest grader she could get for a check flight, according to the student gouge. Not only did he instruct the cones, Hoehl also tested his fellow instructors to see if they were qualified to teach students. He also flight-tested the training planes after they had been through maintenance to ensure that they had been repaired properly. And on top of that he had been an instructor pilot for the Navy's P-3 Orion sub-hunting plane. He knew the T-34C and flight operations like a rabbinical scholar knew the Talmud.

Hoehl wasn't unfriendly. He never got upset with students in the cockpit. His deep voice stayed monotonously even and grim practically all the time over the radio. But his patience was constantly tested by mistakes in the air. And he was uncompromising on grades.

Jenna next worked her way through a stall and spin. She was surprised how nervous she was for this flight. In the students' ready room she had sat in a corner by herself hunched over her flight books. She had that don't-bother-me-I'm-about-to-take-my-check-ride look on her face and other students in the ready room knew to stay clear. Jenna thought Hoehl would be impressed that she talked her way through the spin with no problems.

He wasn't particularly. Though intimidating, a spin was not an extremely difficult maneuver. He expected any student by the thirteenth ride to perform it almost flawlessly.

As the plane neared Brewton, a tiny Alabama town just across the border, Hoehl began the toughest maneuvers the students had to master in familiarization training: "happles and lapples." Short of being shot at in the sky or running into another plane, the biggest problem a military pilot could have during flight was losing engine power. High-performance jets tended to sink quickly without power so unless a pilot could fix the problem and restart the engine, his best bet was to eject. But a prop plane like the T-34C glided well. At 8,000 feet with a dead engine it could stay in the air for as long as ten minutes, which gave the pilot a lot of time to find a landing strip or at least a farmer's field. Southern Alabama had plenty of the latter.

The instructors divided the power losses into two types: those that occurred above 2,500 feet, called a high-altitude power loss (or HAPL), and those that happened between 800 and 2,500 feet, a low-altitude power loss (or LAPL). In each case, the pilot had to quickly find a suitable field for landing and then glide down to it in a short circular pattern even more precise than what it took for a regular touchdown of the plane.

"Okay, you have a simulated flashing fire warning light," Hoehl said abruptly. The warning light was usually the first indication a pilot had that there was a fire in his engine.

"The first thing to do is confirm that the fire exists," Jenna said, checking outside for any signs of smoke.

"There's smoke coming out of the right side of the engine—simulated," Hoehl continued.

"It's confirmed I have fire," Jenna said quickly to demonstrate she knew the emergency procedure. "So simulated, condition lever to fuel off." That was the sliding knob next to the power control lever on the left, which she would have pulled back in a real fire. "Emergency fuel shutoff handle pulled." The striped handle that she would reach back with her left hand to pull would then cut off all fuel to the engine that might feed the fire. Next she had to keep the fire's smoke from filling up the cockpit. "Cockpit environmental control off and cockpit air off," she continued rapidly.

Jenna had performed the emergency procedures correctly. Hoehl pulled the power control lever back to simulate the engine shutting down because of the fire. Since the plane was above 2,500 feet she had to execute a landing for a HAPL, a high-altitude power loss. She searched outside for a suitable field.

"In a real happle, if I had my choice I'd go over to the Brewton field, sir," Jenna said more as a question. There were three remote but paved airstrips arranged in a triangle just south of the town. "Is that okay?"

"Brewton is nine and a half miles away," Hoehl interrupted with a slight edge to his voice. "Do you think you can make it from here?"

Jenna looked outside again and thought for a moment. She was already starting at 7,000 feet. "I think I could make it," she answered.

Hoehl didn't think she could. "Come around to the left here," he ordered.

"Yes sir," she said and banked the plane left. She still thought she could make it but wasn't about to argue.

Hoehl didn't plan to let her think twice about it anyway. "Stay away from the power lines and find a field up here to the north," he said quickly and firmly. The farmers' fields almost underneath them were cleared enough for an emergency landing. That way Hoehl wouldn't have to hold his breath wondering if she would be short of the Brewton runways.

Jenna found a long field to her left. She simulated a distress call on her radio. "Mayday, Mayday, Mayday, this is zero four five [the number of her plane]," Jenna said. "I have a fire and shutdown. I'm located about five miles southeast of Brewton. I'm going to do a happle into a farmer's field."

But the HAPL required quick thinking and precision air work, both at the same time. While she radioed in her Mayday call and checked her equipment to make sure she was ready for an emergency landing, Jenna also had to maneuver her plane to a precise point at the top of the emergency landing pattern, which in this case was 2,500 feet off the ground and near the bottom left side of the long rectangular-shaped field.

Pilots called this point "high key" and they were able to hit it by looking at their altimeter and lining up their plane so the left edge of the field cut under their wing at about one fourth the distance down the wing. From there, the pilot glided the plane forward

In water survival training, aviation students learn how to parachute into the ocean. A pool at the Pensacola Naval Air Station is used for instruction.

Students escape the Helo Dunker from underwater.

A student gets ready to plunge into the water in the Dilbert Bunker.

Stacie Fain
after the Helo
Dunker.

A student practices being lifted by a helicopter hoist.

Students in the hyperbaric chamber being deprived of oxygen.

The T-34C Turbo Mentor.

Jenna Hausvik in the T-34C.

Magnus Leslie after his solo flight.

David Perrin.

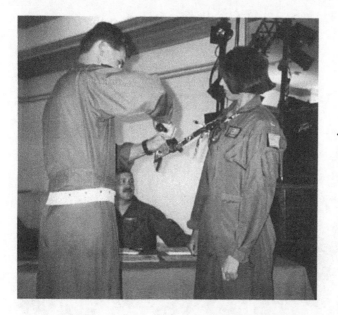

Jenna Hausvik has her tie cut after her solo flight.

David Perrin with his tie cut.

Two T-45 Goshawks flying in formation.

The view from the cockpit when a T-45 dives to a target to drop its bomb.

Mike Sobkowski, foreground, and Jonathan Wise strap on parachute harnesses and survival vests for a flight.

Bill Sigler, left, and another pilot being wetted down after the flight that earns them their wings.

Rob Dunn, right, with Jonathan Wise.

Mike "Mango" Carr, left, on the LSO platform.

Jets launching off the USS *John Stennis*.

The USS *John Stennis.*

Landing signal officers on the LSO platform watch as a Goshawk is about to touch down on the *Stennis.*

Rusty "Wolfie" Wolfard shouts out a grade to his writer for a student who has just landed.

Jonathan Wise and Maria Grauerholz on their wedding day.

K. KREISER

Javier Ball in the F/A-18.

Rich Whiteley suited up before taking off in the F/A-18 Hornet.

Joe Gelardi back from a flight in the Hornet.

Curtis Carroll plots way points on a map for a strike mission.

An F/A Hornet taxies to a catapult on the USS *John F. Kennedy*.

Phillip Clay stands on Vultures Row aboard the *JFK*.

A Hornet prepares for a cat shot at night.

An F-14 Tomcat launches as the sun sets.

Alex Howell mugs for the camera
on the flight deck of
the *JFK*.

Author in the back seat of a T-45 Goshawk before takeoff.

Author on the flight deck of the USS *John F. Kennedy.*

Author interviewing Commander Bill Shewchuck after a flight.

Author in the back seat preparing to launch off the USS *John Stennis*.

Author, second from right, standing on the LSO platform watching jets land aboard the *Stennis*.

Author, center, talking by phone to a jet about to land aboard the *Stennis*.

Author on the bridge of the *Stennis* watching jets land.

Author, right, interviewing Rob Dunn, left, in the squadron ready room after a flight.

Author, right, interviewing an instructor pilot, Lieutenant John Klas, before a flight in the F/A-18 Hornet.

against the wind. Then he banked right at a 15 degree angle so he crossed over the front of the farmer's field to a point wide to the left of it at a lower altitude of about 1,200 feet.

This next point was called "low key" and the pilot, if he was lined up correctly, would see the right side of the field cut under his right wing at about two thirds of the way down. The pilot continued the long loop around, dropping to about 600 feet, until he was lined up behind the clear field about 1,200 to 1,500 feet away. From that point he should be able to land in the first third of the field.

In effect, the pilot had to fly, near perfectly, one rotation of a round spiral. If he began at the wrong point, missing his high key and then his low key by more than fifty feet, he could reach the end of the spiral at the wrong point, which might mean slamming into trees short of the farmer's field.

Speed was also important. The pilot not only had to hit high key at 2,500 feet and one-quarter wing tip distance from the field, his plane also had to be traveling at 100 miles per hour. A variation of more than plus or minus five miles per hour could have him flying too high or too low on the spiral.

Jenna hit the high key at the right point, 2,500 feet above the ground, but her plane was barreling ahead at 130 miles per hour, thirty over the speed limit. Going that fast she couldn't lose altitude quickly enough to loop around the front of the field and hit low key at 1,200 feet. Jenna lowered the flaps on her wings, which would force the plane to lose altitude quicker, but it was too little too late. She was still too high. The problem was compounded by the shape of her spiral down. Instead of taking a wide loop around the right side of the field, which would have given the plane more time to lose more altitude, Jenna had misjudged her wing tip distance and flown a tight pattern close to the right edge of the field.

"I've got my field," Jenna said to herself over the microphone at her lips. "Still looking good."

But the pattern looked anything but good to Hoehl. Because she had flown so close and high coming around to low key, Jenna now had to extend her flight much further out before she made her final U-turn to approach the start of the field. Instead of flying a perfect circle down the spiral, her route looked more like a long jelly bean and still she hadn't given herself space to burn off enough altitude, Hoehl thought.

"I'm a little high and tight," Jenna finally murmured to herself. "I'm going to go ahead and do a little slip."

Hoehl laughed to himself in the back seat. She sure as hell needed to slip, he thought. One way pilots could lose altitude quickly was to tilt the plane to one side while flying straight. This procedure, called a "slip," dramatically lowered the lift the wings provided and could drop the plane at a rate of 2,000 feet per minute.

Jenna finally got rid of the excess altitude as she lined up behind the field, her altitude warning beeper ringing constantly as the feet dropped. She could still easily clear a small pond in front of the open field. But there was another problem.

"What airspeed should you have on the final approach to the field?" Hoehl asked menacingly.

"Ninety knots, sir," Jenna answered.*

"We're at 115 knots," Hoehl pointed out. "We're fast. You have to get your airspeed under control better." He took over the control stick as the plane neared 600 feet and flew it up to end this exercise.

"Even with that big field, because we were so fast and high we still would have had a hard time getting the plane landed before the end of it," Hoehl explained.

Jenna winced. She was having a bad day. The landing patterns had given her trouble during her twelfth fam flight with Consoletti. She knew the correct way to fly the patterns. It was as if something was blocking the directions from her brain to her hands and feet.

But Jenna was still confident. The flight was still salvageable, she thought.

Hoehl wasn't ready to count her out either. Jenna still could pass the flight. But she had to make up for that dismal HAPL with better landings. He tilted the plane's nose up to lift the aircraft to about 1,000 feet, then handed her back the controls. Several seconds later, he pulled back the power lever to idle and told her the engine had flamed out.

Jenna picked another farmer's field not far from the one where she had performed her HAPL exercise. This time she had a low-altitude power loss, a LAPL. She had to race more quickly through

---

* Pilots use nautical measures of speed. One knot equals a little more than one mile per hour. To make it easier for civilian readers, I have chosen to use miles per hour throughout the narrative.

the emergency procedures as she glided the plane to the field. She simulated pulling the power control lever back to idle, pressing the emergency fuel shutoff handle down, and turning the starter on to see if she could get the engine running again. When it didn't, she turned off the condition lever that was next to the power control lever, yanked up the emergency fuel shutoff handle, and simulated a Mayday call, all within a few seconds.

She locked in her shoulder and seat belt harnesses and prepared to enter into the same type of emergency landing pattern she had flown for the high-altitude power loss.

"Which field are you looking at?" Hoehl asked.

"The little one, sir," Jenna said, pointing to a barren patch of land the shape of a circle just below her. It wasn't the best field, but in a low-altitude power loss you worked with what you had, and there was nothing else suitable nearby. Rattling off the procedures over the cockpit microphone, she quickly made her first turn across the end of the field, but, as before, she was still high at low key. She should have quickly lowered her flaps at the beginning to lose altitude, Hoehl realized. Instead, she tilted the plane into a slip to make it descend faster. But even so, she was too high coming into the field. Hoehl took back the controls before she made her final approach.

"Do you think you could land in this field?" he asked.

"I think I would have hit the trees, sir," Jenna admitted. There wasn't the slightest hint of anxiety in her voice. But she was seething inside. Her airspeed had been ten knots too fast and she still wasn't hitting low key at the right spot.

She quickly shook it off, however. Never let them know you're upset, she told herself.

Hoehl had to hand it to her. Jenna was a trooper. Most students would have fallen apart after two disastrous emergency landings. But Jenna didn't lose her head. She was still in this flight, he thought. That's good. But she was still screwing up.

They flew further north into Alabama where the Navy rented two small runways shaped in an L near the town of Evergreen. For the next simulated emergency, Hoehl announced a "chip light." There was a magnet in the bottom of the plane's engine where the oil collected. If the propeller gear box suffered from too much friction and created chips of metal in the oil, the magnet would detect it and alert the pilot that the gear might fall apart. In this emergency, there was still power in the engine so Jenna had more flight time left before she had to land. That gave her more time to find a paved field for what was called a "precautionary emergency landing."

The Evergreen landing strips, which were a little west of Interstate 65, would do just fine, she decided. With power still in the engine, she climbed to the high-key position of 2,500 feet. But her landing pattern was almost a mirror image of the others: too high, too tight, too fast. It was frustrating. Jenna didn't think she had "checkitis," the term students used for choking during their check flights. She still felt calm. She didn't feel even a hint of airsickness. But it was like a basketball game where nothing you threw up went through the hoop.

They practiced touch-and-go landings, landings with the flaps up, then down, emergency landings just after takeoff, and still in all of them she was high, tight, and fast. The landings themselves were safe, but the plane's wheels were touching too far down the runway. That might not be a big deal on a civilian flight, but aircraft carrier pilots had to hit their landings exactly at the beginning of the deck or they would miss the arresting lines. That standard of precision had to be drilled into a student's head from the start of his training.

"You need to land in the first third of the runway or else wave it off," Hoehl said, now not hiding the irritation in his radio communications.

"Yes sir," Jenna said quickly, still trying to exude confidence.

Hoehl was now prepared to give her a down for the flight. Instructors weren't eager to pass out downs. It meant more paperwork to justify the failing grade, reviews by higher-ups of the decision. The student could protest the grade as some now did, which would mean more reviews and second-guessing.

Unstated was the fact that Jenna was a woman. She was just another student as far as Hoehl was concerned. The VT-2 instructors liked to say that the students were all the same to them; all they saw in flight was the back of their helmets. Gender wasn't a factor in the cockpit. The females performed no worse or better than the men. They were all a pampered lot, the instructors grumbled, the women no more so than the men. But still with a woman, the instructor knew that his case for a down now had to be solid. That was because it hadn't been in years past when sexism was still rampant in the aviation community.

But the next part of the flight erased any lingering doubt Hoehl might have had. Even if everything before it goes well, to pass the thirteenth check ride, a student must fly the instructor safely back to Whiting. That meant not becoming lost on "course rules," the precise flight path the student must take when entering various controlled airspaces for the return home. The Navy was picky about the students flying exact routes in those airspaces. The skies over Whiting were too crowded to have planes wandering about aimlessly.

The part of Alabama Jenna had been flying over all day was designated Area Two on her navigation charts. But Jenna had flown little in Area Two and students constantly complained that it was easy to become lost there with nothing for reference points below but brown and green patches of farmland. At least in Area One, another flight training site along the panhandle, the students always had the Gulf coast as a ready reference.

In order to enter the correct flight path in Area Two for the trip back to Whiting, Jenna had to start at a point just south of Evergreen. But she became disoriented. Instead of flying southeast to reach the entry point for the flight path, she aimed the plane northeast and flew past Evergreen. She was heading away from the entry point not toward it, Hoehl knew.

"Sir, I'm looking for Evergreen, but there's a lot of smoke in the way," Jenna said, perplexed.

Hoehl smiled. Even if the smoke from a brush fire on the ground cleared she wouldn't see Evergreen looking in that direction. They had already passed the town. He wondered how long it would take her to realize she was off course.

Two minutes later, Hoehl could see the back of Jenna's helmet twisting from side to side. He knew she now realized the plane was headed in the wrong direction.

Jenna hunted desperately for something she recognized on the ground. Finally, she spotted a sawmill off Highway 31, which she knew was far west of Evergreen. She made a quick U-turn and headed southwest, knowing that she would eventually pick up the entry point.

Hoehl could live with this detour. At least she had recognized her mistake and gotten back on course. But the confusion just seemed to bring on more confusion. Hoehl looked at the altimeter on his control panel. They were flying at 3,900 feet when they should have been at 3,500 feet.

To navigate the correct flight path, Jenna now had to fly parallel to railroad tracks that ran north–south by the tiny village of Castleberry. Then south of Castleberry, where the tracks bent to the right and on into the town of Brewton, Jenna was to bank her plane to the left for a 130 degree heading to the small Conecuh River bridge that was east of Brewton. But there were dozens of creeks and rivers below, all with small bridges crossing them.

"I see the river up ahead," she announced over the intercom. "There's the bridge."

Hoehl lifted his eyebrows. "Where's the bridge, Jenna?" he asked.

"Right there, sir," she said, pointing down. "We're right over it."

"Where's doghouse field?" he asked wearily.

Jenna looked out the cockpit canopy again. "Right there ahead of us," she said. Then it dawned on her.

Doghouse field was a patch of plowed land with a building on it, which from overhead appeared shaped like a doghouse. It was an easily identifiable landmark the pilots used for the point along the railroad tracks where they should begin to bank left for the 130 degree heading to Conecuh River bridge. If doghouse field was just ahead of her, that meant that the Conecuh River bridge was actually far off to the left at the end of the 130 degree heading—not underneath her plane.

"Could that possibly be the right bridge," Hoehl asked sarcastically, referring to the one she had just passed.

"No, the bridge is further down," Jenna said with a sigh. She had just proven that she was lost in the flight path. She now knew she was sunk.

Jenna put the plane on a course for the real Conecuh River bridge. Now she was unraveling. At the bridge, she was supposed to turn right for a 205 degree heading south to Whiting Field. But rough winds had swept through, buffeting her plane.

Five minutes later, Hoehl took over the controls. "What should our heading be?" he asked.

"Two zero five, sir," Jenna answered.

176

"Then why are we heading on one eight zero?" he asked. The plane had been drifting off course to the left and straying into the wrong airspace. He redirected the aircraft back to the right course.

Jenna's next spot for a turn to the left was Point Charlie. From there, she would line up for her landing into Whiting Field.

Please, God, let me at least find Charlie, she prayed to herself. The way point was a farmhouse at the corner of Highways 182 and 89. But there were three road intersections near there and Jenna had always had difficulty picking out the right farmhouse at the corner for Point Charlie. It used to be the one with a large grain silo beside it. But a hurricane several weeks earlier had blown down the silo.

Finally, Jenna found the right farmhouse. She let out a breath. At least one thing in this cursed flight went right.

Hoehl was thankful as well. He wanted the ride to be over.

Climbing out of their cockpit, Hoehl turned to Jenna. "I'm not going to be able to pass you," he said softly.

"Yes, sir," she said just as quietly. Jenna already knew she had failed.

"I don't like giving these any more than you like getting them," Hoehl said. And he really meant it. Jenna would be a good pilot, he believed. She'd recover from this. Some students just needed a bit more time to pick up the flying skills. Hoehl certainly did. He had also received a down as a student on his thirteenth check flight. He hadn't slept the night before because of nerves. And his flight had been even worse than hers. He had fouled up so badly, the instructor had brought him home before finishing the test.

Jenna sat back in one of the padded chairs in the students' ready room, her long legs stretched out and crossed, her fingers laced behind her neck. She stared straight ahead with a blank look on her face. In her mind, the horrible flight churned over and over. Her stomach felt knotted. She had eaten little breakfast before taking off.

The other students in the ready room instinctively knew not to bother her, as if she was a wounded animal that needed time to recover.

Maybe she had been jinxed, she thought. Her solo flight had already been scheduled for that afternoon, presumably after she had passed the check ride. Maybe it was bad luck to put the cart before the horse. She quickly swept aside any thought that the down had been unfair. Jenna knew she had bombed on the flight. No question about that. She would have felt even worse if Hoehl had given her a social promotion. She detested people who were passed to fill quotas. In a way, she was glad how things had turned out today, she decided. She had always been afraid of someone lowering the standards to slide her through flight school. Making her jump over the baby wall at the obstacle course was humiliating enough. To be sure, she wouldn't put up with an instructor harassing her because she was a woman. But she also knew she had no business soloing until she was qualified.

Jenna's stomach relaxed. She began to feel better. She had been treated just like any of the guys. In an odd way, it felt good. Receiving a down wasn't the end of the world. Half the students got them at one point in their training. This just happened to be her time to be part of that half, she decided. You wouldn't be assigned to jets with a lot of downs, she knew. In primary training, a student with three downs might be sent to a review board that would decide if he should be dismissed from the program. But that number was not chiseled on stone. Opponents of women flying combat jets would argue that females were passed in primary training with more than three downs. Sometimes they were. But many of the men got those same breaks as well. Even without favoritism, the number of downs before washing out could go up or down for all students from one year to the next. It depended on how many pilots the Navy needed. If the number of slots was low, the downs might remain at three or lower. If a large pool needed to be filled, more downs would be allowed and the squadrons would spend more time salvaging substandard students.

Jenna flipped through her thick flight folder in the student control room at VT-2's headquarters. The Navy required a stack of paperwork to be filled out for a down, just in case a student didn't feel miserable enough that he had failed. In her folder was a sealed letter from the Navy that she hadn't had a chance to open. She tore apart the envelope now.

It was a glowing note praising her for spending a day with Navy researchers testing how cold-weather suits fit women in the service.

Jenna laughed. A down and a commendation letter for trying on clothes, all in one day. Only in the Navy.

• • •

Mike Hicks stormed into the instructors' ready room. He was so furious his eyes were practically watering. He had postponed a vacation to Disney World with his family just so Perrin could have his twelfth familiarization flight early and then be scheduled for his check ride by the end of the week. His wife, Violet, was hardly delighted with that news. And what does Perrin do in return? He flies like a beginning idiot, Hicks raged.

He vented his troubles to other instructors in the room. Perrin had no "situational awareness" in the air today. He seemed to be flying blindfolded. On spin recoveries he pushed the stick too far forward. Coming out of stalls his power was too low. His emergency landings were all off. If the emergencies had been for real, they'd be scraping him off some field.

And course rules. They were a thrill. At one point Perrin was flying north to Brewton at an altitude at which all the rest of the planes were flying south in the flight path. It was like driving the wrong way down a one-way street. They would have had a head-on collision if Hicks hadn't grabbed the controls and turned the plane around.

He'd been worried that Perrin was too cocky going into this flight. The kid had been nonchalant and joking during the pre-flight briefing. Hicks had been the nervous one, not wanting his pupil to do poorly and then embarrass him on the check ride.

Hicks had decided to sit quietly in the back seat during the ride and not prompt Perrin, to see if he knew the procedures. Let him hang himself if he didn't.

Well, Perrin had practically hanged them both. Hicks had half a mind to give him a down for the flight. He certainly deserved it.

The lieutenant sounded out the other instructors for a second opinion. Maybe this was a fluke, they advised. Let Perrin try his check ride. If this flight hadn't been an aberration, the check ride instructor will flunk him without a doubt.

Hicks thought for a while, then decided to give Perrin a below average. But first he'd have a piece of his ass.

Dave Perrin sat in the line shack's flight briefing room. It was divided into ten small cubicles with chairs in front of them and small blackboards on the back. On some of the cubicles' desks sat wooden models of the T-34C stuck to long sticks that instructors used to demonstrate maneuvers in flight.

Perrin was slumped in the chair, his chin resting on his chest. For the first time in ages, he had nothing to say. He was as depressed and angry as he would ever get. Perrin may have been a wise guy but he worshipped Hicks. He thought he was the best pilot he would ever fly with. He felt horrible that he had let Hicks down.

Hicks plopped into the chair next to him in front of the cubicle. He stared at him for a moment.

"You having girlfriend problems?" he finally asked.

"No," Perrin said quietly.

"You kill someone?"

"No."

"Well, you need to pull your head out of your ass and get your shit squared away," Perrin said with a growl. "The only way you can pass the frickin' check ride is if you pay attention to what you're doing."

Perrin nodded and looked down ashamed.

"Pay attention! It's the little frickin' things that all add up until it's one big old crescendo of shit."

Perrin's piloting skills were solid, Hicks told him. "But you have got to pull it together for tomorrow if you're going to pass the check ride."

Hicks now ticked off the problems with the flight from the notes he had taken: simple turns that were faulty, poorly executed spin recoveries, erratic stall maneuvers, overshooting or undershooting emergency landings, becoming lost on the flight path home. For twenty minutes, Hicks drew picture after picture on the blackboard in front of them to illustrate sloppy maneuvers and patterns. "What's going on here?" he complained. "Where are the brain cells you expended?"

Perrin had no excuses. He shook his head in disgust. He had been to Area Two three times before. "I knew exactly where I was each time," he said dejectedly. "We did spins, everything. I never got lost. Today, I just don't know what happened."

Hicks finally tried to calm Perrin down. He didn't want him too frazzled for his check ride the next day. "Remember each day in the plane is different," he began. Don't dwell too much on today's flight and the bad patterns you flew. "It's the little things that add up to kill you on a ride."

Hicks finally managed a smile and a chuckle.

"Good luck tomorrow," he said. "It's up to you. Have fun, but don't dork it up."

Hicks folded his notes, stood up from his chair and left the briefing room.

Perrin still sat there staring at the scramble of chalk marks on the blackboard.

CHAPTER EIGHT

Redemption

THE snack bar and grill that pilots ate at between flights was just across the wide hangar parking lot behind the west line shack. David Perrin settled down to his favorite lunch of bacon cheeseburger and spicy fries. For his vegetable of the day he tolerated a small salad next to it. The cockiness was back. He had spent the night before brooding over his awful twelfth flight. He wouldn't have argued if Hicks had given him a down. He would have certainly deserved it. But Hicks knew I could fly better than that, Perrin thought to himself. That's why he let me take the check ride this morning.

Smart choice, he decided. Yesterday was a fluke. My brain was in my ass. I knew I could fly better than I did.

He had had a good night's sleep and woke up early this morning for some last-minute studying. One of the items he checked again was the operation of what the pilots called the "naquist," the Naval Air Close Warning System (NACWS). It was a tiny green screen the Navy had just installed on the right side of his control panel, which warned him with a beeping alarm of other planes that might be dangerously near him. Perrin had decided that he would return to Area Two for his thirteenth check flight to prove to himself that he could fly over those fields and not become lost. But just in case he did fly down another one-way street, he wanted to make damn sure he knew how the electronic warning system operated.

But the NACWS never beeped during his flight this morning and Perrin never got lost. He passed his thirteenth check ride with flying colors. It was a breeze, Perrin decided. Maybe Lieutenant Schutt had been a "Santa Claus," the name students gave to instructors who were easy graders. Perrin didn't know, and right now didn't care. Steve "Scooter" Schutt was a handsome lieutenant who flew converted Boeing 707s for the Navy that served as electronic command and control planes. With his blond hair and flashy smile he looked like he'd been cast for a Top Gun movie. Perrin had had far more grueling rides with Hicks, even though Schutt liked to play tricks on him in the cockpit to see if he was paying attention to the readings on gauges. The gouge sheet on instructor high jinks came in handy. Schutt had turned off his generator, screwed with the autoignition test, and slaved the compass off course, all before Perrin had even taxied the plane to the runway.

Perrin was back to bantering with the instructors. Schutt liked the give-and-take. He thought Perrin had his head screwed on tight for this flight. He wasn't perfect in performing his checklists on the ground before takeoff. When he turned the plane in the air, he still kept his head in the cockpit too much watching the angle of bank on his attitude indicator instead of looking outside and using the horizon as his reference. His emergency landing patterns at times were a tad too wide or too tight. But today Perrin made the right corrections in his flight path and nailed the high key and low key. His stick skills had returned. If he stayed relaxed and kept his head on a swivel in the cockpit, he'd have no problems flying safely during his solo flight, Schutt decided.

"Okay, take me home, James," Schutt said after Perrin had finished his last maneuver, recovering from a spin. "By the way, what time did we take off?"

It was another favorite instructor trick. Students often were so excited on their thirteenth flight, they forgot to check their watch to record what time they had taken off from Whiting. In the previous flights they could always rely on the instructor keeping them to within the prescribed time they were supposed to be in the air. But on a solo flight, there would be no instructor backstopping them.

Perrin checked his left palm. He had written the start time on his hand. "Nine hundred hours," he now said, the military time for 9:00 A.M.

Hicks walked by Perrin's table in the snack shop. "Congratulations," he said with a thin smile. Hicks had just heard the news that he'd made it through the check ride. "You realize I paid Scooter $150 to pass you."

"Yeah, right," Perrin said and laughed. It was good to be back on track, having Hicks razz him.

Jenna had hoped for an easier instructor to take her up on her second check ride. Major Barney, who sat beside her now in front of one of the briefing cubicles, was hardly her first choice. Barney was a Marine, which meant he was likely a hard-ass in the cockpit, Jenna guessed, plus he always had this stern, quiet look on his face when she saw him walking around the squadron. He also was VT-2's operations officer and third in the chain of command over the squadron, which meant the instructors, whom she lived in fear of, lived in fear of him.

Jenna had taken two remedial flights after her disastrous ride with Hoehl. The squadron routinely prescribed them for students who had been downed on their check ride so they could work out the problems and catch up with the other students. Scooter Schutt, it turned out, had been her instructor for those two flights. Jenna still flew her emergency landing patterns erratically during the first remedial flight. She now realized her air work was seriously subpar. She wasn't trimming enough after turns. She was fighting the plane too much in the pattern. The check ride had not been a fluke. She was not paying enough attention to her flying skills. Maybe I was too relaxed during that ride, Jenna thought. Schutt patiently corrected every mistake and made her refly the landing patterns until she performed them correctly.

She improved during the second remedial flight. The pattern flying wasn't perfect, but smoother, and she began meeting her high and low keys at the right points. She willed herself to trim the plane more. She still didn't trim instinctively. But at least she was consciously thinking to do it.

Jenna then spent the weekend trying to keep flying cleaned out of her mind. She couldn't live, eat, and breathe aviation every waking moment. There had to be breaks or she'd go crazy. Jenna never cruised the bars as many of the other student pilots did for relaxation. At the Academy, the midshipmen lived to be rip-roaring drunk on the weekends. She hated alcohol. She got drunk once her plebe year to see what it was like. The kamikazes only made her head feel like a bowling ball.

Jenna plunked down on a couch Saturday morning and watched cartoons. She still felt like a kid at heart. She hadn't given up a valuable collection of over 2,000 comic books and she never missed a new movie at a theater in Pensacola that showed them at cut-rate prices. Mike Harris came over and they drove to Petland to look at the animals on sale, then shopped for CDs. That evening she went to a batting cage, whacked baseballs until her arms gave out, then sat in front of her TV set the rest of the night channel surfing.

It had not taken Jenna any time to break out of her funk over the flight. Over the years she had trained herself to be an obsessive optimist. Never let negative feelings overcome you. Always portray a cheery outlook to others, even when you don't feel it. Jenna had a direct and forceful personality. She knew it intimidated many of the men she dated. Basketball had conditioned her to be that way in order to cope with the pressure of intercollegiate athletics. When you talk to yourself on the court after a bad shot, don't say anything negative, the coaches would tell her. Say only positive words.

During her junior year at the Academy, the athletic department had ordered the basketball and football teams to attend a weekend clinic on positive thinking. She sat through encounter groups and was loaded down with feel-good videotapes and workbooks. If you do something wrong, don't put yourself down, the group leaders said. Tell yourself what you need to do right the next time. Say the positive things out loud. Your body will react more to what it hears than to what you think. The clinic had mixed results. The men's basketball team went to the NCAA tournament that year. The football team had another losing season. But Jenna found her basketball playing improved and she was now convinced that talking to herself positively in the cockpit helped.

Loren Barney turned out to be far less intimidating than the leatherneck image he projected. A Mormon, he had spent two years as a missionary in Bolivia before becoming a helicopter pilot for the Marines. He acted almost fatherly with the students, quietly quizzing them on flight operations and gently correcting them in the cockpit. Barney also liked to pick out one part of the plane and have a student dissect every nut and bolt in it. His favorite was the mechanism that spun the propeller.

Barney began drawing complicated line diagrams on the blackboard.

"How is the propeller's speed controlled?" he asked.

"Through two propeller governors, the primary and the overspeed," Jenna answered quickly. The propeller governors controlled the power output from the plane's engine and manipulated the angle of the propeller blades so they would turn at a constant number of revolutions per minute. In effect, the governor was a speed control unit that made sure the propellers provided the correct thrust for the plane speed that the pilot had selected.

"We're out flying and our primary propeller governor fails, what's going to kick in?" Barney asked.

"Our prop overspeed governor, sir," Jenna answered. The overspeed governor kept the propeller from spinning too fast when the primary governor had failed.

"When does the overspeed governor activate?"

"At about 2,332 RPMs."

"And what if the overspeed fails?"

"The fuel topping function engages." This reduced the fuel flow, which would slow down the propeller.

"When does the fuel topping function kick in?"

"At 2,398 RPMs."

"Okay," Barney said and paused, thinking of more questions that might stump her.

"If the primary governor fails and the RPMs are over 2,200 but less than 2,332, what do you do?"

Jenna thought for a quick moment. "I'd try to adjust the RPMs with the condition lever." The condition lever was the knob next to her power control lever on the left, which controlled the revolutions per minute. "If that failed, I'd land as soon as possible."

"How can you tell if the propeller RPMs start to fluctuate?"

"You'll hear it when the pitch of the blades changes."

"What might cause this fluctuation?"

"A faulty overspeed governor test circuit or the primary governor may be malfunctioning."

"What might cause the governor to malfunction?"

"It could be metal particles in the oil system," Jenna said, like a graduate student showing off for oral exams. "If that's the case, I might see a chip light warning soon."

"So what do you do?"

"I'd pull the prop-test circuit breaker. If that didn't work, I'd land as quickly as possible."

Barney set down his piece of chalk. "Very good," he said, impressed. "In fact, outstanding. You'd be surprised. I catch a lot of instructors who don't know what we're talking about."

Jenna had a slight smile on her face.

Just before walking into the line shack for her preflight briefing with Barney, she had corraled another student who'd had him on a check flight for any last-minute gouge on the major. Barney's a propeller fanatic, the student had warned her. Jenna ran back to the squadron's headquarters building where a large chart on the propeller system hung and quickly crammed.

The morning sky was perfect for flying, cool and clear with only wisps of clouds high up. But Jenna hadn't noticed the beautiful weather. A tinge of fear crept over her. Her prebriefing had been flawless, the takeoff from Whiting uneventful. But when Barney gave her the first emergency over Brewton, Alabama, Jenna felt she had approached high key for the landing pattern too wide and had hit low key on the other side slightly low. She was still within the accepted parameters of a safe landing, but then her plane banged down hard on the Brewton airstrip. God, this isn't starting out well, she worried.

Barney had her perform two normal landings. Her pattern work continued to be slightly erratic. She tried to think positive thoughts. But was this going to be a replay of her first check ride, she wondered? Her next emergency landing was adequate, but still she hadn't hit high and low key perfectly.

Jenna gave her first four landings a below-average grade. But

Barney said nothing from the back seat. Jenna forced herself not to let the little mistakes build up in her mind.

"Take me to Evergreen," Barney finally ordered over the radio. Evergreen was about twenty miles north of Brewton.

Her landing patterns over the Evergreen airstrip improved. Barney had her land with the flaps up, then down. They were the best of the day. She touched down as soft as a tissue falling on a table. Jenna began talking to herself more over the cockpit mike.

Barney gave her an emergency as she was taking off the second time from the Evergreen airstrip. She didn't lose altitude quickly enough on the approach to the cleared field just ahead of her and probably would have run into trees at the end of the field if she had actually landed.

Barney ordered her to lift the plane to a higher altitude for spins and stalls. They climbed to 8,500 feet just east of Evergreen. Jenna wasn't impressed with her turn maneuvers, but at least she kept her head out of the cockpit this time and watched the horizon instead of staying glued to the attitude indicator. She recovered from her stalls well, then set the plane up for a spin.

The spins were always her best maneuver, a chance to show off her knowledge of the recovery procedures to the instructor. But this time as Jenna entered into the spin she didn't push the rudder pedal far enough in the direction of the spin or pull the control stick far enough back.

"You're not putting it in full rudder or backstick," Barney quickly warned her as they went through the second rotation. The plane was in danger of entering a spiral, where it would rotate down much faster and disorient the pilot more. Jenna could feel the aircraft begin to whirl more quickly so she made the extra correction.

She cursed herself. A chance to add points to her score, blown.

Fortunately, however, Barney had her repeat the maneuver. She recovered from the second spin perfectly.

By now they were over Texas field, a patch of cleared land southeast of Evergreen that was shaped vaguely like the state of Texas. Barney yanked the power control lever back to test her one last time on an engine failure at high altitude.

Jenna concentrated with all her might on the pattern she would fly. She picked out a clearing south of Texas field for her landing.

She was determined to hit high key at the right spot. She flew several bow tie patterns over it to lose altitude. Still too high. She brought her flaps down early to lose more altitude.

On target. She hit high key exactly at 2,500 feet. Now to the next part of the pattern that always seemed to give her problems: the loop around the front of the field to low key on the other side at 1,200 feet. This time, Jenna picked out a reference point on the left side of the field to fly around so she wouldn't circle too tight, as she had in the past.

"Check speed, check altitude, check field position," Jenna repeated a half dozen times over her mike like a mantra as she made the circle down.

Talking to her hands and feet worked. She hit low key dead on, then banked the plane left and down to 600 feet for a perfect final approach into the front of the cleared field. It was like a ball making a perfect swish through the basket.

"Okay, go home," Barney said quietly before she would have had to make the emergency landing.

Jenna pulled back on the stick. But this time she pointed the plane northwest to Evergreen instead of wandering to the east and becoming lost as she had during the first check ride. She was determined not to miss the entry point for the flight path back to Whiting.

From Evergreen she flew directly over the railroad tracks that went south to Brewton. Jenna banked east when the tracks bent to the west and this time she found the right Conecuh River bridge to make her turn to Point Initial.

Scooter had given her a good tip for not missing Point Charlie next. The windmill may have blown down but just north of Point Charlie was an oil refinery. When the oil refinery came exactly under the tip of your right wing, Point Charlie was directly south, Scooter had told her.

Jenna easily found the refinery and from Point Charlie began her approach into Whiting Field.

Barney sat silently in the back seat. After landing, Jenna began totaling up what she thought were the pluses and minuses of the flight. By now she had a throbbing headache.

She brought the plane to rest at its parking space and pulled back the canopy after shutting the engine down. Barney climbed out of the cockpit and began to walk around the aircraft. She couldn't read anything from his expressionless face. Worse still, he had said nothing about securing the back seat. When a student soloed he had to strap in the parachute pack that rested on the back of the empty back seat so in flight it wouldn't fall forward and jam the controls. Barney hadn't asked her to brief him on how to tie down the parachute for a solo flight. Did that mean he didn't intend for her to solo? Crap, have I failed this again? Jenna wondered to herself.

She climbed out of the cockpit and walked beside him for the long trek back to the line shack. Barney remained sphinxlike. Well, unlike Lieutenant Hoehl, he hadn't immediately said, "Jenna, I can't pass you," she thought. That was a good sign. He hadn't said no. But he hadn't said yes.

They hung up their helmets and survival vests in the line shack's locker area.

"Okay, let's go turn the plane in," Barney said, still with a blank face. The aircraft issue room, where pilots signed the paperwork for the planes they flew each time, was in another hangar a long walk away.

Jenna didn't dare broach the question of whether she had passed. That would have violated student-instructor protocol. She'd just have to wait him out.

Finally, halfway to the hangar, Barney spoke, still looking straight ahead. "Good flight," he said.

Jenna's shoulders relaxed.

"In fact, it was an above-average flight today," Barney continued. She had made minor mistakes. Her flying still needed polishing. But all the students were rough diamonds at this point. If he hadn't been alerted ahead of time that she had flunked her first check ride, Barney wouldn't have guessed from today's flight that Jenna had had problems with her air work.

"You're safe for a solo," he said, finally turning to look at her with a smile. "How does that sound?"

"It sounds great," Jenna said.

CHAPTER NINE

Solo

THE hundreds of T-34Cs were lined up in the aircraft parking lot like rows of corn. Magnus had a long walk to the other end of the lot where his plane, number 287, was parked. It gave him time to think. Am I really going to fly solo, he now said to himself. Until now, he hadn't taken a moment to consider what he was about to do. He had been so focused on passing his thirteenth check ride, so relieved when he did, that it hadn't yet registered with him that he was about to fly alone. There would be no one else in the cockpit to back him up, no one to correct his mistakes.

That reality finally began to sink in. It felt good, or at least partly so. He felt proud that he had made it this far in the flight program. Still, soloing was a big step, a daunting step as well.

Magnus shook his head slightly. Don't get too philosophical here, he told himself. Pay attention to the task at hand, the first

191

step of which was to find your plane. Some students would get so caught up in the moment they'd climb into the wrong aircraft.

Magnus finally tracked down 287 parked at the far corner of the lot. He began to check out its condition. The paperwork Magnus had read on 287 in the aircraft issue room had warned him that the gauge reading for the right wing fuel tank would spike occasionally. Thankfully, the air-conditioner worked. They didn't in many of the planes. But another item caught Magnus's attention. A knob on the circuit breaker for the air-conditioning system had fallen off and the repairmen had never found it. Was it still rattling around in the engine compartment? Magnus now worried.

He took his time walking around the plane for his ground check. On the other fam flights, the instructor had always been there making the checks himself in case the student missed something. Now Magnus made sure he didn't miss a thing. Climbing up on the left wing, he buckled down the parachute pack in the back seat and double-checked his own parachute, then the fittings for his oxygen mask in the front seat.

He hopped off the wing, checked its flaps, then worked his way around to the front of the plane and inspected the landing gear. He opened up the hood to the engine compartment, checked the oil level, and looked for loose wires. He pulled out a penlight from his flight suit and shined it into the compartment. Maybe he'd find the loose circuit breaker knob. Then he felt a little ridiculous. He wasn't even sure what the knob looked like and if the maintenance guys couldn't find it he surely wouldn't.

He ran his hand over the prop blades to see if there were any scratches, then looked around the aircraft for any loose object that might get caught up in the propeller. The rudders and elevators looked fine. The small antennas were all screwed on securely. Seasoned pilots usually spent no more than ten minutes inspecting the plane. Magnus had spent twenty. He didn't want to overlook anything that might embarrass him on the flight.

He climbed into the cockpit, buckled himself into the seat, and began his prestart checks, adjusting the seat and rudder pedals, setting the parking brakes, making sure the emergency fuel shut-off handle was down, the wing flap lever was up, the power control lever was set to idle. A ground crewman driving a yellow generator truck drove up in case he had a weak battery that needed jump-starting. It didn't this time. Magnus then had more than thirty knobs, switches, and handles to turn, flip, or pull in the cockpit.

He closed the canopy over his head and began to crank up the engine. The ground crewman posted himself at the right front of the plane to warn Magnus with hand signals if the engine caught fire when he started the plane. Magnus realized quickly that he hadn't been paying attention to the crewman when he turned on the ignition switch. The instructor had always done that during start-up. Magnus made a mental note to be more careful the next time if he didn't want to barbecue himself in the cockpit before takeoff.

He adjusted the rearview mirror in front of him. It struck him that there was no face and helmet staring back at him from the mirror. The cockpit seemed so empty for a moment. He quickly turned his head left, and pushing the rudder pedals and his power control lever moved his plane out of its parking space and down the taxiway.

But the taxiway was backed up with planes wanting to enter the runway for their takeoff. About half the aircraft were being flown by students on their solo flights. It was like rush hour. Magnus had never seen the airport as crowded as this. It unnerved him slightly. Fortunately the air-conditioner still pumped out a cool breeze.

Fifteen minutes later, Magnus finally made it to the runway. He radioed back to the flight duty officer in VT-2's line shack who kept track of the squadron's students who were in the air.

"Two echo two eight seven solo outbound," Magnus said curtly to signal that he was about to take off.

"Roger, two eight seven," the voice from the line shack acknowledged. "Have a good flight."

Magnus rolled the plane onto the runway. But still he was number twelve for takeoff and had to wait nervously for another fifteen minutes.

Finally, Magnus lined his aircraft up at the beginning of the runway and started racing down it. About 2,500 feet later, the T-34C lifted off from the ground. His eyes darting from outside to his airspeed and attitude indicators inside the cockpit, he quickly raised the landing gear and banked the aircraft to the northeast. Then he radioed the Whiting tower and the Pensacola approach control station that he had taken off. The plane climbed to 5,500 feet and Magnus trimmed it up for a heading north to Evergreen.

He cursed himself. The rules for a solo flight were strict. A student was to fly for from one hour and six minutes to one hour and thirty minutes; no more, no less. He was to perform four touch-and-go landings at one of the outlying airstrips. If he had time, he could wander around a bit. But no funny stuff in the plane. No spins or aerobatic maneuvers. When your time was up, fly back home safely, the instructors had warned him. Even though a student was in the air by himself, other instructors in planes or on the ground at outlying fields watched him like hawks and could give him a down if they saw him flying unsafely. Magnus realized that when he entered the runway he had radioed the flight duty officer that he was taking off. He didn't actually take off until fifteen minutes later, which meant that his recorded flight time was cut by fifteen minutes. He hoped Evergreen field wasn't as crowded as Whiting had been or he'd never get in his four touch-and-go landings before an hour and a half.

With the plane headed on a straight path to Evergreen, Magnus for the first time began to think about being in the air by himself. He looked around outside. God, it was a beautiful day out there, he marveled. He could see forever. For the first time, he wasn't tense or nervous in the cockpit. He thought he would be. But the plane moved when he moved the stick. It acted no differently now than when an instructor was with him. Magnus realized how much he'd learned in these few short months. Thirteen flights and they had him to the point where he could fly by himself and do it comfortably. That made him feel wonderful. It made him feel sure about himself, more sure than he had ever expected to feel.

He saw the town of Brewton ahead of him. He realized how easy the plane was to fly when he wasn't forced to concentrate on responding to emergencies the instructors threw at him. It was even fun to fly.

Magnus kept a close watch for other aircraft in the area. Fifty-five hundred feet was a safe altitude for staying out of the way of planes performing simulated emergencies or aerobatic maneuvers. Every sense in his body seemed more alive and aware of things than it ever had been in his life. He could hear the air flowing over his canopy, like driving on a breezy day in his car with the windows up, the beating air giving off a steady muffled tone. Several students were flying solo to the east of him. Other planes were below him to the west. Magnus was surprised and pleased that he spotted them so quickly, just as an instructor would.

He arrived over the Evergreen runways several minutes later. The runway duty officer, one of the squadron's instructors assigned to sit at the field for the day to watch students land and take off, radioed to him that he could practice his touch-and-go landings on the north–south runway with the large 36 painted at the front of it. But Magnus had to wait his turn in line. He was number five in the landing pattern. More delays, he worried.

Magnus kept his interval from other planes and waited. He finally looped around the field for his approach. Now the nerves returned. He remembered problems he had had with his landing patterns during yesterday's check ride. The runway duty officer on the ground was watching him closely and could send him home if he didn't like what he saw. But Magnus made one of his softest landings ever.

He pulled back the stick and powered up the plane to take off immediately for his next touch-and-go. This time, there were no students' planes in front of him. But a civilian plane had entered the landing pattern. Even though the Navy rented the airfield from the local county, civilian aircraft were given priority rights to use it. Whenever one came into the pattern, the runway duty officer would order the Navy planes to a higher altitude to wait until the civilian plane landed and got out of the way.

This one, however, seemed to Magnus to be taking its sweet time landing and getting off the runway. He looked at the clock on his cockpit panel. He was running out of time. He quickly sped through his three remaining touch-and-go's in the next ten minutes. There was no time left for Sunday driving. He had to make a beeline back to Whiting or he would be in the air more than an hour and thirty minutes.

Magnus dialed up a recorded message on one of his radio frequencies for instructions on which runway to use back at Whiting Field. He never got lost on the flight paths home. Students soloing were warned that if they became lost to admit it and radio for help. But no student wanted to make that kind of call.

This time Magnus banged down on the runway and his plane was slightly off from the centerline. The landing had been ugly, he said to himself. But he didn't care. He had completed the flight safely. For the first time since coming to Whiting Field, he had enjoyed flying. This was why he had wanted to become a pilot. This was what the dream was all about. After all those long miserable nights of studying, those months of being away from Amy, this was the reward. He wanted more of it, more flights where he felt so free and alive. Magnus wasn't exhausted as he had been after other rides. He felt he could turn around right now, take off and fly for hours more.

As he taxied down the parking lot to his space, Magnus became depressed. The flight had whizzed by. It seemed as if he had taken off just seconds ago. The ride had been so much fun, so exciting. No grades. No teacher in the back seat carping at you. Now it was back to the grind of instructional flights.

Jenna was not excited during her solo flight. It surprised her. She thought she'd hear angels singing from heaven, drum rolls, the theme to *Rocky*, anything euphoric for the first time in the air by herself. Instead, her takeoff had been busy with all the flight checks and cockpit control work needed to get the plane out of Whiting. She actually felt more nervous on this flight than the previous two check rides. She was talking to herself more. As her plane sped north, her eyes shifted constantly from the cockpit gauges to the outside. Above all, she didn't want to collide with anything up here. She knew she had the Naval Air Close Warning System that had just been installed in her plane. But other students had warned her that the gizmo wasn't infallible, so she kept a close watch outside.

She was already angry with one mistake she had made at the beginning of the flight. She hadn't twisted one of the knobs far enough on the transponder in her right control panel. The transponder was an electronic device ground control radars used to track planes in the air. After takeoff, Pensacola's approach control station had radioed to complain that her transponder wasn't emitting altitude information on her plane. She turned the knob and the transponder began putting out the signal.

As she approached Brewton, she reached down to the communications box on her left next to the trim wheels and turned the dial to channel 15, the frequency for the runway duty officer at the Evergreen landing field more than twenty miles to the north. She wanted to know what runway the planes were using for touch-and-go landings.

The channel was silent. I still must be out of range, Jenna thought. She turned the dial back to channel 8, the common frequency all the Navy pilots used when they flew in Area Two over southern Alabama.

Five minutes later, Jenna flipped the radio transmitter switch up on her power control lever to alert the runway duty officer at Evergreen that she was approaching the field. She had just passed a power station three miles west of the field, a landmark the students used for the point at which they radioed the duty officer that they were approaching the runways.

That's funny, Jenna thought to herself. She heard some mumbling over the radio. It sounded like the runway duty officer acknowledging that she was coming into the pattern. She flew past the three-mile checkpoint toward the Evergreen landing strips.

Now something was really wrong, Jenna decided. As her plane approached even closer to the Evergreen airstrips she noticed there wasn't the usual chatter on her radio from other students making their landing pattern calls.

She looked down at her communications box.

"Ah, shit!" she said to herself. The channel was still set to 8, the common frequency for Area Two. She had forgotten to turn it back to channel 15 as she neared Evergreen so she could talk to the runway duty officer on the ground.

Jenna quickly turned the dial to 15 and radioed to the duty officer that she was coming around the strip to the break, the point the pilots entered to begin their landing pattern.

Runway duty was one of the more tedious chores an instructor was stuck with, sitting out in the hot sun at a half-deserted airstrip all day, playing traffic cop to students flying overhead. It had been a lousy day for Gary Moe, a Coast Guard lieutenant who was assigned to VT-2 as an instructor. Planes had been cutting in front of one another in the landing pattern all afternoon, even planes with instructors on board. Then there was this crazy civilian pilot who decided to try his own touch-and-go landings, three times.

Moe now had five planes stacked up in the landing pattern, all of them piloted by students on their solo flights. The solos always made him nervous.

And here was this sixth plane barreling toward him out of the blue that he couldn't get to respond when he radioed to it. Who the hell is this, Moe wondered, becoming angrier by the minute.

Finally, he heard Hausvik's voice on his radio speaker. "Navy two echo one six eight solo," the radio call sign for Hausvik's plane. She was at the break and wanted to enter the pattern to land.

The hell she would, Moe thought. A student pilot never entered a landing pattern if he hadn't first established radio contact with the runway duty officer. He could run into other planes. There was no way for the runway duty officer to warn him of danger. Fortunately, Hausvik had kept her distance from the other aircraft in the pattern.

"Navy two echo one six eight solo, negative break!" Moe shouted into his microphone. "Continue upwind and depart the pattern. Go back to the three-mile point and contact me then for entry into the pattern."

Jenna instantly banked the plane right, away from the landing strip.

She knew she was in trouble for this.

Moe was furious. He was ready to give her a down for the safety violation.

Minutes later after finally receiving clearance from Moe, Jenna began her touch-and-go landings. They were all nearly perfect. Moe was impressed.

Jenna had plenty of time for more touch-and-go's than the four required during a solo. After her eighth landing, she wisely decided that she'd probably worn out her welcome with Lieutenant Moe.

Moe decided not to give her a down. Instead he planned to chew her out when they both returned to Whiting.

Perrin had just finished the touch-and-go landings at Evergreen field that were required for his solo. He had a half hour left in the flight. Now it was time for recess, he decided. He hummed to himself in the cockpit, then made funny noises on the microphone pressed to his lips. Driver's ed was over. It was time to tool around in dad's car. He pushed the power control lever all the way forward—max blast, full throttle. Let's see how this drag racer performs. The Turbo Mentor roared to over 240 miles per hour as Perrin bore down east to the town of Evergreen. He couldn't try any aerobatics as he had at test pilot school.

He looked all around outside the cockpit. No one was near him. He could still have some fun.

Perrin lowered the nose and played bomber pilot over Evergreen. He swooped down to 1,000 feet above the town, then pulled the control stick back. Any lower and someone in Evergreen might complain that he was buzzing the city and call Whiting. He pretended to fly strafing runs on the cars driving down Interstate 65.

"Burp, burp, burp, burp, burp, burp," he made the sounds like a kid pretending to fire a machine gun.

After ten minutes of goofing off, he decided it was time to line up at the entry point for the flight back to Whiting.

When Perrin reached the Conecuh River bridge, the Whiting control tower ordered him to bank east on a heading of 165 degrees. In the past, he had always banked to the west and flown a 205 degree heading to Point Initial at the town of Berrydale and from there had headed to Point Charlie at the intersection of Highways 182 and 89 for the landing at Whiting. But the control tower this time wanted him to land from Point Delta, which was on the opposite side of Point Charlie.

Perrin had never banked east at the Conecuh River bridge and taken the alternate route home. He had no clue where he was, so he decided to follow a T-34C flying in front of him that seemed to be flying toward Point Delta. Perrin relaxed. Stay close behind this plane and no one will be the wiser, he thought.

Several minutes passed. Perrin heard over his radio a student in another plane announcing to the control that he had Point Delta in sight and was preparing for his landing.

Perrin thought the voice had come from the plane in front of him. He vaguely recalled that the landmark for Point Delta was a church. He didn't see anything ahead that looked like a church. But what the hell? He flipped up his radio transmitter switch.

"Whiting tower, this is two echo two five five solo," he proudly announced, giving the radio call sign for his plane. "I have Point Delta in sight."

The voice Perrin had heard earlier had not come from the plane ahead of him, but rather from a T-34C much further south. The air traffic controllers in the Whiting tower, who could see the location of Perrin's plane on their radar screens, erupted in laughter.

"Two echo two five five solo, verify that you have Point Delta in sight twenty-two miles away?" one of the air traffic controllers finally said over his radio with a chuckle. Perrin hadn't traveled nearly as far south as he thought he had. If he could see Point Delta twenty-two miles away from the other side of the Alabama state line, the air traffic controllers figured he'd just set a world record for the best eyesight.

"Oh, fuck," Perrin said to himself. He should have kept his mouth shut.

Perrin radioed back and sheepishly asked the tower for directions to Point Delta. He hoped no one else had been listening in on his radio traffic.

A half hour later, Perrin stood in front of a computer terminal at the aircraft issue room typing in information on his solo flight. About a dozen instructor pilots and students milled around in the room filling out paperwork. They were all from squadrons other than VT-2's.

"Hey, d'you hear that radio call from Area Two today," one of the instructors finally shouted to the group. "Some kid had X-ray vision. He could see Point Delta from twenty-two miles away." The other pilots laughed.

Perrin's face turned red. He tried to hide behind the computer terminal.

The banquet room of the Whiting Field officers' club began to fill up with the instructors and students from the VT-2 squadron. In a corner sat several kegs of beer that had been iced down. On long tables were stacked submarine sandwiches and soft drink bottles. Many of the students had brought wives and girlfriends, who pretended to laugh and be interested. The students and instructors swapped aviator jokes and funny stories about flying. But the language was all in flight terms and acronyms so to the outsider it was meaningless and boring.

The squadron tried to have these get-togethers once a month, usually on a day when the weather was so bad no one could fly. The event was called the "tie cutting." No one was quite sure when or why it started or even what exactly the point was behind it. The tradition began sometime in the 1930s when Naval aviators wore ties during their flights. After a student soloed, a ceremony was held and the instructor cut the student's tie. Although the reason had never been written down, most in the squadron today thought it symbolized the student severing his ties with the instructor by flying on his own.

Each student who had soloed wore a tie over his flight suit and presented the instructor with his favorite bottle of liquor. (More and more instructors now were asking for nonalcoholic beverages.) Then the student would try to tell a funny story about the instructor. If it wasn't funny, an emcee for the event would make him cough up a few bucks for the liquor fund.

Next, the instructor would tell a horror story about the student in the cockpit. These tales were always funny. He would then cut the student's tie. If he thought the student was a decent pilot, he'd snip it near the end. If he was a screw-up, the instructor would slice it near the neck. VT-2's instructors always seemed to cut high.

Jenna and Magnus tried vainly to make a few jokes. Jenna had bought a tie with basketball players printed on it. The jokes fell flat and they were ordered to peel off dollars for the kitty.

Perrin stood up. Instead of his flight suit, he wore Bermuda shorts and a polo shirt with a black skull-and-crossbones tie knotted around his neck.

"Perrin learned something on every flight," Hicks began. He was enjoying this.

"On one flight, he learned to push the button that allowed him to talk on the radio." Everyone laughed.

"On the next flight, he learned that if he pushed the power control lever forward the plane would go faster." They laughed again.

"And by his twelfth flight, he'd finally found the horizon." They laughed even louder.

"Oh, God," Perrin said, chuckling and shaking his head.

Hicks pulled out a bowie knife almost a foot long.

Perrin's eyes widened. Hicks looked like he was going to carve up his chest.

The squadron roared and Hicks sliced the tie near Perrin's neck.

Primary training was far from over. Hausvik, Leslie, and Perrin still had more months of learning to fly aerobatic maneuvers and in close formations with the Turbo Mentor. There were cross-country flights, night flights, and more navigating using only the cockpit instruments. But fams were over. The umbilical cord had been cut. They now shared the joy of first flight by themselves.

3

WARRIORS
IN THE SKY

CHAPTER TEN

Pipper to Bull

JONATHAN loved flying solo in the Goshawk jet. He had long abandoned the trembling nerves of piloting alone in a plane. There had been a number of solos since his first one in the T-34C turbo-prop at Whiting Field. Jonathan now felt more comfortable when he soloed. Most students at this point were relieved when they did not have instructors in the back seat of the plane second-guessing them. Jonathan Wise had now been training to be a Navy pilot for two years and had come to realize that each instructor had his own technique for flying the aircraft. Which always differed slightly from that of the next instructor. You couldn't please them all, Jonathan knew, so it was nice to have the back seat of the plane empty. Then you could fly your way.

It was only a few minutes to the Shade Tree bombing range, an inexplicable name since there wasn't a tree in sight on the barren brown dusty plain below. Jonathan was part of a four-plane formation flying more than 250 miles per hour. The Goshawks were arranged in an echelon. In the echelon's lead plane sat the two instructor pilots, their heads turning right and left in the cockpit to make sure Wise and the students flying the two other planes didn't crash into one another during their bombing runs.

Buka flew the second plane about twenty feet to the right and about twenty feet back. Ski flew the third at the same interval to the right while Jonathan brought up the rear with his plane. Ski was the call sign for Marine First Lieutenant Mike Sobkowski, a simple enough origin. Buka had a more complex history. Rob Schroder, also a Marine first lieutenant, had earned it when he singed the inside of his mouth drinking a fiery drink called a Flaming Sam Buka at a bar in Washington, D.C. Wise's call sign was tolerable enough: Zorro. He had been on the Naval Academy fencing team. He also looked somewhat like Zorro, a twenty-five-year-old Zorro, except with blond hair and a light mustache. Dye the mustache and maybe you had Zorro.

The students were just being initiated into the ritual of call signs, in many cases trying to rid themselves of embarrassing nicknames others had pinned on them during the past couple of months. Call signs became serious business, they were discovering. A pilot could be stuck with a humiliating one for his entire career in the Navy. The more ashamed he was of it, the more likely the name would stick. Not until he became a senior officer could an aviator change it on his own and no one junior to him would dare challenge that he was now "Maverick" or "Gunslinger" or something else macho. Trey Sisson, another student in the squadron, thought he might have had to endure being called "Sissy" for the rest of his flying days. Fortunately one of the Marine instructor pilots in the squadron refused to fly with anyone named Sissy. A greatly relieved Trey hoped he had found a convenient excuse to eventually change his handle.

Flying the number four plane the furthest back in the echelon had its pluses and minuses, Jonathan realized. The instructors in the lead plane almost sixty feet forward had difficulty turning their heads back far enough to spot any mistakes he might make in keeping his jet positioned properly in the formation. At the same time, being the final plane in an echelon was like riding the tail end of a

Roller Derby whip. Any turn or jink the planes at the front made became magnified for the number four plane, which meant Wise always had to race to keep apace with the formation.

The four planes flew north of El Centro, a dusty California town near the southern border with Mexico, then looped east around the Superstition Mountains. Todd Vaupel, a thirty-year-old Marine Corps captain piloting the lead plane, eased his stick back and the aircraft climbed to 8,000 feet. His back seat partner for this flight, Navy Lieutenant Ed Miller, thirty-two, twisted his head to the right to make sure the three other planes behind them—his three "chicks," as he called them—followed suit. Both officers were leaving the service. Newly married, Vaupel had flown Harriers, the Marine attack jet that could land vertically like a helicopter on a carrier. He wanted no more of long tours at sea. He was tired as well of the grind in the training command where instructor pilots were expected to work six days a week cranking out graduates for the fleet. Miller was a snuff-pinching good old boy from North Carolina who flew the S-3 Viking sub-hunter jet off a carrier. He planned to go to Harvard University, earn an MBA, make a bundle in investment banking.

The entire formation now was pointed northwest, on a straight line to the Shade Tree target. No sooner had the planes reached the 8,000-foot altitude than Vaupel pushed the stick forward so his jet dove down toward the target. Shade Tree was actually a collection of hundreds of old tires arranged to form three circular lines, spaced fifty feet apart, which created a giant bull's-eye target the length of a football field in diameter. At the center of the bull's-eye were the charred remains of an old Army tank. Spotters in a nearby tower on the ground radioed to the pilots where the bombs struck. Hitting a bull's-eye meant having the bomb drop inside the first inner circle, which measured forty feet wide. A ditch also had been cut for a mile leading up to the bull's-eye target, which from high above appeared to the pilots as a long line that they used as an aid to align their approach.

About fifty seconds away from the target, Jonathan reached up to the upper left of the instrument console in his cramped cockpit and turned a selector knob on his armament control panel to BOMBS. The bombs he would drop weren't real ones, but small devices painted blue that looked like what World War I pilots threw out their cockpits. Made of cast iron, they had a small smoke-producing charge that detonated when they struck the ground so the spotters could mark where they hit.

Perched on top of the front instrument console in Jonathan's cockpit was a heads-up display—a flat piece of glass shaped in a rectangle that operated much like a speech TelePrompTer. It now lit up with a confusing array of yellow numbers and symbols, as well as a two-ringed bull's-eye bobbing up and down or right and left on the glass. Wise next punched the left button underneath the selector knob, which would allow him to drop first the four simulator bombs hung on his left wing. When he was done with them, he'd push the right button to drop the same number of bombs from his right wing. The toggle underneath the knob and buttons, called the master arm switch, Jonathan left on safe for the moment. He would flip it up to MASTER ARM, which would allow the release of the bombs, only moments before he dropped his ordnance.

Jonathan still had butterflies. He always did before every flight. He hoped that would never change. If the butterflies went away, flying would become boring or, even worse, he would become lazy in the cockpit and make mistakes. In a combat jet flying near the speed of sound that could be deadly.

Jonathan pushed the throttle forward with his left hand. Outside, the Goshawk's engine roared, propelling the jet to 450 miles per hour. Inside the cockpit, Jonathan heard only a moderately loud hum; the jet's noise was always far behind him in the fast-moving aircraft. That was different from the puttering T-34C turboprop, which was a noisy rattletrap inside the cockpit.

Jonathan trimmed the plane one more time for the higher speed at which he was now flying. That was another thing different from the T-34C, he had quickly learned. No more clunky trim wheels on the side. To trim the elevators and ailerons on this aircraft, he merely lifted the thumb of his right hand holding the stick to push a tiny ridged button located on the top of the stick. A knob on his left console trimmed the rudder. So much more convenient, particularly since Wise had only thirty seconds in this part of the approach to select bombs, punch in the left wing release, flip the master arm switch to safe, trim the plane, and scan his instruments to make sure the plane remained on course.

The pace quickened even more. Miller glanced behind him at

the three planes trailing Vaupel's. "Pretty decent formation comin' in here for the Mach run," Miller radioed in his slow North Carolina drawl. None of the jets was too far apart or bunched together too closely. As Vaupel's jet raced over the target below for the Mach run, he radioed back to the three other planes: "Okay lead's at 2,700, four four five on the true."

Vaupel was reading off the altitude (2,700 feet) and airspeed (445 miles per hour) for his jet. Mach was a speed measure, Mach 1 being the speed of sound. The Mach run was a first pass over the target when the formation checked its speed and altitude. Jonathan glanced down at the readings on his cockpit instrument panel. Sobkowski and Schroder did the same. Their instruments all showed the same readings for airspeed and altitude. The cross-checking was important. The jets would be circling and coming down on the target at 450 miles per hour in rapid-fire succession. If the airspeed and altitude readings in any one of the jets was miscalibrated, the pilot could be at the wrong place at the wrong time in the bombing formation, perhaps causing a catastrophic midair collision. This time they were all on the same sheet of music.

Wise rechecked his radar altimeter to make sure that if he dove below 2,000 feet in the bombing run an irritating "beep-beep-beep" would blare into the radio receiver of his helmet, warning him that he had to immediately break out of the dive. If he didn't, within a few hundred feet there wouldn't be enough airspace between him and the earth to pull up. He'd end up in a smoking hole in the ground.

Past the bull's-eye, Vaupel radioed again: "Lead's breaking." In the formation he was the lead and his plane was designated number one. This was the signal he now gave to tell the three other planes that he was banking his jet sharply to the left at a 60 degree angle and soaring up to 8,000 feet. Eight seconds later Schroder was supposed to jerk his number two plane to the left in the same maneuver. But nervous, he jumped the gun and broke left just after six seconds, announcing over his radio: "Two's breaking."

"That looked like a quick eight seconds there," Miller said to Vaupel over their intercom.

"Yep," Vaupel agreed.

Miller made a note on the notepad strapped to his leg. He'd bring up the fast start with Schroder after the flight. The break maneuver was performed to space the planes out at exactly eight-second intervals when they flew the bombing pattern. Two seconds off here, two seconds off there, and planes would soon be banging into each other, Miller knew.

"Three's breaking," Sobkowski radioed at the proper eight-second interval.

When it came Jonathan's turn, he shoved the stick with his right hand smartly to the left as if he was taking a forehand volley shot in tennis. The jet snapped into a sharp bank, throwing his head to the side. In the next second, Wise pulled the stick to his chest to make the Goshawk soar up to 8,000 feet. The hard left turn along with the pull-up plastered him against his seat. This was a three-G maneuver, meaning he was experiencing three times the force of gravity on his body. Wise in effect now weighed almost six hundred pounds.

He immediately felt his G suit inflate. The suit was connected by a tube to air that came from the jet engine. When the aircraft's weight sensor detected an increase in Gs, the air from the engine would be rammed into the G suit, inflating it, to squeeze the lower half of a pilot's body and push blood back up to his brain. For Wise, it felt like giant hands were wrapped around him from the waist down squeezing him like a tube of toothpaste.

"Did you hear four call breaking?" Miller asked Vaupel over their intercom.

"No," Vaupel answered.

Wise indeed had forgotten to radio that he had made the left turn. Miller looked back to make sure Wise had broken left, then made another note on his pad. Another mistake Miller would bring up later.

Jonathan leveled off at 8,000 feet. He had only several seconds more to prepare for the next phase of the bombing run. His concentration had to be intense. This was "dynamic flying," the seasoned pilots liked to say. A euphemism for maneuvering the warplane violently through the air. But this was what Jonathan had been waiting for. Waiting for years, in fact. The fun stuff. Until now, he and the other students had been learning to take off, to land, and to fly in a plane, which any civilian pilot learned. Now he was learning something few pilots ever would. He was learning to use the plane as a weapon. He was taking his first step toward becoming an air warrior.

This was why he had joined the Navy to become a Naval aviator. The thrill he now felt as he prepared his jet to drop bombs was almost overpowering.

Jonathan had always taken a long-range view of his life. At age eleven, he had marched into the kitchen to announce to his startled father that he planned to attend the Naval Academy. And he did. His plebe year, upperclassmen would scream at him hysterically because he always kept a smile on his face, as if he was enjoying the hazing. He in fact began enjoying the abuse when he willed himself to take the long view. Plebe hazing would not last forever.

The Academy grind seemed to never end, however. Jonathan majored in aerospace engineering, where the instructors had what he thought was a crocodile mentality. They ate their young. Each year got worse. He subsisted on four hours sleep a night. The other twenty hours he spent studying and drilling. Even weekends were consumed with football games.

Fencing had been a relief. He became engrossed as well with other martial arts, such as firing high-powered rifles in competition. Fencing, Wise now discovered, was not dissimilar to the aerial combat he was beginning to learn. As with flying in combat, in fencing the swordsman needed quick reactions. He had to practice certain maneuvers with the foil until they could be performed reflexively without thinking. Then a fencer had to be creative in putting together combinations of maneuvers to overcome an opponent.

Jonathan liked finding other analogies. Flying military planes in formation, he would say, was like driving on a crowded highway during rush hour. He enjoyed looking for holes in the traffic several cars ahead, tried to weave his vehicle around others to exploit the holes and get ahead. All the while he had to keep his distance from other cars so he didn't rear-end them, just as in planes. Even today, Jonathan and a roommate would find a deserted road and race their cars along it with their bumpers only inches apart to simulate the precision flying needed to keep their six-ton Goshawk jets hurtling through the sky only furlongs apart.

Less than a year after graduating from Annapolis, Jonathan decided he was in love. He wanted to marry the woman who had become his best friend.

Maria Grauerholz had the same single-mindedness as Jonathan. At age fourteen, Maria had visited the Marine Corps Museum in the Washington, D.C., Navy Yard, where she spotted a gold statue of an F/A-18 Hornet jet. "I want to fly that," Maria told her father.

She was vaguely aware at the time that the military barred women from flying combat jets. That didn't matter in the mind of an adolescent. Maria's goal was set. The Hornet. The combat exclusion law? She'd worry about that later.

Her parents at first did not think she was really serious about becoming a fighter pilot. When four years later she presented them with an application to the Naval Academy that she had filled out on her own, they realized Maria was.

Maria and Jonathan met the summer of 1990 after their plebe year, while bobbing up and down in small patrol boats the Academy sailed along the East Coast. They soon became good friends, and by their senior year the best of friends. Everyone at the Academy needed a close friend to help them endure the four dreary years. Maria never found it unusual that hers was a young man. There were so few women at Annapolis.

As they worked their way through preflight indoctrination training at Pensacola, Jonathan and Maria finally began dating. By the time they reached Whiting Field for primary training they were engaged. It was the best of all worlds for them. They were both excited about flying. It seemed to be all they talked about when they were together. And now they were in love.

Jonathan had decided he wanted to fly helicopters. He became airsick in the prop plane and wasn't particularly interested in filling up paper bags the rest of his life, pulling Gs, or spinning around in a combat jet. Maria remained determined to fly Hornets. She graduated from Whiting with honors and won her first choice, jet training. By then she had already been stuck with a call sign. During a flight, an instructor warned her that she had to be assertive over the radio when she dealt with air traffic controllers. It had been a family joke that her father had wanted to name her Helga because of her Germanic background.

"I can use my Helga voice," Maria chirped up over the intercom.

Bad mistake. From that day on, her call sign for better or worse was Helga. The nickname clung even harder when classmates realized Maria had two hobbies. A soprano in the Academy glee club, she had begun singing with the Pensacola Opera. Her favorite opera: *La Traviata.* Maria also was a power lifter. Her top bench press: 168 pounds. Squats: 303 pounds. Dead lift: 330 pounds. She'd never be rid of Helga.

Jonathan's flight grades at Whiting were slightly lower than Maria's, but good enough that the Navy ordered him to take jets instead of choppers. He glumly moved to Meridian, Mississippi, to begin training in the aging T-2C Buckeye, a twenty-five-year-old, twin-seat jet the Navy was phasing out as a trainer.

To his surprise, Jonathan fell in love with jets. The T-2Cs were old, beat-up Edsels but very forgiving for students with "snakes in the cockpit"—the saying pilots had for nervous rookies who jerked the stick and stomped the rudder pedals. Jonathan discovered he could easily cope with pulling Gs in the jet. His airsickness bag stayed in his flight suit pocket.

There was even something comforting about sitting on top of a rocket that would eject his seat if he had trouble. He had always been worried when he flew the T-34C turboprop that he wouldn't have enough time to climb out of its cockpit before it crashed in an emergency.

Jonathan now never looked back. He wanted to be a fighter jock. He wanted to fly the F/A-18, just like Maria.

Maria headed for Kingsville, Texas, a cattletown along the Gulf Coast whose Naval air station had received the newer T-45 Goshawks the service had just bought. A hybrid of the British Aerospace Hawk jet, the Goshawk was typical of most Pentagon programs. The $16 million plane cost the Navy $6 million more than it was supposed to and took two years longer to build than the contractor had promised. Instructor pilots liked the Goshawk nonetheless because it had more radios and navigation equipment in the cockpit than the old Buckeye and was more fuel-efficient.

Beginning jet flying—whether in the old Buckeye or new Goshawk—was like being on fast-forward for Jonathan and Maria. The T-34C turboprops at Whiting had seemed overwhelming enough at first, all that math they had to perform in their heads on altitudes, speeds, distances, fuel flows, and navigation way points while steering the plane. Now the calculations had to be performed three times faster. Instead of flying 150 miles an hour, they were flying 450 miles an hour. The T-34C flew no higher than

10,000 feet; the jets routinely reached 25,000 feet. Checklists had to be performed more quickly. Jets, Jonathan found, were far more maneuverable and sensitive to steering in the air. Students had to think far more ahead in them.

The prop plane was also roomy compared to the cramped cockpit of a jet. Students now had to squeeze into a G suit as well as a torso harness that strapped them tightly to the ejection seat. The oxygen mask felt like a suction cup on their face. Straps were wrapped around their legs and ankles and attached to the seat. (If the pilot had to eject, the straps yanked his legs to the seat to keep his legs from flailing and being chopped off by the cockpit rim as he blasted out.) It gave Maria a secure feeling, like a swaddled baby. For other students it could be claustrophobic at first.

At Meridian, Jonathan learned the basics of jet flying in the Buckeye, how to handle the aircraft in the air, how to fly aerobatic maneuvers such as loops and rolls, as well as flying in formation with other Buckeyes. Maria was learning the same but in the newer Goshawk. In October 1995, Jonathan joined Maria at Kingsville for advanced jet training in the Goshawk. They both were assigned to the same training squadron, VT-21, nicknamed the "Redhawks."

The mood in a jet squadron was different as well from that of the squadrons in primary training. To qualify for jets, all the students had to have top grades from the primary phase. Jonathan and Maria were surrounded by overachievers. They were now part of the cream, the smartest, the most aggressive, the most competitive, the best sticks among the Navy's future pilots. Failure was the most painful for this group. In primary training, instructors encouraged students along like a father teaching his son to ride a bike for the first time. Jet instructors were more critical in the back seat, far pickier about the ride. VT-21's instructors were bland but unforgiving and quick to form a bad opinion about a student at the first sign of weakness. At least that was the students' view.

The squadron's instructors thought they had a unit full of sensitive young egos. No rocks among them—except for the occasional reckless student who would give an instructor a heart attack in the jet. Those were afflicted with what the instructors called NAFOD, for "no apparent fear of death." But all of their pupils were thin-skinned, the instructors complained. Jet flying wasn't a business for the fainthearted, yet gripe at these prodigies and they let you know their feelings had been hurt. Some would even talk back like spoiled brats.

Because of scheduling delays, students moved through the jet training at different paces. Maria had jumped ahead of Jonathan in the training pipeline. Jonathan also had had to learn to fly two jets, the T-2C at Meridian, then the T-45 at Kingsville, which put him behind his fiancée, who only had to learn the T-45. By the time Jonathan began bombing targets, Maria had already finished her advanced training.

Maria had also realized her dream. She had graduated second in her class and had been rewarded for it by being assigned to fly the F/A-18 Hornet, the Navy's top fighter and attack jet. If Jonathan made the cut for F/A-18s as well, they would be the first husband-wife combat fighter team in the Navy.

Whether the Navy was ready for husband-wife fighter pilots remained to be seen. The service was still feeling its way with single women flying combat jets. The squadron in Kingsville took this couple well. But that was because both Maria and Jonathan were good students. The instructors at VT-21 complained behind closed doors that other women with poor grades that would have washed out a man were being given second chances to complete jet training. The Navy was still desperate to put more females in jets and hadn't yet accepted the fact that some had to be allowed to fail, they argued.

Maria was torn on the subject. In the past, women had been harassed in jet training. It annoyed her that a lot of below-average male pilots were helped along in jet training and nobody raised a peep about them. But there were very few women training for combat jets so far. Each was under a white-hot spotlight. If any one of them had below-average grades, the news spread through training units like wildfire. And if a woman was helped along, the whole world seemed to know about it.

The double standard was unfair. But what made Maria just as mad were the women who had been helped along but who had grades so poor they didn't belong in jets. Or women who received bad grades they deserved, but hid behind cries of sex discrimination. It only made the job of the good female pilots like Maria more difficult. Everywhere she went she'd have to first prove that she wasn't a quota queen.

But there was nothing Maria could do about it, she realized. More females had to go through jet training before the Navy calmed down and the poor female performers were allowed to quietly fail just like the men. Until then, all Maria could do was her best. The male pilots who would never accept her were idiots she ignored anyway. The open-minded ones, like Jonathan, would realize she was qualified and treat her as an equal.

But Maria did feel the pressure to do well. And it was not just the pressure she put on herself because she was a competitive person by nature. For better or worse, she was one of the leaders of a new breed of pilots. She felt pressure to do well because she was a woman.

Maria would never forget the day she was about to make her first solo landing on an aircraft carrier. An enlisted woman had rushed up to her all excited.

"Oh, you're such an inspiration to us," she said admiringly.

"Thanks," Maria answered with a thin smile. As if she didn't have enough worries just flying a jet onto the carrier. Now the future of women rode with her in the cockpit as well. It seemed that every time she had feminism pushed to the back of her head so she could concentrate on the mentally grueling task of combat aviation, somebody reminded her that she also carried the weight on her shoulders of being a pioneer for her gender.

If Jonathan was going to join Maria in flying F/A-18s, it was crucial that he perform well the next two weeks. This was what flying Navy combat jets was all about: bombing targets on the ground, shooting enemy planes out of the sky, taking off from and landing on carriers. To earn a seat in an F/A-18 Hornet or F-14 Tomcat, the Navy's premier combat jets, Jonathan had to have above-average grades in these three skills. He would learn dogfighting with enemy planes and landing on carriers later. But first he had to master what the Navy called the "weapons phase," or "weps" for short. Putting bombs accurately on targets. It was the beginning of the make-or-break phase of his training.

The VT-21 squadron conducted most of its instruction in the muggy hot skies over Kingsville, Texas. But routinely the squadron packed up its equipment and flew its Goshawks, students, instructors, and maintenance crews 1,400 miles west for two-week stays at El Centro. Nestled in the agriculturally rich Imperial Valley, which depended on irrigation canals cut in from the Colorado River, the 2,289-acre Naval Air Facility northwest of the town was ideal for learning to fly combat jets. Vast desert ranges ringed by mountains were available for practice bombing. Relatively few airports were nearby so the Navy pilots had to worry less about pesky civilian planes straying into their airspace. Students only had to avoid flying more than ten miles south and over Mexico. (They learned quickly that the dark and pointy Signal Mountain just below the Mexican border was a good reference point to stay north of.)

But most important was El Centro's perfect flying weather year round. The Imperial Valley's warm, dry climate recorded more days of sunshine with crystal clear skies than any other part of the United States. Jonathan was grateful they were here in April when the temperature during the day was a bearable eighty degrees. From June through September the thermometer would rise to 100 degrees during the day, making the cockpit an oven when taxiing on the ground with the canopy closed. Pilots called it "El Sweato" then.

The harried officers who grappled with the squadron's daily training schedule also appreciated the two weeks at El Centro because its instructors and students were far away from "snivels," the military's chauvinistic slang for family excuses—sick kids, angry wives, demanding girlfriends—that kept pilots out of the cockpit. A detachment to El Centro meant concentrated and undistracted training with students flying two flights every day.

"One's abeam, two point five," the radio crackled in Jonathan's earpiece. It was Vaupel in the lead jet letting the three planes behind him know that he had just reached an altitude of 8,000 feet, a mile and a half southwest of the target below. The pilots called this the "abeam position" and used it as a reference point for entering into the bombing pattern. ("Two point five" stood for the 2,500 pounds of fuel Vaupel's jet still had in its tanks.) When the other three jets crossed the abeam point they would radio the same message. That way, each jet would know where the other is in the pattern. From the abeam point, Vaupel would now fly almost two and one half miles to the target's rear before making a sharp left turn to line up on the target from the southeast.

The flight pattern the jets flew to bomb the target looked like the path of a giant roller coaster—more than one and a half miles high, over two miles long, and about two and a half miles wide. The jets circled the target at 8,000 feet. When they were about two and a half miles southwest of the bull's-eye the jets made a hard left turn, then dove at a 30 degree angle toward it. Near the bottom of the roller coaster—about 5,000 feet from the target when the jet had dropped to 3,000 feet in altitude—the pilot released the bomb. In aviation slang the release was called the "pickle." Just before the jet streaked over the target the pilot pulled the plane up and to the left in a neck-snapping maneuver so the aircraft soared almost vertically back to the abeam position. Then the roller-coaster ride was repeated for the next bombing run.

Jonathan had practiced the bombing runs for five days in the simulator to prepare for the real rides. He had performed well in the hundreds of bomb drops in the simulator. He had delusions in the back of his mind that he would do just as well when he began the real bombing runs earlier in the week.

The real hops were a rude awakening, however. An instructor was in the back seat during the early flights to correct his mistakes. Jonathan made many. He realized quickly that the simulator could never replicate the winds buffeting his plane up and down and from side to side as he struggled to line the aircraft up on the target. He felt like he was driving a car down a steep hill, tossing a water balloon out the window, trying to hit a fire hydrant on the sidewalk.

In the simulator, Jonathan also didn't have to worry about keeping a precise distance from four other planes in the bombing pattern. There weren't all these communications that had to be made so the pilot of each plane knew where the other aircraft were. Critical communications. Without them, two jets might end up diving toward the bull's-eye at the same time or colliding in midair. Radio calls had to be made crisply and clearly. One student stuttering or garbling his transmissions could dangerously bottle up the entire formation. In the early rides, instructors worried not so much about accurate hits as they did about the students not running into each other in the pattern.

The communications, turns, pulls, lineups, bomb drops happened in rapid-fire order in the pattern. At first, Jonathan hardly had time to think about what he was doing. His mind had become saturated with procedures and calculations. On early hops he would lose count during the flight of how many bombs he had dropped. He was just "slinging bombs," as the instructor called it. The hits were all around the bull's-eye target.

By the end of a week of bombing at El Centro, the hits were expected to be more tightly bunched near the bull's-eye. Bombing was an acquired skill, much like throwing darts. Some students mastered it quickly. Others never did. The key was having a quick, nimble mind, Jonathan found. Some students were unable to quickly analyze bomb hits and correct for mistakes as they raced through the pattern. At first, Jonathan was happy just to be flying in the right place in the pattern, making the correct radio communications along the way, then dropping the bomb from the plane. He seemed too busy doing all this to have any time left for analyzing his drops and making adjustments.

There were dangers to look out for as well. The more dynamic the flying became, the more risks were taken. The week before Jonathan's squadron arrived at El Centro, an F/A-18 had plowed into the desert sand during a bombing run over one of the ranges. The T-45 Goshawk had fewer bugs than most new aircraft, but so far the squadron had lost three jets in accidents. One student had to eject because his Goshawk had two blown tires when he landed. Two other students collided their jets in midair.

Running into another Goshawk wasn't the only danger. Civilian aircraft off course, sometimes even drug trafficking planes from Mexico, wandered through the bombing pattern. Several days earlier while Jonathan was flying with an instructor, a Marine Cobra helicopter, apparently lost, flew through the range airspace while the Goshawks were swooping in on the target. Another four-jet formation returning from a bomb run had to scramble out of the way when a Cessna plane flew head-on toward them.

"Two's abeam, two point five," Jonathan heard over his radio. Schroder had reached the abeam point.

"One's in hot," Vaupel radioed to the other pilots four seconds later, the signal that he had rolled his plane to the northwest and was bearing down on the target with the master arm switch turned on, ready to drop a bomb. The instructor pilot flying the lead plane always dropped along with the students. It kept the instructors' skills honed.

"One's off safe," Vaupel radioed after he had dropped his bomb and begun pulling his plane up. That let the formation and the spotters on the ground know he had turned his master arm switch to safe to prevent the jet from accidentally dropping ordnance.

"Two's in hot," Schroder radioed a second later. He was now making his dive to the target.

"Two's off safe," Schroder radioed after his first dive.

"Three's in hot," Sobkowski radioed as he dove down just behind Schroder.

Jonathan brought his wings level at the top of the roller coaster moments before he was supposed to make the left turn for the run northwest into the target. It had to be a sharp left. Just banking the jet left at a 90 degree angle so its wings were perpendicular to the ground wouldn't be enough. The jet would still swing too wide.

To make the sharp left turn, Jonathan had to perform what aviators called a "roll-in." With his left hand he moved the throttle back to medium power and in the same instant shoved the stick smartly to the left with his right hand. With his left hand he pushed the power setting forward, then with his right pulled the stick back to his chest and even further to the left. That caused the plane to roll over almost upside down and the aircraft's nose to swing around sharply to the left. Jonathan felt himself straining for an instant against his shoulder and lap belts as he hung suspended upside down in the jet.

In almost the same second, Jonathan remembered to toggle the master arm switch to ARM with his left hand, then to flip the transmitting switch on the throttle in order to radio to the rest of the formation: "Four's in hot."

Being inverted now also caused the jet's nose to dive down to the target as the plane lost altitude. With the Goshawk's nose now pointed at a 30 degree angle at the bull's-eye target, Jonathan in the next instant quickly backhanded the stick to the right so the aircraft would roll back to being right side up. The entire maneuver—the roll-in along with arming the plane and radioing to the formation—took just over three seconds.

The jet's wings wobbled up and down as Jonathan in the next

breath struggled to line up the plane for the dive to the target. He squinted at the two-ringed bull's-eye bouncing around on the heads-up display, the HUD, in front of him. In the center of the rings was a small dot. The pilots called this the "pipper." Jonathan had precalibrated the heads-up display so that all he had to do was steer the plane toward the target and drop the bomb exactly when he saw the HUD's pipper cross over the bull's-eye on the ground. The bomb would land on the bull's-eye. Simple enough, at least in theory.

But to make it actually happen, Jonathan now had to juggle more than a half dozen calculations in his head during the next eight seconds. Whether the bomb hit the bull's-eye depended first on the altitude at which Jonathan released it. The higher the altitude, the more time the bomb had to fall—also the more time gravity had to bend the bomb's trajectory further down. Therefore, if Jonathan released the bomb too high as he dove to the target, the bomb would fall short of the bull's-eye. If he released at too low an altitude during the dive, the bomb would fall beyond the target. At the 30 degree angle that his plane dove toward the target today, the manual told Jonathan he had to drop the bomb exactly at 3,000 feet to hit the bull's-eye. Every 100 feet that he was off in altitude translated into fifty feet off from the target. If he released at 3,100 feet, the bomb would fall fifty feet short of the target. If he released at 2,900 feet, the bomb would fall fifty feet long.

Next Jonathan had to consider his speed. If he flew the dive too fast, he would fling the bomb ahead of the target. If he was too slow in the dive, the bomb would fall short. The speed he had to fly at for a thirty-degree angle of dive was 450 miles per hour. Ten miles per hour too fast and the bomb would fall fifty feet long; ten miles per hour too slow it would be fifty feet short.

The angle of his dive was also important. If he was 1 degree too steep in his dive angle, the bomb would fall fifty feet long. A degree too shallow and the bomb would fall fifty feet short. He had to keep the angle of his dive exactly at 30 degrees.

There were other factors to consider, such as the Gs his plane was pulling. For the 30 degree dive angle his G force had to be .87, or slightly under the normal pull of gravity. Any more and the bomb would fall short; less and the bomb would fall long. The plane's wings had to be level during the dive. If they were banked slightly to right, for example, the bomb would fall short and to the right. If the aircraft yawed right or left it would throw the bomb off to one side or the other. Finally, the wind could push the bomb off target.

It was rare for any pilot to perform everything perfectly in a bombing run—that is, fly at exactly the right speed, dive at the right angle, and release at the correct altitude. It was just as unlikely that there would be no wind to blow the bomb off course. For that reason, Jonathan had to learn how to quickly compensate for errors. For example, let's say Jonathan's dive angle was 1 degree too shallow and his airspeed was ten miles per hour too slow. If he released the bomb when the pipper in his HUD crossed the bull's-eye on the ground, the bomb would fall 100 feet short of the target. To compensate, Jonathan must allow his pipper to drift 100 feet past the target before he released the bomb. If he did that, the correction would result in the bomb hitting the bull's-eye.

On the other hand, if Jonathan happened to be 1 degree shallow in his dive but ten miles per hour too fast, the two errors would cancel out each other. If Jonathan dropped the bombs when the HUD's pipper crossed the bull's-eye, the bomb should hit the bull's-eye.

The combination of different types of errors could be endless. Today, a light crosswind blew from northwest to southeast. That meant Jonathan had to aim his pipper above and to the left of the bull's-eye to compensate. Barreling down in a bumpy jet with dials in the cockpit spinning, the HUD's images dancing about, and the radio crackling, Jonathan would go crazy if he tried to calculate all the Newtonian physics of free-falling ordnance in those few seconds. Or, worse yet, he might fixate on the target and forget to pull up before his plane dipped below 2,000 feet. (Below 2,000 feet, Jonathan would quickly run out of space between him and the earth to lift the jet before crashing.) No, to bomb accurately he had to have a combination of video arcade skills in the cockpit and a sixth sense developed over many runs about when to release the bomb to compensate for errors.

Jonathan, however, would rarely drop bombs this way when he finally reached a combat squadron in the fleet. This was the "Red Baron method," as the instructors liked to joke, having a student learn to throw a dumb iron bomb off his plane to a target below based on calculations in his head of where it would hit. In an F/A-18 Hornet, fancy radars and high-speed microcomputers would determine distances, speeds, altitudes, G loads, and wind speeds, then automatically release the bomb at the right time. All the pilot had to do was place the diamond in his heads-up display over the target and pickle. The bomb often was not a hunk of iron and explosives but rather a precision-guided munition crammed with electronics that steered it with pinpoint accuracy to the target. But the students needed first to learn to bomb as their grandfathers and great-grandfathers had done before they could rely on the gizmos. Plus there was always the chance that computers would fail and they would have to drop manually like the Red Baron.

Aerial bombing had become a specialized art in the military. In the post–Cold War world, pinpoint attacks with precision-guided munitions were becoming an even more critical form of warfare. Neither the Navy nor Air Force were likely to indiscriminately carpet-bomb cities or troop concentrations in the future. More likely one smart bomb would be guided to an important target in a conflict short of total war in order to decapitate an enemy's command center.

Within the Navy, pilots who had flown the service's A-6E Intruder attack bomber had specialized like doctors in dropping certain types of ordnance. Some were experts in stand-off, land-attack missiles, others dropped only laser-guided bombs, still others specialized in bombing only enemy air defense. But the Navy had retired the aging Intruders and the pilots who flew them were leaving in droves for the airlines. The admirals could well rue the day they gave up the Intruders. An entire generation of the service's most skilled bombers was being wiped out. In their place, the pilots who flew the F/A-18 Hornet would have to be skilled not only in dogfighting with planes in the sky but also in bombing targets on the ground that the Intruder normally struck. It might be too many jobs to heap on one pilot in future wars.

For the moment, Jonathan had enough on his mind with the one job at hand. He was determined to make his first bomb drop a good one. Get off to the right start. Jonathan struggled to keep the plane diving exactly at a 30 degree angle and on course so it followed the long black line on the ground leading up to the bull's-eye. It seemed impossible to juggle all the factors affecting the bomb's trajectory in his head and correct for errors. His brain couldn't process it fast enough.

Jonathan had developed a procedure from flights before. As the Goshawk rattled down toward the target, Jonathan first checked the top reading on his HUD to make sure his wings were level. In previous hops, he had been overcorrecting for his wings dipping one way or the other.

The wings were level this time. He was flying the aircraft much smoother than before.

He next looked at his pipper as it crawled quickly up the line on the ground toward the target. On line so far.

He made a quick check of his altitude.

Then glanced back to the pipper.

He made another quick correction in the angle of his dive.

Back to the pipper again.

On course.

Jonathan's grip on the stick tightened as he tried to keep the plane steady. He watched his altitude reading. The plane was now racing down the chute at about 450 miles per hour, but he ignored the airspeed reading on his HUD. He ignored the dive angle and G readings for the moment as well. Too much to absorb. The Gs he could feel in the seat of his pants.

Four thousand feet.

Thirty-eight hundred feet.

Thirty-six hundred feet.

Suddenly Jonathan realized he had a problem.

His jet was about to drop to 3,400 feet, 400 feet higher than the required release altitude, but his pipper was just about to cross the point on the target that he should be releasing the bomb. In aviation jargon, this was called an "early sight picture." His bomb sight had the target lined up before his plane had reached the proper 3,000-foot altitude to drop the ordnance.

If he dropped his bomb when the pipper crossed the target it would fall short because his altitude was still too high. But if he pickled the bomb at the correct altitude the pipper would have already passed the target and his bomb would drop long.

Jonathan winced. In less than a half a second, he decided how he'd correct for the error.

He pushed his stick slightly forward, which caused the nose of his jet to dip down. That tilting movement had the effect of moving the pipper in his HUD back from the target. It bought him more time so he could release the bomb both when the pipper was on the target and when the plane reached the proper altitude.

It was a neat fix, Jonathan thought. His heart pumped faster. His flight suit felt sticky inside. His mouth became dry from the cool oxygen he was sucking in.

Three thousand feet.

Jonathan punched the red button on his stick with his thumb to release the bomb.

"Beep," the sound Jonathan heard in his earpiece. An X flashed beside the BOMB reading on his heads-up display, indicating the aircraft had released the ordnance.

More beeps in the next second, warning that his plane was now dropping below the 2,000-foot mark and if he didn't pull it up soon the aircraft would bore into the ground.

Jonathan tugged the stick slowly but forcefully back to his chest. The Gs quickly set in. His G suit inflated as his shoulders sagged and he felt squished to the bottom of the seat.

One, two, three, four, five Gs. The HUD's G reading moved up quickly but stopped at five as the jet recovered from its dive and began to gain altitude. Jonathan had to remind himself to pull up smoothly from the bomb drop so the five Gs he had to endure came on slowly. Students who jerked the stick back could easily pull six Gs or more. One student yanked back so hard, the plane pulled eight Gs, leaving the instructor, who was caught off guard by the sudden movement, blacked out for a moment in the back seat.

Jonathan reached up with his left hand in the next second to flip the MASTER ARM switch down to the SAFE position. His arm felt like a bowling ball was slung from it.

"Four's safe," he radioed hurriedly to the ground spotters and the planes in the formation.

When he saw the nose of his plane rise up to the horizon Jonathan relaxed his pull of the stick slightly, which lessened the Gs he was experiencing.

In the next instant, he snapped the stick to the left. That caused the Goshawk to make a sharp roll to the left.

Jonathan then yanked the stick back to the center and again

pulled hard to his chest. The Gs returned. The Goshawk roared up to the left. Five thousand. Six thousand. Seven thousand feet.

Jonathan quickly scanned ahead to make sure he wasn't coming up too quickly on Sobkowski's tail in plane number three in the pattern. He wasn't.

Then he glanced down and to his left quickly to see that jet number one, flown by Vaupel and Miller, the instructor pilots, was diving down near the target behind him. It was. Jonathan's plane was in the right position.

Jonathan flipped the radio switch with his right hand. "Four's abeam," he said hoarsely, letting the other jets know he had reached the abeam position in the pattern.

No sooner had he said it than the spotter on the ground radioed to jet number four, his jet: "Four, you were one four seven at one o'clock."

The spotter used the points on a clock to tell the pilot where the bomb had fallen. "One four seven at one o'clock" meant the bomb had struck 147 feet beyond the bull's-eye and slightly to the right where the one on the clock would be.

He was long and slightly to the right.

Why? Jonathan wondered.

He had thought he'd made the proper corrections to have the bomb fall shorter.

But he actually hadn't. When Jonathan dropped the nose of his jet during the dive, it had had the effect of lowering the G force of his plane below the .87 G it was supposed to maintain for the 30 degree angle. At the same time, the incorrect G load threw off his dive angle. The combination of the two effects made his bomb still fall long instead of on target as he expected.

Jonathan should never have dipped his nose. As for the altitude, he should have split the difference—allowed the pipper on his HUD to move past the target point, then dropped the bomb when his jet reached 3,200 feet instead of 3,000 feet.

Jonathan was supposed to use the trip the jet made circling around the target for another bomb drop to think through the proper corrections. But there was precious little time for thinking. Everything happened so fast. A pilot had only about one minute and fifteen seconds between bomb drops. Take away the time he had to spend in that interval with flying the pull-ups, the turns, and the rolls, checking where he was in the formation, watching his fuel level, altitude, and speed gauges, making the correct communications calls along the way, plus writing down his bombing score on the pad strapped to his knee, Jonathan had all of about fifteen seconds between each bomb drop to ponder what he had done wrong and consider a correction.

He shoved the stick to the left, then jerked it to his chest, then back to the left again, for another roll in to line himself up behind the target.

"Four's in hot," he radioed again.

Jonathan jiggered the stick ever so slightly to make final corrections in his 30 degree angle of dive and as before focused on his pipper, then the altitude, then the pipper, then altitude, altitude, altitude.

Again, he was too high as the pipper crossed the target.

Again, Jonathan dipped the nose of his plane down to make a steeper dive angle and give his pipper more time on the target until he reached the right altitude.

He punched the red button on his stick.

"Beep." An X flashed beside the BOMB reading on his heads-up display.

Jonathan had learned he had to develop a cadence in the bomb drop. Pickle the bomb, pause for a few brief seconds as it dropped, pull the jet back up the roller coaster at a high angle of bank, roll to return to the abeam position. Pickle, pause, pull, roll. Pickle, pause, pull, roll.

"Four, you were one nine six at twelve thirty," the ground spotter radioed as Jonathan banked left into the abeam position.

"Shit," Jonathan muttered.

He was still long. In fact longer than before. The corrections he was making were supposed to work, he thought. Why the hell weren't they?

Two lousy scores. He still had six bombs left but this wasn't a good beginning, he worried.

The grading system for the bombing exercise was complicated. For the eight drops, the instructors threw out the three lowest scores and the three highest scores. Then the two middle scores were averaged for what was called a "circular error probable," or CEP. In the first nine flights, the instructors didn't expect the students to record a CEP much lower than 150 to 300 feet. That was average. They weren't marksmen yet. By the tenth through thirteenth flights, the average was 125 to 250 feet. For the final two flights of the weapons training—numbers fourteen and fifteen—average was considered seventy-five to 150 feet. A student who scored seventy-five feet or less was given an E Award for excellence.

The students and instructors waged three bets for each hop: the best first bomb dropped, the best CEP for the day, and the one bomb dropped closest to the bull's-eye. The payoffs were slips of colored paper called beer chits that could be redeemed from the losers, who bought the winners free beer. Jonathan's drops so far were average, but it was hardly comforting. Average wouldn't get him F/A-18s.

Rob Schroder lowered the nose of his jet for his dive to the target. This was his tenth flight dropping bombs, one more solo run under his belt than Jonathan. Average on a tenth flight was considered at somewhere between a 125 and 250 CEP. But Schroder was well on his way to breaking out of that pack. He hadn't dropped a bomb during his first two dives. Schroder had a game plan for this flight. Forget winning a beer for the first bomb drop. He was determined to win an E this hop, so he flew the first dive for practice to calm his nerves and get a feel for the pattern. In real combat, a bomber pilot wouldn't have the luxury of making practice runs before dropping. The first bomb would usually be the one that counted. But in training, the students were allowed to make passes and not drop every time.

Schroder's strategy paid off. His first drop, which he finally made on the third dive, struck 200 feet directly below the bull's-eye at the six o'clock position. Nothing to write home about.

But Schroder learned from his mistakes. During the next drop, his bomb struck just eight feet from the center of the target. A bull's-eye!

Drop number three was just as good. Twenty-nine feet from the center at ten o'clock. Schroder was scoring better than the instructors. That was what he expected. Schroder was a perfectionist. That was why he had joined the Marines. They were a cut above the Navy, he thought. The Marine instructors at the Naval Academy were the most impressive men he had ever met. Thank God somebody was still maintaining standards in the military.

Twenty-five years old, Schroder was a tall and muscular young man reared in Hattiesburg, Mississippi. But only a trace of his Southern accent remained. Much of it had been lost during his four years at the Academy. It wasn't the only thing Schroder thought he'd lost at Annapolis. His parents were furious with how cynical and unexcited he was on graduation day. But he just wanted out of the institution. The Academy was supposed to be a pinnacle of high standards with a code of honor just like the old South's. But it had been corrupted by political correctness, as Schroder saw it. Standards had been lowered. Women, minorities, every special-interest group was being catered to. Probably because the school was too close to Washington, he figured. The politics rubbed off.

The drugs, cheating, sex scandals, it was too much for him. Hell, he'd read in the paper this week that a midshipman was involved in a car theft ring. The place was going to pot, Schroder believed. All because there was too little attrition in the institution. Not everybody was cut out for the Academy, but they kept too much trash in that school. He couldn't change it, so he was glad to be rid of the place and in the Marines.

Schroder's next three drops were seventy-five feet, sixty-one feet, and sixty-five feet from the target. Five terrific hits. He had an E cinched. He could relax now. He'd lost a beer for the best first bomb dropped, but had won two for the closest hit and the best CEP.

Mike Sobkowski was one happy Marine. Certainly not about this flight. He was spraying the target with bombs high or low by about 160 feet. Sobkowski was happy he got in the Naval Academy. He began wanting to be a submariner, but he was happy he had changed his mind and joined the Corps. Sobkowski, twenty-five, looked like a Marine. His lineage was a mixture of Polish, Irish, and German. He was square-faced with a GI crew cut and a bright smile that never seemed to leave. Mike wanted to be an infantryman, but he was happy with aviation. He wasn't the best pilot but he made up for it by working harder than the other students, he thought. Not a quick learner but a thorough learner. He intended to be as gung ho as they come. The instructors liked that. He hadn't yet mastered the art of pipper placement. He was no dead eye. But he'd make up for it with enthusiasm.

Jonathan became more frustrated. His third drop he overcorrected for being long. The bomb fell 105 feet short at five o'clock, not much better than being more than a hundred feet off in the other direction. The fourth drop he was back to his old spot around the twelve o'clock position, but this time even further out at 259 feet. It seemed to Jonathan that he couldn't escape being long.

Miller looked back and noticed the same thing. Wise seemed to be putting his bombs into one hole. Either he didn't realize he was making the same mistake or he was doing nothing to correct it, Miller concluded. Miller didn't have time to dwell on it now. He'd grill Wise after they had landed. He turned his head in the other direction to make sure none of his chicks were flying too close in the pattern. All pumped up, pulling Gs and dropping bombs, these kids could become sloppy with their air work.

Jonathan was becoming slightly fatigued. He felt like he had been pumping iron for the past half hour from the roller-coaster ride's constant G loads. His fifth and sixth bomb drops were excellent, seventy-five and forty-nine feet from the center of the target. But they would be tossed out as part of the best three. His final drop returned him to his sinkhole: 189 feet at twelve o'clock.

Fortunately for Jonathan, his worst three scores—and they were lousy ones, he thought—were thrown out as well. The two middle scores averaged to 149, his CEP.

It was one point above average for this flight. But Jonathan had begun the hop with his sights set on scoring a CEP of seventy-five. It was probably naive to think he could bag an E on his first solo flight, he now realized. But no student took off with his goal being to score average. Not in this hyperaggressive group. As it was, Jonathan had fallen behind just a bit. The CEP on his bombing flight before this one was 148, one better. Maria had done well throughout the bombing training. It was one of the reasons she had won F/A-18s. Jonathan wasn't competitive with his future bride. That was kid stuff. He was proud that he had been mature enough not to let testosterone come between them. But he felt an ever so slight tension as he walked away from the Goshawk that day. He had to start scoring Es, soon. Or he wouldn't be joining Maria in flying the same plane.

CHAPTER ELEVEN

Dogfighting

ZIGGY and Nibbs pointed their T-45s toward the Quail operating area. Quail was the name given to a giant chunk of airspace—on the map it was shaped like a trapezoid—which the Navy reserved for air combat maneuvers seventy-five miles northeast of its El Centro airfield. Quail was perfect for aerial dogfights—a flat valley below with no one living on its dusty brown mesa. It was Saturday afternoon and the skies were bright blue. A gray haze surrounded the Little Maria Mountains that bordered the western stretch of the operating area. The Plomosa Mountains poked up about 1,200 feet on the east. The Colorado River snaked south through the center.

Fifty miles south of Quail, the two jets began limbering up for the fight.

"G warm right," Ziggy radioed Nibbs and shoved his control stick sharply to the right, then down.

Ziggy's jet, never dipping below 300 miles an hour, turned almost over and dove 1,000 feet in what amounted to a 90 degree bank to the right. Ziggy's G suit quickly bloated as the force of gravity weighing on his shoulders increased. One, two, three, four times his body weight. The pressure of four Gs made Ziggy feel as if he weighed almost 800 pounds.

Ziggy performed the maneuver to test his body's reaction to the blast of gravitational pull. The G warm-up was mandatory for what he was about to do. How well the body resisted excessive Gs varied from day to day. A pilot could lose his tolerance for it if he had been away from the plane for months.

Ziggy took the Gs well, no graying from the sides in his vision, one of the early signs that too much blood might be draining from his brain. He could feel his heart beating faster. But he was in good shape. He barely had to take more than a few extra breaths because of the pressure on his chest.

"Two," Nibbs radioed back and performed the same maneuver almost in the next instant. For the exercise, Ziggy was designated the lead plane for radio communications. Nibbs was "Dash Two," or simply "two" for short.

"Six, clear," Ziggy radioed to Nibbs, letting out a breath as his plane leveled behind Nibbs's and the gravitational pull returned to normal.

"G warm left," Ziggy radioed and performed the same maneuver to the left.

The jets turned sharply and violently each time. They had to in aerial combat. Milliseconds were precious. The quicker a pilot snapped his jet one way or the other, the more advantage he might gain over his opponent. Position was critical in a dogfight. A pilot had to seize it instantly. Crisp movements. The slightest delay could spell the difference between killing or being killed.

This was the eighth flight Bill Sigler had made to practice "air combat maneuvers," the jargonesque term for aerial dogfights. Kevin Nibblelink, a thirty-one-year-old Navy lieutenant with more than a thousand hours' experience in the F-14 Tomcat jet fighter, had been his instructor for many of the flights.

Ziggy, the call sign Sigler used when flying, had found aerial combat different from anything else he had learned up to this point. Landing a plane, taking off, piloting in formation, even dropping bombs were all paint-by-the-numbers skills. Cut-and-dried. A student had to memorize thousands of details and procedures, master the precision of flying to set points. But aerial combat had no clear-cut rules, no set way to fight, relatively little to memorize. Aerial combat required thinking on your feet. The battle was fluid. A good dogfighter was like a football running back who could juke and weave to avoid tacklers, who could instinctively make the right choices in an instant, find holes in the defense, run through them.

A dogfighter had to be aggressive, always thinking about staying on the offensive. Every turn of the jet, every maneuver had to be performed with one thing in mind. Attack. Attack. Attack. It was the only way to stay alive.

Top Gun glorified jets spinning around one another for the kill. But in modern air combat, dogfighting had largely gone the way of the bayonet charge. The Vietnam War was the last time American warplanes engaged in true aerial dogfights. During the Persian Gulf War, Iraqi jets did everything they could to avoid combat with U.S. warplanes.

Today's jets were designed to fight from long distances with sophisticated radars that spotted the enemy miles before the pilot did and smart missiles that electronically locked on to a hostile plane from afar. The fighter aces of World War I and II were athletes, race car drivers, expert marksmen with quick reflexes and strong arms to pull the stick. Today's fighter pilots were piano players with radar and weapons system buttons under each finger. A dogfight was supposed to become a small part of air-to-air combat. If a modern jet jockey found himself in one it was probably because he made a mistake and hadn't detected on his radar the enemy plane a hundred miles away.

That was the theory behind modern air combat. But as with most forms of warfare, theory rarely matched reality. The jet's expensive missiles, which could take out enemy aircraft that the pilot could not see, often proved worthless. The reason was disturbingly simple. For all the advances in aircraft technology, jet fighters still didn't have reliable electronic hardware to differentiate between a friendly or enemy plane. Despite the fancy missiles hanging from their wings, most jet pilots still did not fire on an adversary until they had gotten close enough to confirm with their own eyes that it was the enemy. In modern combat, with politicians terrified of any casualties, even more so of friendly-fire casualties, generals and admirals refused to allow pilots to shoot unless they could see the enemy. That meant Ziggy still had to learn to battle in close quarters.

If he was stuck in a dogfight, the pilot wanted to finish it in a violent burst of maneuver and fire. In the movies, the close combat might last for a half hour. In real life, modern jets will spend no more than sixty seconds battling one another. Air combat studies have found that if the pilot hadn't killed the enemy plane, or "bogey," in the first minute, his odds of surviving the fight dropped sharply. He was better off running away after a minute and living to fight another day.

In the previous training flights, Sigler had been flying set-piece maneuvers with Nibblelink, whose call sign was Nibbs. One time Ziggy would start out on the offensive attacking Nibbs. The next time Ziggy would be on the defensive with Nibbs attacking. Today the two men would fight from a neutral start. Either could become the attacker depending on who gained the advantage first.

The two jets climbed to 15,000 feet and continued flying to the northeast.

"Ziggy's fenced in left," Sigler radioed to Nibblelink. Glancing to his right he could see Nibbs's jet about 1,000 feet away. Ziggy flipped the MASTER ARM switch on his left console, then turned the knob above it clockwise to the air-to-air gun selection. That activated the simulated machine gun in his plane. He selected a tracking option for firing at enemy planes. His heads-up display glowed with a dancing bull's-eye and numbers sprinkled about it.

"Nibbs is fenced in on the right," Nibblelink responded.

They were preparing for the snap guns exercise. Ziggy had no idea where the term "fenced in" came from. It was the prearranged signal he gave Nibbs to tell him the plane's armament system was activated and he was ready to begin the mock battle. Fighter pilots used quirky lingo in the cockpit. Many of the sayings had obscure roots. For example, an F-14 Tomcat pilot who finally spotted an enemy plane and could take over its tracking from his radar operator in the back seat, would radio "Judy" to his partner. "Roger, punch," the radar operator would radio back. It came from the *Punch and Judy* puppet show.

"Ziggy's speed and angels on the left," Sigler next radioed.

More slang. This time the preparatory command for executing the snap guns exercise. The drill was simple. The two planes flew S patterns toward one another. When they crossed at the middle of their Ss, the nose of Ziggy's jet—if he had positioned it correctly—would be aimed at the side of Nibbs's plane. Ziggy would try to fire off a burst of simulated machine gun rounds at Nibbs's fuselage. He would have less than a half second—the snap of a finger—for the kill. But before they had climbed into their planes, Ziggy and Nibbs had agreed they would perform the exercise at an altitude of 15,000 feet and at a speed of 300 miles per hour. "Speed" meant Ziggy was flying at 300 miles per hour. "Angels" meant he was at 15,000 feet.

"Nibbs speed and angels on the right," Nibblelink answered. He was ready.

For the first maneuver, Sigler was the shooter. Nibblelink played the target. Ziggy jerked his stick to the right, recentered it in the next second, then pulled the stick back to his lap. That banked the Goshawk sharply.

Ziggy felt his shoulders sag from the force of three Gs. He wanted to maneuver as hard to the right as he could so his jet would loop inside the left turn he expected Nibbs to make. The pressure on Ziggy's shoulders relaxed briefly as he glanced up through his cockpit bubble. For an instant he could see Nibbs's jet streaking toward him.

Nibbs had made a smooth looping turn to the left. He had intentionally made himself an easy target so Ziggy could practice shooting at him. This was a warm-up so Ziggy could calibrate his eyeballs to pick off Nibbs with his jet's machine gun.

Ziggy tried it now. He pushed the stick forward, then sharply to the left. The three Gs returned as his jet banked violently to the left. Ziggy performed the tight S maneuver so the aircraft's nose was pointed perpendicular for a brief second toward the flight path he expected Nibbs's jet to fly. If Ziggy was lucky, he could squeeze off a hundred simulated rounds from his machine gun, ten of which might stitch their way along the fuselage of Nibbs's jet as it whizzed by.

Ziggy was about to squeeze the gun trigger on his stick, but held up. All day, Ziggy would try to fly as close as he could to Nibbs. But both pilots had agreed ahead of time that their jets would get no closer than 500 feet during the dogfight to avoid midair collisions. As Ziggy banked left to line up his shot, his Goshawk was dangerously close to invading the 500-foot bubble protecting Nibbs. If he kept up the hard left bank, he might ram the side of Nibbs's jet.

Ziggy instead eased up on the turn. Nibbs's jet shot past him to the left.

Ziggy next banked hard to the right for a wide circular turn much like what Nibbs had just made. They reversed roles. Ziggy would now be the target in the snap guns exercise. Nibbs would be the attacker.

Nibblelink glanced quickly to his right as he spotted Sigler far off in that direction. He banked hard enough to line up the Goshawk's nose to fire at Ziggy's side. Pilots divided an enemy plane into quarters. Anything forward of the wings was the front quarter. Behind the wings was designated the rear quarter. Nibbs and Ziggy had agreed that forward-quarter hits wouldn't count in the dogfight. They were always difficult to pull off because there was far less time that the enemy plane was in the gunsight. The purpose of these maneuvers was to fly to the rear quarter of the other's plane where the pilot had a few more seconds for the kill.

Nibbs eyed Ziggy's jet through the gunsight of his heads-up display. His only shot was at the front quarter. Nibbs decided to take it anyway and squeezed the gun trigger. The machine guns did not actually fire. For training flights, the pilot instead heard in his helmet's radio earpiece a *brdrdrdrdrdrdrdrdrd*, a static sound to simulate machine-gun fire when the stick's gun trigger was squeezed. In real combat, there would be more than just static in the ear. When an F-14 Tomcat's 20-millimeter cannon machine gun spewed out 6,000 bullets per minute, the jet vibrated as if it was belching.

In this exercise, a pilot made a simulated kill if the opposing jet was about to pass through the gunsight of his heads-up display as he was squeezing off machine-gun rounds. There was rarely any dispute between the pilots over whether a shot hit. Like master chess players they usually knew long before the last move if they had been defeated.

"Trigger down," Nibbs radioed to Ziggy, adding quickly, "forward quarter, invalid." Nibbs wanted Ziggy to know he was taking a shot he shouldn't take. Ziggy knew it already.

The jets crisscrossed at 300 miles per hour.

"Nibbs in as the target, maneuvering," Nibblelink radioed. They switched roles again.

"Roger, Ziggy's the shooter maneuvering," Sigler radioed back.

This time Ziggy took a wide turn to the right, craning his neck to the left all the while so he could keep Nibbs in sight almost a mile away on the other side.

The pilots fattened or thinned the S turns they made to try to get the jump on each other when they crossed. Just as Nibbs began his turn to the right, Ziggy banked hard to the left, then right, the G force slinging him from one side to the other in the cockpit. The violent turns paid off. Ziggy was in perfect position to blast Nibbs's jet at the wings.

Scoring a hit was no easy feat. Firing at a moving aircraft from another moving aircraft was like taking a running jump shot at a moving basket. Ziggy had to make three quick calculations in his head.

The first was how much to lead his target. Ziggy's jet was about a thousand feet away from Nibbs's, which was speeding along at 300 miles per hour. From that distance, it took about a second for the bullets from Ziggy's machine gun to reach Nibbs. In one second, Nibbs's jet would be long gone if Ziggy aimed directly at him. Ziggy therefore had to fire the bullets in front of Nibbs's jet as it zipped by so Nibbs would run into them.

Distance posed another complication. The maximum range for Ziggy's machine gun was 1,000 feet. His heads-up display had been precalibrated to tell him if he was within range to shoot as Nibbs's jet flew through the pipper displayed on the HUD. If Ziggy saw that the hash marks of his HUD's pipper fit neatly over the outlines of Nibbs's jet, he knew he was no more than 1,000 feet away and could squeeze off a shot. But the further away from Nibbs's jet that Ziggy was, the more gravity dragged down his bullets. Again, if Ziggy fired directly at Nibbs, the bullets would follow an arcing path and dip underneath Nibbs's plane. Ziggy therefore had to aim the pipper on his heads-up display a bit high of the target to correct for the sinking bullets at the longer range.

Calculation number three: fighter pilots called it "the plane of motion." An aircraft, no matter how it was moving, always flew at an angle, or plane of motion. It could be a vertical plane of motion or horizontal, or a combination of both. Whatever the angle, to pump

Nibbs's jet full of lead Ziggy had to fly on the same plane of motion as Nibbs when he pulled the trigger. Otherwise, Ziggy would be lucky if he got off just a few potshots that might scar Nibbs's jet but not bring it down. For example, if Nibbs was climbing sharply to the right and Ziggy was climbing sharply to the left, their paths would cross but Ziggy might only land one bullet on the fuselage of Nibbs's jet. Far better if Ziggy was also climbing to the right as he met Nibbs—which he was doing in this case. Flying on the same plane, Ziggy would have a better chance of sawing Nibbs's jet in half from nose to tail with the bullets.

The game could get even more complicated. In the split second that Ziggy had to calculate the lead, range, and plane of motion, Nibbs was also maneuvering to foul up Ziggy's arithmetic and spoil the shot. As Ziggy flew to the center of his S, Nibbs had two choices. He could pull up the nose of his jet so Ziggy's bullets would fly under him. Or he could push the nose down and force Ziggy to miss high.

Nibbs couldn't telegraph his movement too early or Ziggy would shift the angle of his jet again to compensate.

Both jets played cat and mouse as they neared the crossover point in the snap guns exercise.

Ziggy had a hunch that Nibbs would raise his nose.

He guessed right. Nibbs went high and Ziggy had done the same so he was on Nibbs's plane of motion and in line for the shot.

Ziggy could see the tiny white and red jet—Nibbs's aircraft as it appeared a third of a mile away. It was speeding to the heads-up display he was now looking through.

Ziggy squinted and pulled the gun trigger.

Brdrdrdrdrdrdrdrdr!

A bright yellow X flashed on his HUD to indicate the machine gun was firing.

Nibbs's jet zipped through Ziggy's HUD. It fit like a glove within the pipper's hash marks.

"Trigger down, snap," Ziggy radioed quickly, pleased with himself. "Good shot."

Kevin Nibblelink smiled. He hadn't wanted to come to the training squadron to wet-nurse students in these puny little Goshawks.

Nibbs was opinionated, not afraid to speak up when he disagreed with superiors. That had gotten him in trouble with bosses in the past. Being assigned to the training squadron instead of a better job with F-14s was punishment for not keeping his mouth shut, Nibblelink guessed. But after a few months of flying Goshawks he was surprised at how much he enjoyed teaching.

Ziggy had been his best student. In fact, Ziggy was in a class all his own, perhaps the best pupil to ever fly a Goshawk, Nibbs thought. His scores were so high other students had begun to grumble that he was blowing the curve for them. There were the good students, and then there was Ziggy. No one came close to him. No one worked as hard as he did or had such natural ability in the jet. Ziggy looked and acted like a warrior. He had a Prussian face: pointed nose, piercing gray eyes, short cropped blond hair, a handlebar mustache, and a deep voice that commanded instant respect. Ziggy would be an admiral one day, Nibbs was sure of it.

Ziggy was also what the pilots called a "retread." His left eye was weaker than the right. Even with glasses to correct the vision problem, the Air Force wouldn't accept him as a jet pilot. But the Navy would take him if he rode in the back seat of an F-14 as a radar intercept officer. The Navy also held out hope to him that he could one day switch to the front seat and become a pilot. The chances of moving were slim. Fewer than a handful of radar operators were allowed to retrain as pilots each year. But it was the only door even slightly opened for him to become a jet pilot so Ziggy chose the back seat of an F-14.

One month after finishing radar operator school in Oceana, Virginia, Ziggy was in the middle of the Desert Storm war, a terrified young ensign who hadn't known who Saddam Hussein was before he packed off for Saudi Arabia. The third night of the air war, Ziggy sat in a Tomcat roaring over northwest Iraq, electronic alarms blaring into his ear warning him that Iraqi surface-to-air missile radars below were trying to lock on to his jet for a shot. White flashes of antiaircraft fire streaked past his cockpit. He couldn't believe it. Two years before, he was a dumb farm boy from upstate New York. Now he was dodging SAMs in a $40 million jet. It took more than a half dozen missions for him to shake himself out of his daze and concentrate on the combat.

After the war, Ziggy applied to the Navy's personnel bureau to transfer to the pilot program. He was turned down the first year he sent in the paperwork. The radar intercept officer who beat him out for the last slot in the transfer program had been an admiral's aide. Carrying coat bags and running errands for an admiral was about the dullest job an officer could be stuck with in the Navy. But if that's what it took to be a fighter pilot, Ziggy would do it. He became a bag carrier for two years and this time his admiral's recommendation was enough to get him into pilot training.

Now Ziggy was as terrified as he had been during the Persian Gulf War. This time the fear was of failure. He was a thirty-year-old Navy lieutenant taking two years out of a promising career to retrain as a jet pilot. His contemporaries in the service would meanwhile move ahead of him on their career tracks. He was starting the service over as a rookie aviator and he didn't even know if he could fly. Sure, he had combat experience and basic air sense. The radar operator was an important part of the Tomcat operation. But the pilot up front flew the jet. Ziggy had never done that and he didn't know now if he could.

He had even more on the line than a freshly minted ensign starting flight school. Ziggy's instructors expected him to do well. He was already seasoned. He had even served with some of his instructors when they were young ensigns. Now they were teaching a retread new tricks. Failure for the new kids coming through flight school was painful enough. In Ziggy's case, he already had a career in the Navy that would end if that happened. Ziggy was deeply religious. God had a plan for all of us, he felt. Ziggy didn't know what that plan was. But he prayed that it now included him flying jets.

Ziggy had no time to relish the shot he had just taken at Nibbs. This was still just the warm-up. They were basketball players taking practice shots before the real game—in this case one-on-one aerial combat. No sooner had Ziggy pulled the trigger than he quickly began scanning his navigation instruments to make sure he and Nibbs were on course to enter the Quail airspace. They raced over the snaking Colorado River below.

Entering Quail, they set themselves up for their first fight: the low angle to hard counter. Ziggy was on the right at 16,000 feet and a mile away from Nibbs's jet. He would be on the offensive. Nibbs, who was flying at 15,000 feet, would pretend to be the enemy plane on the defensive trying to elude Ziggy. The low-angle-to-hard-counter fight would give Ziggy practice invading Nibbs's "control zone."

The theories of aerial dogfighting had changed little since World War I. Ziggy's goal was simple: position his plane behind Nibbs's to shoot it down. In more sophisticated combat, jet fighters such as the F-14 Tomcat or F/A-18 Hornet could knock out a plane traveling head-on toward them. But in this combat, Ziggy maneuvered to place himself at Nibbs's six o'clock, the term fighter pilots used, based on the position of a clock hand, for being at the rear of a bogey.

Reaching the rear for a kill could be complicated. To shoot an enemy aircraft with a machine gun or send a missile up its exhaust pipe, Ziggy had to position his jet behind the bogey's in an area the pilots antiseptically called the "control zone." Imagine the enemy pilot dragging behind his jet a giant cone thousands of feet long and wide. To make his kill, Ziggy first had to maneuver his jet inside that giant cone. Outside the cone, Ziggy would be at the wrong angle or distance to make the shot. For the past seven flights, he had been learning the basic tactics to reach the enemy's control zone. That was what air combat maneuver training was all about.

But the imaginary cone was constantly moving in different directions as the enemy plane twisted and turned. For Ziggy, two Goshawks maneuvering in aerial combat seemed like a couple of dogs skidding around on a slippery linoleum floor. High-speed jets didn't just take sharp turns like cars at a stoplight. If Ziggy followed Nibbs too closely and Nibbs suddenly banked hard to the right, Ziggy (like a dog skidding on linoleum) would overshoot forward and not have enough time or maneuver room to stay on Nibbs's tail.

Ziggy had to follow from a greater distance. But not too far. If his jet was more than a half mile away, Nibbs would have enough time to make a U-turn and come barreling at Ziggy head-on. Positioning a jet into an enemy's imaginary cone required a razor-sharp eye for distance and angle. A pilot had to have almost a sixth sense for where his jet should be miles ahead in the dogfight—much like a quarterback gauging where a long pass should fall in order to hook up with a receiver downfield after he has run a complicated pattern. But being in the right place at the right time was far more complicated in the sky than on a football field. An enemy plane on the run was constantly making turns to elude its attacker or looking for ways to turn the tables and counterattack.

To stay behind Nibbs, Ziggy had to think in circles. All rudimentary fighter maneuvers were based on circles. Each time a bogey jet turned, it was beginning a flight path into an imaginary circle. If the plane kept up the turn at the same angle of bank, it would fly in a circle. The faster the pilot flew as he banked the plane, the wider his turn circle. The slower he flew, the smaller the circle.

If Ziggy wanted to make a kill from behind, he had to maneuver his Goshawk so that he would fly to at least some point along Nibbs's turn circle. It was the only way he would get inside Nibbs's imaginary cone behind him and take a shot.

Ziggy was thinking circles now as the two planes streaked northwest. The pilots called the maneuver a low angle to hard counter because Ziggy was now trying to outwit Nibbs with sharper turns.

Sigler took a quick glance at Nibbs's jet to his left.

"Fight's on," Ziggy radioed. He was beginning his attack.

Ziggy tilted his jet left to close in on his bogey.

Nibbs immediately dove his jet down to try to evade.

The Colorado River valley came racing toward Nibbs as he headed for the ground.

Ziggy dove down too as he banked to the left. He thought he could take a shot at Nibbs, not with his machine gun, but with a missile that could be fired from a longer range.

"Fox two," Ziggy radioed. Fox two was the signal that he had let loose with a heat-seeking infrared missile such as the Sidewinder against Nibbs's tail. If the pilot radioed "Fox one," it meant he had fired a missile that used a radar to home in on its target, such as the Sparrow with a thirty-mile range. If he said "Fox three," it meant he had fired a long-range radar missile such as the $477,000 Phoenix, which could down a plane from more than 100 miles away.

Nibbs simulated that he had easily defeated this missile shot.

"Chaff, flares, continue," he radioed, the signal that his jet had dispensed the decoys to throw off the heat-seeker in Ziggy's missile. Keep fighting, Nibbs ordered.

Nibbs jerked his head to the right to spot Ziggy behind him.

Then he yanked his jet to the right as it dove. The pressure on his body became a crushing five and a half Gs. With his head turned right, it felt like another man had wrapped his hands around his neck and was hanging from it. Flying practice dogfights with students was always perilous. Nibbs had to be constantly on guard, always looking over his shoulder to watch for students who might run into him. He only had time for quick glances ahead or at his cockpit instruments to make sure he stayed on course before he had to turn his head back to keep the student in sight behind him. It felt like piloting the plane almost sitting backward.

Nibbs was now trying to make what amounted to a U-turn. If Ziggy continued flying northwest, Nibbs would be able to shake him off his tail and pass by him nose to nose.

Ziggy meanwhile had to make his hard counter to Nibbs's turn. It was the circles game again. To reach the imaginary cone behind Nibbs, Ziggy had to eventually make an even harder right turn than Nibbs in order to fly into Nibbs's turn circle.

Ziggy first pointed his jet to the ground and banked it slightly to the left. That edged him closer to Nibbs's flight path.

As Nibbs turned left, Ziggy banked even harder to the right at 300 miles per hour. The pressure hit him like a ton of bricks. Six Gs. More than Nibbs had experienced. Ziggy felt his cheeks peel back against his face.

Nibbs immediately saw that Ziggy had banked right, and at a sharper angle than the turn Nibbs had made. Nibbs forced his jet to bank right even harder to close in on Ziggy. Nibbs wanted to tighten his turn so much that it would force Ziggy outside his circle, like the dog skidding across the linoleum.

Ziggy hung on. His nose continued to dive to the ground. His wings were almost perpendicular to the earth as the Goshawk strained in its right bank to keep inside Nibbs's circle.

The hard bank paid off. As Ziggy's jet swung around, he could see Nibbs just ahead and to his left. Ziggy was nearing Nibbs's imaginary cone, his control zone. He slipped his right forefinger around the gun trigger on the stick and peered into the heads-up display.

The tail of Nibbs's plane in the next second crossed the left corner of the HUD screen.

Nibbs banked right again to shake Ziggy off his tail.

But not enough to foil Ziggy from getting off a quick shot against the side of his jet.

"Trigger down, snap!" Ziggy radioed, his voice sounding like a gurgle because of the pressure of the Gs on his throat. "Good shot."

Nibbs's jet had passed through the pipper of Ziggy's HUD as the jet simulated burping out bullets.

Ziggy was pleased. Another hit.

But suddenly he began to feel queasy.

The snap shot had been too easy. Nibbs usually flew aggressively in these exercises and was hard to catch. Something didn't seem right, Ziggy thought to himself. The turns had been violent, but Nibbs usually played hardball in the dogfight. Normally Ziggy didn't get off a successful shot against him in the opening go-round of the circle. If you zapped an instructor in the first 180 degrees of a turn, that was about as good as a student would ever be at this stage of his rookie career, Ziggy knew.

But Nibbs had taken the punch. He could be a cunning dogfighter, "a good stick," as fighter pilots put it. Nibbs never seemed to become disoriented in air combat. He always knew where both he and Ziggy were in the sky. Skillfully like a lightweight boxer, Nibbs could pick away at an opponent with fancy footwork in the sky and well-placed jabs.

Now Nibbs had a surprise up his sleeve for Ziggy. A snap shot usually didn't put enough bullets into an enemy jet to down it. What the incoming fire usually did was scare the hell out of the enemy pilot and make him angry like a wounded animal.

Nibbs smiled and proceeded to flush out his attacker.

Ziggy barreled down toward Nibbs like a ton of bricks. Three hundred and twenty miles per hour. Three hundred and fifty miles per hour. Three hundred and seventy. Down, down, down. The light brown ground appeared closer and closer. He had to watch that he didn't plow into it.

Just as Ziggy was about to close in, Nibbs slammed on the brakes. He yanked his stick back, sending the nose of his jet high up into the sky. At the same time Nibbs flipped out the aircraft's speed brakes, square flaps attached to each side of the rear fuselage that looked like corrugated steel screens, which, when they slammed against the onrushing wind, quickly slowed down the plane because of the air drag. Nibbs's Goshawk was like a galloping stallion that had come to a screeching halt by rearing up on its hind legs. The jet shuddered all over as its wings buffeted the wind. Within seconds it had slowed to about 200 miles per hour.

By the time Ziggy realized he'd been outfoxed there was nothing he could do. He desperately yanked the stick to his lap to pull up his jet and slow down as well. The aircraft shook violently. Six Gs slammed against Ziggy's chest as the aircraft pitched up. He forced a breath out and strained the muscles in his stomach as his G suit squeezed his hips and legs. But there was no way to reduce his speed as quickly as Nibbs had or stay inside Nibbs's circle. Like the skidding dog, Ziggy's Goshawk raced underneath Nibbs's jet to the left and now outside his circle. His aircraft had overshot its quarry.

Whoa, this is no fun, Ziggy thought to himself as he slid past Nibbs. Twenty seconds ago, he had been the attacker ready to pounce on Nibbs for the final kill. Now with the overshoot, Ziggy had lost most of his advantage in position. He was no longer chasing Nibbs's tail. Instead, Nibbs was flying south and Ziggy was rushing north in the opposite direction. And if Ziggy didn't watch out, Nibbs would make another sharp U-turn and close in on his tail. The hunter would be the hunted.

Nibbs chuckled to himself. It was fun playing with the beginners. "Oh, shit," Ziggy muttered. Now he realized how badly he'd screwed up. Ziggy had been too eager to take that first shot. He'd thrown airspeed and geometry out the window, and charged at Nibbs like a raging bull instead of perhaps waiting a few more seconds, then closing in on him slower so Nibbs couldn't fake him out as he had. Ziggy had been too greedy.

Now he needed to salvage this mess. Nibbs wasn't yet in a position to hurt him. But he was a far better dogfighter and soon would.

Ziggy turned his head up and to the right. He could see Nibbs flying south still at the higher altitude, his jet now traveling no more than a lazy 180 miles per hour, Ziggy guessed.

There was no way Ziggy could point the nose of his jet up in order to shoot at Nibbs. The best he could do was maneuver somewhere to a point behind Nibbs. He had to get back to the imaginary cone trailing Nibbs.

Ziggy banked right and put the plane into a slow dive. He kept the speed to no more than 150 miles per hour, hoping he might be able to loop back up and maneuver again behind Nibbs, but this time with less airspeed so he didn't overshoot.

But Nibbs was no fool. He realized immediately that Ziggy was trying to recover from a bad situation. Nibbs wasn't going to let him off that easy. He banked his jet left so it would travel north and pass Ziggy's Goshawk going in the opposite direction.

But as he jerked the stick to the left, he took a one-second intermission to set up the ground rules for the next part of the combat. The instructors and students just couldn't race about the skies in free-form combat. They faced the constant peril of midair collisions with all this dogfighting. The instructors wanted the maneuvers and countermaneuvers to be realistic. But it wasn't worth getting someone killed.

"ROE?" Nibbs asked over his radio before he made the head-to-head pass with Ziggy. What were the rules of engagement for this next round?

"Ziggy's high," Sigler answered.

"Nibbs will be low," he responded.

The fight was back on, but Ziggy had allowed the roles to switch. He had started out on the offensive but very quickly had overshot his target and flown ahead of Nibbs.

Instantly, Ziggy began what the pilots called a "flat scissors." In a real dogfight, Ziggy would avoid it like the plague. A pilot flew into a flat scissors because he'd made a mistake as Ziggy had and was now trying to shake a bogey off his tail. Or better yet, put Ziggy again behind Nibbs.

In a flat scissors, the two jets looked as if they were weaving over and under each other like two strands of yarn in a cord. As they now crossed head to head, Ziggy banked his jet down and to the right while Nibbs banked his jet up and to the left. Then the two pilots began flying their jets as slow as they could as they circled around to cross again in the weave. Ziggy pulled his throttle back and pointed the nose of his jet up as far as he could before aerodynamics would force it to stall. Nibbs did the same. They were running an unusual race at this point. The winner was the plane that finished in second place.

The whole point of the flat scissors maneuver was to put on the brakes so the other guy would jump ahead of you. If Ziggy traveled forward at a slower rate than Nibbs did as the two planes weaved back and forth, Nibbs would eventually be flushed out in front. Ziggy would be back on his tail and could get off a shot.

But by the time they crisscrossed a second time in the weave, Ziggy realized he'd never get behind Nibbs. Nibbs was playing the same game he was, only better. Nibbs had managed to fly the scissors at 170 miles per hour. Ziggy couldn't seem to stay below 220. He was getting ahead of Nibbs, exactly what he wanted to avoid in the flat scissors.

This isn't working, Ziggy said to himself. Nibbs was far more skilled in manipulating his jet speed and the turns during the weave. Every time he had sucked Ziggy into a flat scissors in past dogfights, Nibbs had ended up behind him. Now it was happening again. Ziggy cursed himself. He was flying a sloppy jet. Nibbs was a wily bandit. Oddly enough, all the instructors seemed to become better dogfighters when they trained students. In a regular Tomcat squadron, Nibbs might fly no more than twenty-four hours a month and perhaps no more than a half hour of it would be in air combat maneuvers. Fuel was the problem. Dogfighting burned too much of it so practicing became expensive. But in the training squadron, Nibbs could fly as much as fifty hours a month in dogfights. It was against beginners but the instructors found it still honed their skills. After Nibbs crossed underneath him the third time in the weave, Ziggy quickly decided not to bank left again to continue the scissors. Instead he pushed his jet down and made a long arcing turn to the right. But he didn't plan to run away. Ziggy wrenched his stick more sharply to the right, hoping to scoot back behind Nibbs, who was flying north of him.

But Ziggy was like a boxer taking a roundhouse punch against a quicker opponent. He gave Nibbs plenty of time for a countermove.

"That's a big commitment there," Nibbs radioed smugly. He could see Ziggy's jet off his right shoulder.

"Yep, it was," Ziggy radioed back glumly. He couldn't have telegraphed his move more if he had sent Nibbs an engraved notice.

Aerodynamics had defeated Ziggy again. As he arced his jet down and to the right, Nibbs had turned right as well to try to stick to Ziggy's tail and fire a shot.

Looping around behind Nibbs was now out of the question. The best Ziggy could hope for was that Nibbs's jet would inadvertently overshoot him and barrel ahead.

But that wouldn't happen either. Nibbs had plenty of room to make his right turn and still stay behind Ziggy. There wasn't much Ziggy could do about it. His jet was now diving toward the ground gaining speed every second.

Options raced through Ziggy's mind. He could yank his stick back, pull up the nose of his jet. That would slow his forward travel. Maybe it would squeeze out Nibbs so he didn't have room to make his right turn and still stay behind Ziggy. Maybe it would force Nibbs to overshoot him.

But the problem now was speed. Ziggy's eyes darted to his airspeed indicator. The jet was still crawling at 150 miles per hour. Once Ziggy had decided to drop his nose to make his right turn he was stuck. He couldn't pull the jet back up unless he increased his airspeed beyond the 150 miles per hour it was traveling. If he pulled the Goshawk up at 150 miles an hour, he would stall the aircraft because its wings would not be able to provide enough aerodynamic lift at that speed and angle. But if he increased his airspeed now, he gave Nibbs more room to make his right turn and get behind him.

Ziggy kept his nose pointed to the ground. Nibbs dove his plane down as well, then banked it even harder to the right so he remained on Ziggy's tail.

Ziggy tried to dodge and weave as best he could to shake off Nibbs as the two jets sped toward the ground. Nibbs had the upper hand, but he still wasn't in position to take a shot. Ziggy shoved his throttle forward so his jet raced to 300 miles per hour as he dove. He finally had the speed to pull up. Decisions now had to come in split seconds. There was no time to second-guess. Only to act. Then worry about the consequences in the next second. Ziggy had to pull his plane up, shoot it vertically into the sky like a rocket. It was the only way to escape Nibbs.

He turned his head back over his right shoulder to try to spot where Nibbs was on his tail.

In a flash, he was blinded. Ziggy was squinting directly into the sun. Nibbs had used an old fighter tactic. To camouflage his plane from the enemy, an aviator could fly in front of the sun so the other pilot couldn't see him. The sun trick usually blinded the other pilot for only a brief moment. But that might be just enough to maneuver for the advantage. Ziggy now lost Nibbs in the sun for a few seconds. It gave Nibbs more time to roll in behind Ziggy unobserved.

Nibbs now was ready to blast Ziggy from behind.

The little bastard, Ziggy thought and smirked. He relished this type of fighting. If there was honor in combat, it was in the air, one-on-one. Two warriors matching wits and daring. Death to the defeated. To the victor, another day to live and fight.

Dropping bombs was the ugly side of war, Ziggy felt. He had never been able to completely stomach it during the Persian Gulf War. There was no such thing as a surgical air strike. The bombs always killed innocent men, women, and children on the ground, people Ziggy didn't hate, whose only crime was being in the wrong place at the wrong time. Ziggy realized he was merely a tool of war. He had to follow orders. But why would God allow such barbarism to happen? he wondered. Ziggy hadn't resolved that in his mind as yet.

But battling another jet pilot in the sky was different. There was no innocence up here, Ziggy thought. If someone was in a jet armed to the teeth with missiles and guns, he was there to kill you. Taking another's life could never be completely right. But Ziggy had a much easier time sorting through the morals of this kind of combat to make the kill. Virtue among the whores, he reasoned. Aerial combat was the purest form of warfare he would ever see.

Ziggy had shared his feelings about the morality of war only with a few other pilots. Most aviators didn't want to discuss it. Being too introspective could be misread as vulnerability. Sensitivity was ridiculed in this business. A jet pilot never dared reveal that side of his personality, if he had it. His mission was to find an enemy's vulnerability and exploit it. Even within a squadron, pilots probed each other for weaknesses.

But Ziggy had to think this through. He was a Christian and some day he would have to answer to a higher being for what he had done on earth, even if it was under orders. Ziggy would never forget the seven-year-old girl who had written to him as a class project during Desert Storm. They exchanged dozens of notes.

One day the letters stopped. Ziggy wrote a final note to ask why. The girl's father answered. She had seen a photo of an F-14 and had asked what the bombs underneath it were for. "To kill people," the father had answered. The little girl was repulsed. She could never bring herself to pen another letter to him.

Though Nibbs was now bearing down on him, Ziggy wasn't about to give up. He had enough airspeed now. Ziggy yanked his stick back and to the left. His helmet slammed against the cockpit headrest, the G force pressing his chest against the back of the seat. He was forcing his jet to soar up, then bank to the left. Ziggy was trying desperately to swing high in a wide circle to force Nibbs to jump ahead of him. Imagine two skaters racing down a straight line, but suddenly one veers off to the left, then back in, to slip behind the other. That was what Ziggy was trying to do to Nibbs, only in a three-dimensional maneuver.

Nibbs also pulled his jet up in the next instant. But he only turned slightly to the left. Go ahead, Ziggy, make your left sweep and cross in front of me, Nibbs thought.

Nibbs was thinking two moves ahead, calculating how he would still end up behind Ziggy.

As Ziggy rocketed high into the sky and to the left, he crossed over Nibbs's jet streaking by underneath.

But as he made the crossover to the left, Ziggy kept his stick back to his lap and flipped the jet upside down to force it into a barrel roll. The horizon he now saw twirled around like a baton. In a barrel roll, the Goshawk looped up and quickly down to the left like a roller-coaster ride where the passengers are turned upside down to make the vertical turn.

The barrel roll should have forced Nibbs to overshoot so Ziggy could end up again behind him, or so Ziggy thought. But as any roller-coaster passenger knows, the downhill trip is always faster than the ride uphill. Coming out of the barrel roll, Ziggy had kept the stick pulled back to his lap for too long. The nose of his jet pointed down, racing for the ground. His airspeed continued building and building.

Ziggy couldn't get the damn plane back up to slow down!

"Oooh, that was a big lead turn," Nibbs radioed, recognizing instantly the countermaneuver he needed to make.

"Yeah, I didn't mean to do that," Ziggy said. The barrel roll had been poorly executed. Ziggy hadn't wanted to end up nose down with too much speed. He was in trouble again.

"Yeah, your nose went too far down," Nibbs radioed, agreeing. He had seen Ziggy's problem instantly.

"You're going to get it this time," Nibbs said with a chuckle. He readied for the kill.

Nibbs yanked his stick back so his Goshawk rocketed vertically into the sky. His forward travel stopped on a dime.

Ziggy could only watch helplessly as he continued his dive to the ground.

As Nibbs saw Ziggy's jet slide ahead, he pitched his stick forward. Like a roller-coaster car reaching the top, the nose of his Goshawk jackknifed down. From bright blue sky he saw the earth in an instant. Nibbs's stomach felt as if it was being squeezed like a balloon up into his throat, the same feeling a roller-coaster rider felt pitching up, then down on the tracks. In this case, the sharp jackknife down created a violent, negative G force on Nibbs's body. Pilots often threw up at this point; it felt as if a hand was reaching down into their stomachs to pull out their insides. If he hadn't been strapped to the seat, Nibbs would have flown out of the cockpit. His legs and arms floated up sharply like a rag doll's.

Nibbs now dive-bombed to Ziggy's tail. On his HUD screen he almost had the back of Ziggy's jet lined up in the pipper.

Not quite.

"No shot, continue," Nibbs radioed. It signaled he was diving in for the kill, but Ziggy was still at an angle too far below him to make the hit.

Ziggy was now on the run. He drove his jet even further to the ground to try to escape Nibbs's pursuit. Don't try to be a hero, Ziggy and the other students had been taught. If in an instant you were at a disadvantage, bug out. Run away. Live to fight another day. That was what Ziggy was now trying to do, diving down further so Nibbs would be at too high an angle above him to take a shot.

The two jets were falling like rocks. Nibbs finally pulled his stick back to ease out of the dive so his nose would be lined up on Ziggy's tail for a shot. The pressure on him from pulling up became severe, past six and a half Gs. His jet felt as if it was being buffeted by the onrushing air as the wings strained to maintain their lift.

Ziggy looked behind him. Nibbs continued to close in. Ziggy's uniform was soaked with sweat by now. He could have wrung water out of his helmet liner. The harnesses that strapped him into the cockpit were soggy and smelling sour. He always had to wash his flight suit after every dogfight.

Ziggy was disgusted with himself. This was not turning into one of his better dogfights. He checked his altitude readings. His Goshawk was racing down at 400 miles per hour and nearing 10,000 feet. They had agreed before the flight that they would dogfight no lower than that level for fear of hitting the ground.

"Knock it off, deck!" Ziggy radioed. It meant they were at the 10,000-foot point. It was time to stop the fight before it became too dangerous.

"Roger, let's knock it off," Nibbs agreed, somewhat relieved. Any further and they'd be scraping him off the ground. "Recommend west, two twenty."

"Concur, coming right to west," Ziggy answered.

Both pilots eased off the throttles, banked their jets up, then to the right for a U-turn that would have them flying west along the Quail range.

They set up for the next fight. This time, Nibbs would begin on the offensive. Ziggy would try to regain the advantage.

They started with what the pilots called a "rolling scissors." It was a version of the flat scissors. Both pilots would try to outmaneuver each other by slowing their downrange travel in crisscrossing flight paths. But in a rolling scissors, the pilots would take wider turns and rolls as they wove back and forth. From afar, the patterns the jets flew looked like Chinese dancers twirling long ribbons around and around as they ran across the stage.

Both planes now traveled west at 250 miles per hour; Ziggy at 15,000 feet, Nibbs at 16,000 feet. Ziggy pushed his throttle forward to position his jet slightly ahead of Nibbs and to the right. That put him on the defensive.

"Fight's on, cleared in," Ziggy radioed Nibbs and made a hard, four-G bank to the left.

"Nibbs is cleared in," Nibblelink responded, pulling his stick back and banking his jet up and to the right just as hard.

They began the roller to see who could sneak behind the other.

Nibbs kicked his nose up, flipped his jet to the right, then upside down in a barrel roll to try to swoop left and down behind Ziggy.

But Ziggy knew what to do to keep from jumping too far ahead of Nibbs. As Nibbs kicked up his nose to begin his barrel roll, Ziggy struggled to stop his downrange travel. He jerked his stick back, then to the right so his Goshawk also rolled up, to the right, and finally flipped upside down into its own barrel roll.

The two jets were now trading barrel rolls to force the other to jump ahead. As Ziggy bottomed out, Nibbs rolled upside down over the top. As Nibbs bottomed out, Ziggy rolled upside down over the top. Around and around they went.

The game now was speed and the tightness of the rolls. Whoever was slower at the top of the rolling scissors and kept his nose pointed up further could slow his downrange travel more and win the fight by coming in second. Like a coiled spring, the more Ziggy could keep the rolls in his pattern tight and his speed down, the less he traveled downrange. Ziggy pulled the stick back as hard as he could each time to lift up his jet for a tight roll-over.

But he found he was traveling too fast. Ziggy wanted his speed around the rolls to be about 250 miles an hour. But he was traveling at 280 miles per hour. That forced him to have a wider radius in his turns.

Nibbs, on the other hand, flew at a slower speed and kept his turns tighter. Ziggy again ended up flushed out ahead of him.

Nibbs tried to line up his jet to shoot Ziggy as he bottomed out at the end of the roller. He tugged back his stick to line up a shot on the side of Ziggy's jet.

Nibbs pressed the gun trigger.

Brdrdrdrdrdrd.

He missed.

Instantly, the tide of battle changed.

Like a roller coaster coming out of the bottom of the ride, Ziggy's jet now raced at more than 300 miles per hour—enough for him to rocket vertically up into the sky and briefly escape being on the run in the rolling scissors.

"Very nice right there," Nibbs radioed a compliment in a strained voice. The Gs forced air out of his lungs as he spoke.

Nibbs's jet hadn't built up enough airspeed to follow suit. He couldn't chase Ziggy into the sky or take any more shots.

Ziggy had dodged the bullets. But as they returned to the rolling scissors once more, Nibbs still had the advantage.

Ziggy banked up and left in the roller, but he did so much too early. As he drove his plane to the bottom of the roller, Nibbs pulled his Goshawk up to tilt it over to the left as well. As Nibbs bottomed out, he had Ziggy in his sights again.

He simulated firing a missile.

"Fox two, invalid," Nibbs radioed.

Ziggy had jerked his jet up and looped it around to the opposite direction, evading the missile in the process.

He was still alive, but just barely.

"Very nice," Nibbs complimented again over the radio. "There's no way I'm going to follow you through that."

Nibbs quickly pulled up the nose of his jet and looped it back so it was flying in the same direction as Ziggy.

Ziggy dove to the earth. His Goshawk picked up speed as it sank like a stone. With enough speed he hoped to outrun Nibbs.

But Nibbs was far too savvy a dogfighter for Ziggy to shake off. Nibbs dove his jet as well. He raced south to catch up with his inexperienced quarry.

Within seconds, Nibbs had closed the gap. He fired a simulated missile.

"Fox two, half a mile, three twenty," Nibbs announced to Ziggy over the radio, letting him know the distance and speed (320 miles per hour) when he took his shot.

"Roger, four fifty, chaff, flares," Ziggy radioed back. He was flying faster and dispensing the simulated chaff and flares as decoys. Ziggy calculated that his jet had dodged Nibbs's missile.

Ziggy began a neck-snapping break turn to the right. It was the hardest turn he had tried all day. His G suit ballooned immediately. Ziggy began the hook breathing so he wouldn't pass out. He squeezed his chest and neck muscle and forced air out of his throat, making a "hook" sound, so the blood would stay trapped in his brain. The pressure built to seven Gs. The skin on his face sagged and his jaw dropped. His arms weighed down like lead.

Nibbs could see Ziggy a mile ahead of him making the violent break turn to the right. He wanted to get back on Ziggy's tail, back to his control zone.

Nibbs could try to do it in one of three maneuvers. The first was the simplest. He could just aim the nose of his jet in the direction Ziggy was taking and go after him, making the same hard break turn Ziggy had made. Pilots called this "pure pursuit." But that wouldn't end up doing Nibbs much good. Even if he could make the same turn Ziggy had, he'd still be too far away from him.

The other maneuver was what the pilots called "lead pursuit." Nibbs could take a shortcut. He could turn right now, in effect cut across the circle Ziggy's jet was making, then try to intercept Ziggy at the other end. Lead pursuit was much like what a marksman performed to shoot at a moving target. If the marksman aimed his rifle directly at the target, he'd miss because by the time the bullet arrived the target would have long moved ahead. But if the marksman led the target, fired ahead of it, the target would run into the bullet. Simple enough.

But in fast-banking jets, a pilot flying lead pursuit faced the same problem as the slipping, sliding dog. Nibbs's Goshawk could end up overshooting Ziggy's jet, flying off in one direction while Ziggy was turning in the other. For that reason, the best maneuver Nibbs could perform was what pilots called the "lag pursuit." It meant chasing Ziggy with more finesse and patience. Instead of racing after him—the first impulse a beginning dogfighter had—Nibbs would slow his jet slightly, delay for a few seconds before he began the same right turn Ziggy had made. That way, he would fly into Ziggy's turn circle in a more controlled fashion. He'd be less prone to overshoot.

But to pull off a successful lag pursuit, Nibbs first had to be able to accurately gauge miles ahead of him where Ziggy's plane would be flying in its turn circle. Then with seat-of-the-pants calculations for distance, speed, and banking capability, he had to maneuver his jet so it would eventually line up in that same circle.

Nibbs could have done that. He'd had years of experience practicing finesse maneuvers like the lag pursuit. But this time he decided not to. Instead he would intentionally overshoot Ziggy with a lead pursuit. Nibbs wanted to see how well Ziggy could shake off a pursuer if an enemy plane like his had made a mistake.

"Here comes the overshoot," Nibbs radioed, letting Ziggy know that he planned to fly poorly to see how Ziggy would react.

Nibbs banked his jet quickly to the right of Ziggy to take the shortcut that would intercept him at the other end of the circle Ziggy was flying.

Ziggy came around from his right turn and could see Nibbs flying toward him just off to his right.

Ziggy waited two seconds, then pulled his jet to the right even more, hoping to circle around and maneuver behind Nibbs again when he overshot.

But Ziggy should have waited two seconds more. He had turned sharply to the right too soon.

"Ooh, that was a little early," Nibbs cooed over his radio. His rookie had given Nibbs another break.

Nibbs banked right sharply. He wouldn't overshoot Ziggy to the left as much as he had planned. In fact, now he had a chance to plant some bullets on Ziggy's tail.

Nibbs peered into his heads-up display. He could see Ziggy's jet approaching his pipper. From this distance it looked like a red and white toy plane.

He squeezed the trigger on his stick.

Brdrdrdrdrrd.

"Trigger down. Ahh! Unassessable!" Nibbs radioed, frustrated. Ziggy's plane had shifted slightly to the left eluding his shot.

The hunt was now on. Nibbs wouldn't give Ziggy any more breaks. He wanted to finish him off once and for all.

There were other ways to invade Ziggy's control zone. The two men were fighting three-dimensional combat. Two Army tanks maneuvering against each other on a battlefield did so on one plane, a horizontal one. But jets could fight both horizontally and vertically. To gain an advantage over an enemy pilot in his turn circle, Nibbs could perform what aviators called an "out-of-plane maneuver."

The best way to visualize why a pilot would do it is to imagine both jets flying horizontally and Nibbs's jet lined up in Ziggy's turn circle. Nibbs's jet is on one side of the circle and Ziggy's jet is on the other side. If neither plane could fly faster than 300 miles per hour and both jets kept chasing each other at the same speed along the same horizontal plane, they would just fly around and around in the same circle. Neither would catch up to the other's imaginary cone.

But Nibbs could break this never-ending merry-go-round by putting his jet on another plane of flight. He could pull his aircraft up, so it soared high into the sky, then quickly aim his Goshawk back down into the horizontal circle Ziggy was flying to close in on him. Pilots called this a "high yo-yo." Nibbs could also loop down if he wanted to. That was a "low yo-yo." Either way, it was a maneuver designed to be a shortcut for Nibbs to take through Ziggy's circle to line up behind his jet without overshooting.

Nibbs began with a high yo-yo. He pulled the nose of his jet up steeply so the Goshawk vaulted high into the sky and then rolled down as Ziggy made a wide circular sweep to the left.

That got him closer to Ziggy. But Nibbs wasn't yet in a position to take a shot. He tried a low yo-yo next. Ziggy reversed course and made a wide turn to the right. Nibbs dove down in an even sharper turn to the right to try to come under Ziggy and on his tail.

Nibbs succeeded up to a point. He was now flying directly behind Ziggy. But Ziggy had dodged another bullet and driven his jet toward the ground so Nibbs was at too high an angle for a shot.

But flying down the roller coaster again meant that Ziggy would be gaining speed and flying ahead of Nibbs.

Ziggy pulled his stick back and to the left so his Goshawk roared up and banked into a turn to the left. That would slow him down.

"Ziggy's high," he radioed to Nibbs. Ziggy was forcing another flat scissors maneuver on Nibbs and wanted to set the rules of the game so the two jets wouldn't collide. They would begin crisscrossing again over and under each other. Ziggy hoped he would be able to flush out Nibbs so he would finally be forced to fly in front.

But not for the first crisscross. As Nibbs banked right and down, he could see Ziggy coming the other way from his right.

Nibbs pulled the gun trigger again.

Brdrdrdrdrdrd.

"Trigger down, unassessable," Nibbs radioed. Ziggy had dived low at the last minute so Nibbs couldn't tell if his bullets would have struck him.

Ziggy pulled his jet back up in a steep bank, turned hard to the right, and then pointed his nose back to the ground. He was flying up and down like a bucking bronco. Ziggy didn't have any earthly idea whether it was doing any good. But he was desperate. Anything to keep Nibbs from peppering his jet with bullets. His chest heaved as he sucked in oxygen from his mask.

"Ooh, that nose is coming up," Nibbs radioed as he saw Ziggy bobbing up, then down to his left. "Very nice."

This time, Nibbs tried the lag pursuit. He delayed his right turn by half a second to have a better angle on Ziggy for another shot during the crossover.

Brdrdrdrdrd.

"Trigger down, unassessable," Nibbs radioed after Ziggy dove under him in the crisscross.

Ziggy knew Nibbs was flying his jet better than he. He expected that. Flushed out as he was, the shots would be coming at every crossover. He'd juked the jet up and down and right and left to avoid the hits. He felt like he was in a Western movie with a gunslinger firing at his feet making him dance.

Ziggy had to become more unpredictable. Nibbs was looking at a long fat target every time he fired at the side of Ziggy's jet fuselage during the crossovers. Ziggy quickly decided to make his jet skinnier in Nibbs's sights. In real combat, fighter pilots instinctively scrunched their shoulders together hoping that the enemy's bullets would miss them. It didn't do any good, but it made the pilots feel better.

Instead of banking to the left in another crossover, Ziggy swept further right and shoved his stick forward so the jet again dove to the earth. That way, Nibbs had a smaller tail target to aim at. Ziggy also hoped his tail would be too low for Nibbs's machine gun.

It was. So Nibbs let loose with another simulated missile that might have a better chance of catching up to Ziggy's nose-diving jet.

But no sooner had he fired it than Nibbs realized Ziggy was still at too low an angle for the missile to reach his jet.

"Fox two, invalid, angles," Nibbs radioed quickly to signal the miss.

Ziggy remained in the dive for a few more seconds frantically hoping to gain airspeed so he could cross in front and under Nibbs before Nibbs could get off another shot.

One second, two seconds, Ziggy counted to himself, then pitched the jet back up and banked to the left.

It worked. Nibbs didn't have enough maneuvering room to bank sharply to the left and remain behind Ziggy.

"Here comes the overshoot again," Nibbs radioed. He fired his machine gun but he missed. Ziggy was too low.

"Yeah, better timing," Nibbs radioed. "Missed high." It was a compliment. Ziggy had timed his turn to the left perfectly so it forced Nibbs to overshoot.

Nibbs looked out the right side of his cockpit to see what Ziggy would try next.

Ziggy was flying more than a mile south. "Well, he's doing something," Nibbs said to himself, but he couldn't tell what it was.

Ziggy in fact wasn't sure what he would try next. He had slowed his plane as much as he could in the crisscrosses of the flat scissors.

That had foiled the potshots Nibbs had been taking at him. But Nibbs was too experienced a pilot for Ziggy to flush out.

It was time to escape. Ziggy decided to begin "pitchbacks and extensions." On the ground it would be the equivalent of a car racing forward, coming to a quick stop, then racing forward again. In the air, the pilot used speed and altitude to create the neck-snapping effect.

As the two jets crossed again, Ziggy pointed the nose of his Goshawk down and gunned the engine to 220 miles per hour. That was the "extension" part of the maneuver. But as the aircraft neared 10,000 feet, he pulled back on the stick and made a hard right turn to the south, which made it slow down suddenly. That was the "pitchback."

Nibbs pulled up and cut to the east to intercept Ziggy as he made his southern turn. But Ziggy unloaded on altitude again as they crisscrossed and came barreling down once more, this time at 260 miles per hour. They were flying what looked like lopsided figure eights. Ziggy was now trying to use airspeed to get further and further away from Nibbs.

"Nice unload," Nibbs complimented over the radio. He flew east for a second more, then banked his jet down and to the right to try to plant a missile on Ziggy as his jet pitched back again, this time to the left.

"Fox two, invalid, angles," Nibbs radioed after firing the simulated missile. Ziggy was too high for it to home in on him.

Ziggy pointed the nose of his jet down for a third time and this time revved up its engine to 320 miles an hour.

Nibbs looped around to the left and dove as well, but his jet was only traveling 180 miles per hour, not nearly fast enough to catch Ziggy for another shot.

Ziggy could escape if he pulled the right trick out of his bag.

Nibbs decided to give him a hint. "Go right over the top, Zig," Nibbs radioed. "I'm unloading right now trying to catch you."

Going over the top meant performing an Immelmann, a backflip maneuver named after the World War I German ace Max Immelmann, who first perfected it as a way to quickly reverse a plane's direction.

Near the end of his roller-coaster ride down, Ziggy pulled the Goshawk straight up for the beginning of what was almost a vertical loop. The G force slammed him back to his seat. As the jet turned almost upside down to point in the opposite direction, Ziggy tilted his head back and looked down to get a fix on the horizon below, then jammed the right rudder pedal with his foot and flicked his stick to the right.

The Goshawk flipped back on its stomach so it was now flying in the opposite direction, headed toward Nibbs.

Nice move, Nibbs thought. "I would unload off the top of that and bug," Nibbs advised over the radio. "There's really not much I could do right here. I can't follow you. That's a good bug."

Ziggy did just that. He pointed his jet down one last time, raced it to more than 400 miles per hour.

Nibbs made a U-turn and dove down to try to catch him, but Ziggy was far ahead. He'd escaped.

"Knock it off," Nibbs radioed, the command calling off the fight for the moment.

"Roger, knock it off," Ziggy answered, relieved that he'd finally managed to elude Nibbs, even if he'd done it with a little coaching from his teacher.

Sweat was streaming down Ziggy's back. His neck ached. He had spent the entire time in the defensive maneuvers looking over one shoulder or the other watching Nibbs sneak up on his tail. With his head always turned sideways the whiplash from the Gs slamming him back had been painful. In a five-G turn, his head weighed about a hundred pounds when he had to move it.

If dogfighting had taught him anything, it was that being on the defensive was far more miserable than taking the offense. In the early turns, Ziggy had been angry with himself over how horribly he had flown in trying to evade Nibbs. He had redeemed himself by finally bugging out successfully. But the fight had been ugly until then. It had taken him three tries at the flat scissors maneuver before he actually performed it correctly.

Through all the twists and turns, Ziggy had to also keep watch that his jet didn't stray outside the Quail operating area. That was no mean feat. They had 900 square miles in which to maneuver but the space was quickly consumed at the speeds they flew. Ziggy felt like a boxer in a tiny ring with Nibbs punching him from corner to corner. But learning space management wasn't just important for keeping the Los Angeles air traffic controllers off his back. If a fighter pilot became too engrossed in combat and didn't notice that he was straying over hostile territory, he might outmaneuver his opponent in the air but lose the fight to antiaircraft fire from the ground.

Ziggy was angry. All he had been flying for the past two weeks were air combat maneuvers. He should have been sharper. But dogfighters weren't born. It was an acquired skill, extraordinarily perishable, he was beginning to realize. There were so many subtleties involved in the maneuvers. Seasoned fighter pilots who stayed away from air combat maneuver training for several months always felt rusty when they climbed back into the cockpit. Reaction times slowed. Ziggy had flown two dogfighting hops a day for a week, then had taken a break for the weekend. It was startling to him how much his skills had deteriorated when he returned to the jet on Monday morning.

The two pilots reversed direction again and began flying northeast. The next battle would be what they called "neutral one-versus-one." It was the ultimate test of fighting skill. This time, neither plane would have the advantage at the beginning. They would fly at each other nose to nose like two medieval knights jousting on horseback to see who ended up the winner.

"Level three for me this time," Ziggy radioed Nibbs as they set up for the duel. "I'm not fighting too well today."

Before they had taken off, the two men had agreed that on a scale of one to five—five meant the instructor gave the student the best fight he could in the air—Nibbs would fly a number four dogfight. Ziggy wanted to see how he would stack up against Nibbs flying nearly his best dogfight. Not too well on the offensive and defensive sets, Ziggy discovered. He decided to dial back the neutral one-versus-one to a three from Nibbs.

The two jets set themselves up at the northeast corner of Quail. When they zoomed past each other head to head and a thousand feet apart, Ziggy was on the right, Nibbs on the left. The fight had begun.

Nibbs banked left and pointed his nose down.

Ziggy banked left and pointed his nose up.

But Ziggy's left bank would end up being a feint.

"We're going two-circle and I'm nose low," Nibbs radioed. He thought that Ziggy would make a wide circle to the left and he would make his wide circle in the opposite direction by turning left also. It was called a "two-circle maneuver" because after each jet crossed nose to nose they made what looked like two separate circles side by side. Both pilots would then try to fly the circles at different speeds or angles hoping that it would give them an advantage over the other at the end of the turn.

But with his nose pointed down, Nibbs's jet couldn't help but gain airspeed. That meant that as he traveled around his circle at the faster speed, Nibbs was forced to take a wider barnyard turn.

Ziggy looked back and spotted it instantly. He realized he had to change tactics. Ziggy saw quickly that where he'd end up in this maneuver was being shot. He had pointed his nose up when he began his turn to the left, which meant his jet was rearing up and gravity was forcing it to slow down. By the time he banked around to the left, Nibbs, who was traveling faster in his turn circle, would be behind him and on his tail.

Ziggy decided to change directions. He wouldn't play Nibbs's game. Instead of banking left he'd make a hard break turn to the right.

"We're going one-circle and I'm nose high," he radioed Nibbs. In a one-circle maneuver, after the two pilots crossed nose to nose, one banked left and the other pilot banked right. From above it looked like the two circles the pilots flew overlapped each other, in effect making one circle.

"Ahh, very nice reversal," Nibbs radioed admiringly. His student was being crafty, changing tactics midgame. Ziggy was now trying to use his slow speed to his advantage. He was able to make a far tighter turn to the right than Nibbs's wide turn to the left.

The two jets were about to cross nose to nose at the south end of the one-circle maneuver. But Nibbs was at a critical disadvantage at this point. He was roaring toward Ziggy at a much faster speed than Ziggy was coming at him. Ziggy's jet was flying slow enough that it could easily pirouette in a U-turn, then come up behind Nibbs, who had no choice but to forge ahead. Nibbs was flying too fast to reverse course that quickly.

But there was one problem as Ziggy was about to cross Nibbs nose to nose. Ziggy couldn't find him.

Nibbs, who had pointed his nose down in the maneuver, was 3,000 feet below him. Ziggy looked all over. He couldn't see Nibbs coming underneath him.

"Ah, shit! I lost you," Ziggy radioed, frustrated.

"I'm right beneath you," Nibbs answered, deciding to help him out a little. "Do a split S right now!"

A split S was a nifty maneuver for a pilot to make a quick U-turn, then race off against a bogey lower down once the turn was completed. In the sky the U-turn was taken vertically instead of horizontally as a driver would in a car. The jet looked like it was flying in the direction of a hook pointed down. The pilot first had to flip the aircraft upside down, then loop it down. At the bottom of the loop he straightened out the plane so it was flying in the opposite direction.

Like a diver about to bounce off a springboard, Ziggy pulled the Goshawk up just above the horizon, which he could see in front of him, then relaxed the pressure on the stick. Quickly he flipped the stick to the left and pushed in the left rudder pedal so the jet turned upside down. Then, making sure his wings were upside down but level, he pulled the stick back to his stomach so the inverted jet would swoop down. At the bottom of the loop, Ziggy relaxed the back pressure on the stick so the Goshawk was traveling in the opposite direction right side up.

Ziggy was behind Nibbs but only for an instant. He had gained the advantage but Nibbs wasn't planning to let him have it long enough to take a shot. Coming out of the loop at the bottom, Ziggy's jet naturally built up more speed from traveling downhill. As he bore down on Nibbs from behind, Nibbs yanked his stick back so his jet banked up steeply. That forced Ziggy to overshoot and swoop below him, missing any chance for a shot.

In order to chase Nibbs, Ziggy pulled his jet up so it was flying vertically as well. Both Goshawks were rocketing straight up.

Ziggy knew Nibbs would very quickly have to loop back down and plunge to the earth. He would have to do the same. But Ziggy decided to delay his loop down for several seconds. The lag turn, he hoped, would again put him behind Nibbs as both jets dove down.

Nibbs, who was flying much slower, did a tight back flip with his jet.

Ziggy soared up past him. One second. Two seconds. Ziggy pulled the stick back so the Goshawk would loop back and down.

With his jet upside down in the next instant, Ziggy tilted his head back and to the left to see if he could spot Nibbs below.

He was blinded by a flash of light.

It was the damn sun again, he realized. Nibbs had positioned his jet between Ziggy's line of sight and the sun to the east. It bought him another precious few seconds to fake out Ziggy.

"Lost you in the sun," Ziggy radioed angrily.

"Roger, horizon, left at eight o'clock," Nibbs radioed Ziggy to explain how he'd just been snookered.

After Nibbs back-flipped his jet, Ziggy had expected him to fly it in a straight line toward the ground. If he had, the lag turn Ziggy had made to loop over would have positioned his jet to come in right behind Nibbs.

But when Nibbs saw that Ziggy would be blinded by the sun, he decided to take a detour. Instead of diving down, Nibbs rolled his jet right side up, then pointed it left on a straight line parallel to the horizon. Before Ziggy could realize it, he was flying pointed down while Nibbs was flying to the east.

Then in the next instant Nibbs looped down as well in a lag turn so he was flying in the same direction Ziggy was and practically underneath him.

"I'm almost right below you," Nibbs finally radioed to him.

For Ziggy, it was like having a fly buzzing all around him and not being able to see the pesky insect.

He finally looked down and spotted Nibbs. "I understand. I have you visual," he radioed, somewhat embarrassed. In effect the two jets were back where they started. Neutral, one-versus-one. Neither had the advantage.

Nibbs pulled up his jet again.

Ziggy did the same.

The two pilots began flying in circles, but this time vertically instead of horizontally. Their flight paths were like two Ferris wheels. Only in this case the jets were racing end over end in giant loops at 350 miles per hour, chasing after one another.

Each pilot had to carry out the chase with precision. If either Ziggy or Nibbs didn't fly his circle perfectly, the other might have a chance to sneak up behind. Speed and angles, how the pilots traded one for the other, became critically important as the two jets circled. Ziggy could slow up his jet, fly a tighter circle than Nibbs, try to take a shortcut, then get off a shot as Nibbs edged in front of him.

But if Ziggy slowed too much, Nibbs might fly fast enough in his wider circle to sneak behind him. Or if Nibbs slowed to a tighter circle, Ziggy might widen his with more speed and fly more in an egg-shaped pattern. Looping outside of Nibbs, he could take a shot at Nibbs below while he was at the top of the egg. Or, Ziggy and Nibbs could fly the Ferris wheel patterns so the two jets didn't travel parallel. Where the two circles crossed they would try to get off shots at one another. The combinations, the feints, the thrusts could be endless. The subtleties in the way a pilot flew his plane—how hard he pulled the plane up at the bottom of the loop, how much he relaxed the G force at the top—could make the difference between having the advantage or being on the run.

Nibbs decided to make his first circle tight, pulling his stick back more forcefully so he would loop upside down and back quickly like a diver performing a back dive.

Ziggy decided to make his loop at the top wider, hoping to catch Nibbs from behind with his faster speed when they came out of the bottom of the loop. As he began to turn upside down, Ziggy tilted his head back at the horizon on the other side to get his bearings.

He felt the weightlessness of being upside down and pointed to the ground. The jet was only traveling at about 140 miles per hour as it looped over the top.

Then came the crushing six Gs as Ziggy hit the bottom of the loop at 340 miles per hour. His ribs now ached. It felt as if his chin was going to come down and kiss his pelvis. His G suit inflated and squeezed his chest to force the blood back to his brain. The bottoming out could press as much as seven Gs on the plane. Ziggy didn't want to apply much more force. The wings would fall off at ten Gs.

As he came up out of his loop and back to being right side up, Ziggy spotted Nibbs above him in the bright blue sky racing toward the top of still another loop. Ziggy was too far away for his machine guns so he let loose with a simulated missile.

"Fox two," he announced on the radio, his voice raspy from the G force. But Ziggy was too wide in the angle of his loop for the missile to reach Nibbs. Nibbs had planted his stick in his lap and pitched his nose back around too quickly. He jinked away, escaping the heat-seeking missile.

"Invalid, angles," Ziggy radioed in the next half second. The missile had missed, he realized.

Nibbs arced his jet up and over so it back-flipped at the top of the loop. If Ziggy had been playing it conservative, he would have pointed his in the same direction and followed Nibbs. But he probably would not have caught up to Nibbs's tail at the top of the circle and the two planes would have just looped around on the Ferris wheel one more time with neither gaining the advantage.

Ziggy decided to be bold. As his jet soared up to the top part of the loop he pulled his stick back sharply so it made a smaller circle than Nibbs's jet had. Ziggy was chancing a lead turn, taking a shortcut underneath the top of Nibbs's circle to try to catch him on the other side. If he was lucky, he might manage a shot as the circle his jet made crossed the path of Nibbs's circle. Ziggy didn't worry about overshooting his target this time. Nibbs's jet was pointed down now. There was no way, upside down and plunging toward the ground, that Nibbs could reverse course and fly horizontally to counterattack Ziggy.

"Ah, nice lead turn," Nibbs complimented, his voice almost breathless from the pressure of the Gs on his lungs.

But Nibbs pulled his jet in and down as it back-flipped over the top of the loop so Ziggy was at the wrong angle to take a shot before they crisscrossed. Ziggy suspected that Nibbs was ratcheting up his fighting intensity to a four. It didn't seem to Ziggy that Nibbs was pulling his punches. Who could expect him to? That went against the instincts of a fighter pilot.

Seven times Ziggy and Nibbs circled one another in the high-speed Ferris wheel. At the top of the circle, at the bottom, Ziggy would take shots whenever he cut across Nibbs's flight path. Nibbs tightened his circle, widened it, pitched back sharply at the top of the loop, shifted his path to blind Ziggy with the sun behind him, anything to avoid being shot at or to regain the advantage.

They were like a dog chasing its tail—only vertically. Either pilot could have given up, flown away, tried another maneuver. But the circles almost became addictive to both men. Neither dared to leave the Ferris wheel. It only took a minor deviation in the flight paths, just the slightest misstep, a subtle change in angle, and in an instant you could be lined up for the kill, they both knew. The circle fight—around and around again—became too tempting to pass up.

Ziggy sure as hell wouldn't. He was on the offensive at the moment, in the driver's seat. He had spent most of the day being shot at. Now he was doing the shooting. He wasn't about to give up the offensive just because he was in a dizzyingly repetitive maneuver. It was fun being the guy squeezing the trigger for a change. It was exhilarating.

Ziggy willed himself to stay on Nibbs's tail hoping the instructor would make a mistake. Be patient, be patient, he told himself. Don't be greedy. He realized that every move he made against Nibbs forced Nibbs to react, forced him to bleed his airspeed to dodge Ziggy's shots.

The Ferris wheel in fact was whittling away at Nibbs's maneuvering room. With each circle, Nibbs had to jink more to avoid Ziggy's shots. That forced his jet to slow down more as it made the circles. Nibbs's circles were becoming smaller and smaller. His flight was becoming more predictable.

Finally, Nibbs's airspeed fell below the 300 miles per hour he needed to soar vertically. His jet strained to reach the top of the loop, then began to fall off to one side. Nibbs's jet was traveling too slow to make another loop. He had to escape.

"Trigger down, snap," Ziggy radioed as Nibbs's jet limped over the top of the loop for the final time. "Unassessable, sun." The sun behind Nibbs had blinded Ziggy again as Nibbs's jet crossed his flight path at the top of the loop.

Nibbs wasn't about to hang around for another shot from Ziggy. As his jet reached the top he flipped it over so he was now right side up instead of falling back upside down. Nibbs gunned the engine, pushed the Goshawk's nose down, and fled the circle.

Ziggy immediately understood that Nibbs was trying to escape. Ziggy wasn't about to let him out of his grasp. He quickly flipped his jet right side up as well, pushed the throttle forward and the stick down, and went after him.

Ziggy could see the tail of Nibbs's jet. He punched the button for the simulated missile.

"Fox two," Ziggy radioed. But Nibbs, almost from a sixth sense, had banked his jet up at a 40 degree angle before Ziggy fired. That foiled the missile, which couldn't make the same hard turn to reach the hot tailpipe of Nibbs's jet.

"Invalid angles," Ziggy radioed back, acknowledging that he'd missed. But Ziggy wasn't about to give up. He smelled blood. He was now closer than ever to Nibbs's tail. Just a few more turns and he would bag him.

Ziggy banked up to chase Nibbs. He flew above him to come back down to take another shot at the instructor. The two jets were in another flat scissors, only this time Nibbs was on the defensive and Ziggy was boring down on him.

Brdrdrdrdrd.

"Trigger down, snap," Ziggy radioed as Nibbs's jet neared his HUD sight. Not quite. Nibbs was still too far away.

"Missed, long," Ziggy radioed.

But Nibbs was running out of space and ideas for eluding Ziggy. He banked the jet to the left. Ziggy banked right. This time, Ziggy had the sun behind him.

"Ah, I've lost sight," Nibbs groaned as he looked back to his right to try to find Ziggy. Nibbs was blinded for a change.

Nibbs banked his jet up, then to the right, and then down in a corkscrew maneuver hoping it would be enough to escape. But he still couldn't spot Ziggy because of the sun.

Ziggy meanwhile gently turned first left, next right, then flew up in order to cross over the top of Nibbs.

Carefully, carefully, carefully he flew now. He was within striking distance. He didn't want to make any mistakes. One more maneuver and he'd have Nibbs.

That maneuver was now. The moment Ziggy crossed underneath Nibbs's flight path, he rolled his jet into a back flip, then to the left in a quick U-turn. As the nose of his Goshawk pointed down in the direction of Nibbs, Ziggy pulled his throttle and slammed on the aircraft's speed brakes to let Nibbs jump ahead. He could see Nibbs flying toward the ground on his left. Ziggy fought off the temptation to bank left and point directly at Nibbs. He might fly past Nibbs again and miss the shot.

Instead, Ziggy finessed it. As Nibbs banked to the left to evade him, Ziggy did as well, but at not as sharp an angle. He rolled his jet to the inside of Nibbs's turn. Nibbs was forced to fly across his gunsight.

This was finally it. The best shot Ziggy had had all day. Hell, it was practically the only shot, he realized. Ziggy tightened the trigger grip on the stick. He eyed the pipper on his HUD. He could see Nibbs's jet approaching in his heads-up display.

One second. Two seconds. Nibbs's jet was barely outside the HUD's box that Ziggy was viewing.

Ziggy squeezed the trigger.

Brdrdrdrdrdrd.

"Trigger down, snap!" he radioed excitedly.

Nibbs flew directly into Ziggy's line of fire.

"Great shot!" Ziggy shouted.

It was, Nibbs realized. If the bullets had been real, Ziggy would have sawed Nibbs's jet in half. His student had finally beaten him. Nibbs was pleased.

"Recommend knock it off," he radioed. Time to end the fight. They were running out of fuel.

"Roger, knock it off," Sigler agreed. After being on the defensive all day, the direct hit felt good.

"Ziggy has the lead," Sigler said, finally relaxed. He would guide the two planes home.

Ziggy glanced about the cockpit at its navigation dials, then at the map clipped to his knee. He wanted to make sure he didn't get the two jets lost on the trip back.

The joy of nailing Nibbs faded quickly. Ziggy was now feeling miserable about the flight. Pilots stayed alive by dwelling on the negatives instead of the positives, he knew. Of the hundreds of flights he had flown as a radar intercept officer, he had performed no more than a half dozen perfectly, as far as he was concerned. In the back seat of the Tomcat, Ziggy had committed the first cardinal sin only once—allowing a bogey he was supposed to be watching sneak up on him—and the second cardinal sin never—failing to warn the pilot in the front that he was about to hit the ground. Today Ziggy had sinned.

As soon as he collected his wits after the dogfight, he began to isolate the maneuvers he had performed poorly. He had lost sight of Nibbs six times and twice dipped below 10,000 feet (the simulated ground level the pilots had agreed to). A first-timer was expected to make these mistakes during training. But a retread like Ziggy should have more air savvy than a green ensign going through flight school. In the real world, mistakes weren't allowed. Losing sight of the enemy or flying too low got you killed. That was the standard Ziggy would eventually have to live up to. He might as well begin living up to it now, he had decided.

Nibbs was delighted with Ziggy's performance. Nibbs wasn't supposed to make him an air ace at this point. Ziggy was only learning the fundamentals here. The dogfights would become far more complicated later. The two pilots today had been flying Goshawks with the same performance capabilities. In real combat, Ziggy might be flying against a foreign jet built to bank and climb differently than his aircraft. There would be exceedingly more difficult one-versus-two jet fights, two-versus-two fights, air combat with five or more jets where the skies become a spaghetti of planes weaving over and under one another.

But Ziggy had the potential to become one of the best dogfighters in the business, Nibbs believed.

Nibbs pulled his Goshawk into a parking place on the tarmac of the El Centro air station. A ground crewman outside signaled him to begin shutting down the engine.

Too bad about Ziggy, Nibbs thought as he flipped switches to power off the plane. It didn't look like he would become a combat pilot. In fact, Ziggy's Navy career was probably over. Such a waste, Nibbs said to himself.

Ziggy had sat on a plastic drink cooler in the squadron ready room, his shoulders slumped, elbows resting on his knees, his head bowed. It was two hours before his flight with Nibbs that Saturday. The ready room was a makeshift one the squadron occupied in a vacant hangar during its two weeks at El Centro. It had a dingy linoleum floor, bare whitewashed walls, a few metal chairs scattered about, and dusty worn couches the instructors and students lounged on during breaks between flights. A folding table behind Ziggy was piled high with cups, lunch meats, loaves of bread, and bags of chips the pilots snacked on during the day.

Students and instructors buzzed about Ziggy as he sat quietly on the cooler. No one dared go up and talk to him. He was damaged goods. A man who was seeing his lifelong dream vanish before him.

The telephone call from another student had awakened him early that morning with the news. *The San Diego Union-Tribune*, which covered naval aviation heavily because of the air bases in its city, had the story on the front page of its Saturday edition. The Navy had been snakebit with mishaps in its F-14 Tomcats. So far, three had crashed that year. One had exploded off the San Diego coast killing its two crewmen. Another crashed in the Persian Gulf. But it was the third accident that spelled the end of Ziggy's career.

On January 29, 1996, Lieutenant Commander Stacy Bates made too steep a climb taking off from the Nashville, Tennessee, airport. Crash investigators guessed he had been showing off for his parents, who watched the takeoff at the airport. Becoming disoriented when the jet flew into a bank of clouds, Bates plunged the plane into a fatal dive. Bates, his radar intercept officer, and three people in a house the Tomcat hit died in the fiery crash. Bates had been a marginal student in flight school, then accident-prone when he climbed into the F-14. He had crashed a Tomcat the previous year and two radar operators in his squadron had refused to fly with him.

But Bates became Ziggy's problem now because of one other thing. He had been a retread, a radar operator like Ziggy who had convinced the Navy to let him retrain to fly combat jets.

Among the Navy's fighter pilots, the resentment toward retreads ran deep. The Navy spent hundreds of thousands of dollars training a radar intercept officer. Then it turned around and spent another million dollars retraining him to be a pilot. It was a waste of money, the regular pilots grumbled. You couldn't teach an old dog new tricks. The retreads were thirty-year-old officers like Ziggy whose reaction times were not as quick as new ensigns in their twenties. The retreads usually returned to flight school with attitude problems. The instructors were their contemporaries. The retreads weren't fresh sponges soaking up the training like the new kids. Sure, the retreads often did better in training. But it was smoke and mirrors, the regular pilots argued. The retreads were seasoned officers who already knew their way around the Navy. They could talk a good game and instructors tended not to grill them as hard as they did the newcomers. In squadrons the retreads generally were not the best pilots. The record showed they had a slightly higher accident rate than pilots who had never been radar operators.

The Navy was under tremendous pressure to stem the mishaps in F-14s. Admiral Mike Boorda, the chief of naval operations, decided to throw the babies out with the bathwater. No more radar operators would be allowed to return to flight school. As for Ziggy and the ten other retreads now in training, they "will be allowed to finish, but they will be assigned to aircraft that carry a pilot and co-pilot, such as transports and patrol aircraft," according to the *Union-Tribune* story he now read.

The retreads were the problem, the Navy conveniently decided. Get rid of them and the accident rate would drop. Those already in the training pipeline could fly as long as it was in a prop plane with another pilot watching them. It was typical of the way the Navy made decisions. Mindless. Faced with a complicated problem, the service could always be counted on to overreact with all the subtlety of a battleship firing sixteen-inch guns. The Navy's bureaucracy was one of the most rigidly stultifying in the entire federal government. Decisions from the top, particularly the stupid ones, were rarely questioned, the chain of command never challenged. No matter that Ziggy was the best student in his class. An exemplary exception to the norm. Grades that few other students would match for a long time. No matter that there was little chance he would experience the same problems other retreads had. It all didn't matter now. This was guilt by association.

Ziggy stared at the floor, clutching the newspaper in his hands. He felt as if he'd been mugged. The article hit him like a shotgun blast. He had had no inkling that the Navy was considering a measure as drastic as this. All those years of hard work, of hoping and praying and maneuvering through the system so he would have a chance to become a fighter pilot. He had devoted his entire adult life to chasing this dream. All of it now down the drain.

Ziggy had telephoned the Navy's bureau of personnel Saturday morning after reading the *Union-Tribune* story. He would be graduating from flight school in just a couple weeks. Air combat maneuvers was the last phase of his training before he received his gold wings and was officially a jet pilot. Was there some wiggle room? he asked the bureau's detailers who made job assignments. The Navy was a big, cumbersome bureaucracy. It might take weeks before the bureau's pencil pushers began moving the paperwork to carry out Boorda's orders. Could he sneak through the program before that? Ziggy asked the detailers.

Not a chance, they responded. The bureaucracy had moved quickly in this case. Boorda's order instantly froze the job assignments. Ziggy was left with two humiliating choices. He could finish his jet training. He would be sent to another school to learn to fly a cargo plane. Then, according to service regulations, he would owe the Navy eight years on boring flights as a "trash-hauling" pilot—with another aviator baby-sitting him in the cockpit because the Navy no longer trusted him to fly safely by himself. Or, he could drop out of jet school now, before graduating in a couple of weeks. The Navy would transfer him back to an F-14 squadron, where he would have to serve three more years as a radar operator.

Ziggy decided to drop out early. He would fly this last flight with Nibbs. Dogfighting was fun. He wanted to go up one last time. He would compartmentalize his thoughts, block out of his mind during the flight the fact that it might be the last, enjoy the pure excitement of aerial combat. Then he would stop. If he couldn't fly jet fighters he didn't want to fly at all. He would return to an F-14 squadron, serve out the remaining three years of his contract. Then he would resign. The quicker the better. There was always a chance he might be able to convince the personnel bureau to grant an exception to Boorda's order and let him become a fighter pilot. He would be on the phone all day Monday pleading with Washington. But Ziggy knew the chances of that happening were slim.

If he couldn't get a waiver to Boorda's order, Ziggy wanted out. He had played by the rules all these years. He had done everything the Navy had demanded of him so he could fly combat jets. Now the service had changed the rules midstream. He no longer wanted to be part of an organization that broke its promises.

CHAPTER TWELVE

Wedding Day

IT was a warm, sunny Sunday afternoon in May for Hempstead, New York. A hundred and twenty people filled Saint Paul's Greek Orthodox Church for the wedding. Jonathan Wise wore his dress white uniform. Maria wouldn't have been caught dead wearing a uniform to her wedding. They both may have been Naval officers but Maria Grauerholz was the bride after all. She wore an off-white wedding dress to match the train and veil her mother had worn on her wedding day.

Maria wanted the wedding ceremony to be Greek Orthodox, her family's religion. So Jonathan would share a piece of his heritage, his two best men wore Scottish kilts. A bagpipe was played at the reception.

Jonathan was beaming as Maria's father walked her down the aisle to him. He had wiped out of his mind the fact that their wedding day had already presented them with the first strain their marriage would face in the years to come.

Who would have ever guessed that his training squadron would be working during the Memorial Day weekend, Jonathan thought. Both their schedules had been so crowded. Learning to be a combat pilot left little time for a personal life. But Jonathan and Maria thought this holiday weekend was the safest of all to be married. Jonathan was scheduled to take the second most important step of his life four days after the wedding—his first landing in a jet aboard an aircraft carrier. If he qualified in the solo landings, he would become a carrier pilot. But ships have a knack for changing course and not making appointments. The carrier Jonathan's class was supposed to land on had been diverted to another location. The training squadron had to choose another ship for the students' first landing, the USS *John C. Stennis*, which would be steaming off Key West, Florida, on Memorial Day weekend.

For Jonathan there was no choice. He couldn't put off the wedding. His classmates would have to make their first carrier landings without him. Jonathan would have to delay qualifying for at least a month, when the squadron booked the next nearby carrier for its training flights. It would also delay his graduation from jet school.

There would be many more times when their careers collided with their personal lives. They were sailing in uncharted waters as possibly the first husband-wife team in Navy combat jets. So far, they had been separated six of the eighteen months they had been engaged because they were on different training schedules. The Navy prohibited a husband and wife from sailing on the same ship. They also wouldn't be able to serve in the same squadron. About the best Maria and Jonathan could hope for was that both would be assigned to units on the West Coast. Jonathan would be in a squadron assigned to one carrier. Maria would fly in a squadron for another carrier. When one was on land the other would likely be at sea. If they were lucky, they calculated that they would be together six of the next eighteen months.

If they were lucky. Ship schedules didn't always cooperate, as Jonathan discovered this weekend. They might be separated even more. After flight school, they both owed the Navy eight years in the cockpit so the service could recoup the investment it had made in their training.

Jonathan and Maria weren't complaining. All Navy families endured long separations. They knew what they were getting into. They had thought through everything. Life would be hard. But both were about to embark on an adventure. Both wanted desperately to be fighter pilots. God was giving them that chance, they believed. The chance to soar through the clouds. To tumble and roll in a multimillion-dollar jet. To pull Gs. To make the most exciting landings and takeoffs a pilot could make. They weren't about to let their marriage stand in the way of this dream. Neither wanted to shortchange the other on this opportunity. At least through their first couple years at sea, Jonathan and Maria would make do with seeing each other for brief moments when both were in port.

Later in their tours, the Navy would be more agreeable about assigning them to jobs where they could be together. But usually that meant one spouse would have to accept a less desirable desk job while the other got the choice flying assignment. Inevitably one member of a husband-wife team would have to sacrifice his or her career so the other could move up the ladder.

But that was years away. For now, Jonathan and Maria were sharing a romance, for each other and for flying.

CHAPTER THIRTEEN

Carrier Qual

THE view was spectacular. Behind him and to the west only wisps of clouds danced across the light blue sky. In front, billowing dark thunder clouds stacked in stairsteps to the morning sun. But they were far to the south, no threat to the dangerous landings Rob Dunn would make today. He felt so serene. If there was a heaven, it had to be off Key West with the rippling dark blue Florida Straits below him. Streaks of aquamarine from crosscurrents interrupted the dark blue, along with the white speckles of cigarette boats and trawlers sailing near the coast. He tried to make out Cuba, which lay fifty miles to the south. Dunn had naively asked how he would know if his jet accidentally strayed into Cuban airspace. "The missile coming off the rail of a MiG ought to be the first clue," Mango had said with a laugh.

Dunn had been circling his Goshawk at almost a mile high. He was with a group of four jets, led by an instructor piloting one of the aircraft. Below him, the aircraft carrier looked so peaceful as it steamed sixty-nine miles from Key West toward the east. From his altitude, Dunn could just make out the trickle of a churning light blue wake the behemoth left behind. Everyone had told him the ship would look like a postage stamp from this altitude. To Dunn, it seemed larger. But he was amazed at how narrow the vessel appeared, a thin dark tube on the sea almost like a submarine. How the hell could anybody land on that pencil, he wondered.

"It's show time!" a voice on the radio broke his thoughts. The carrier's air traffic controllers had taken command of his flight and ordered him to drop his altitude to enter the landing pattern.

Dunn banked the jet to the left and down, then lined up behind the carrier at 1,200 feet. He could feel his heart thumping in his chest. For every other new maneuver, the student had an instructor sitting in the plane's back seat the first time. But not for the first carrier landing. The student landed on his own. Dunn was flying solo. No instructor would dare sit in the back seat. Too nerve-racking. The instructor would be too tempted to grab the controls and pilot the aircraft himself. Landing on a carrier was considered one of the most dangerous things a student did. It had to be performed alone.

Mango and Wolfie stood on the landing signal officer platform rubbing sunscreen lotion all over their faces. It would be a hot one today. Almost two o'clock and the Caribbean sun beat down on the carrier deck. The steel platform jutted out from the right side of the carrier at the rear where the jets landed. Two large gray consoles sat on the platform with television screens that played videotapes of the jets coming in. Large white dials on the consoles gave readings on the wind speed across the deck and on the rocking of the ship. (Today the ship sailed smoothly on glassy calm seas and a headwind blew an ideal twenty-two miles per hour from the bow to stern.) Four black phones hung from the console to patch the landing signal officers—LSOs—to the planes above or to the ship's captain perched in a padded chair on the bridge. Mango tried to avoid talking to the skipper, who never called the platform except to complain.

Stretched out to the left of the platform was a large padded basket made of black vinyl that the LSOs could jump into if a landing plane veered off course and was about to crash into them. Only once in his Navy career had Mango edged close to the basket when he thought a jet would hit him.

Mango adjusted his black Blues Brothers sunglasses and straightened the white canvas vest LSOs were required to wear to identify them on the flight deck. Some wiseass had written "I like boys" on the back of Mango's vest. His wavy black hair blew back in the wind. Mango and Wolfie refused to wear the cranial helmets with their Mickey Mouse ears that were required for other deckhands. No LSO would be caught dead in those goofy things. Fuck the carrier's "safety Nazis." A landing signal officer had to look cool.

Lieutenant Mike "Mango" Carr and Lieutenant Rusty "Wolfie" Wolfard were a breed of aviators becoming extinct in the Navy. Mango—he earned the call sign in a politically incorrect era because his mother was Filipina—was a civil servant's son from Washington, D.C. He was proud of the fact that he was nothing like these obsessive yuppies coming into flight school today. Mango picked a major in college that would graduate him the quickest, joined the Marines briefly on a lark, transferred to the Navy, then didn't give a damn about grades in Navy flight school. Wolfie was a tall country boy from Sheffield, Alabama, who managed country music songwriters before deciding he wanted to become a Naval pilot.

Mango and Wolfie met in flight school and had been close friends for eleven years. They both served in Desert Storm. Mango flew E-2C Hawkeye planes, the airborne command posts directing the carrier's air war. Wolfie piloted A-6E Intruder attack bombers.

For Rusty Wolfard, flying combat missions over Iraq was the most intense experience he had ever been through in his life. Carrier deckhands would rush up to him with tears in their eyes before closing the canopy on his cockpit, yelling at him to "Kick ass!" Nights bombing oil complexes in Iraq became surreal with antiaircraft fire and SAMs whizzing over, under, and around his jet. Like the Fourth of July. He'd catch himself becoming mesmerized by the light show for brief seconds, then terrified by the reality that he could die in an instant. It brought out the best and worst in the men around him. Wolfie saw some pilots so stricken with fear they turned in their wings and refused to fly. Other friends paid with their lives. After the third night of low-level bombing, the carrier crew presented the Intruder squadron with a barrel of brass balls because they flew so close to the gunfire. Before a combat mission, the pilots would play Wagner's "The Valkyrie" full blast in the carrier ready room. Then one aviator would pull out his penis and crank it like a Model T as the others cheered. It would take Wolfie six hours in the ship's gym to unwind from the fear he felt after each mission.

The Navy was now all fucked up, Mango thought. He was getting out. Wolfie already was out. The only reason Wolfie stood on the LSO platform was because he loved flying and had stayed in the Naval Reserves after quitting. When the Navy retired the A-6, it tried to stick Wolfie on a ship. He refused and took up training students in the Goshawks as a reservist so he could stay in the air.

It was all this political correctness that had finally gotten to them, all the ass covering, all the goddamn pencil-pushing admirals in Washington who let Congress lead them around by the nose. The warriors were leaving. The good aviators were fleeing in droves to the airlines. A generation of fighters lost, and left behind were the careerists more interested in punching tickets than flying.

Mango had been passed over for lieutenant commander. Because of Tailhook, he was sure of it. He had been at the 1991 convention. What he saw was one big drunken frat party, not the assaults. But those Gestapo agents from the Pentagon's inspector general's office interrogated him as though he was a prisoner of war. Even if he had seen anything he wasn't about to rat on his brothers to these guys. Mango was labeled "uncooperative" with the investigation. It drove a stake into his Naval career. The brass wouldn't admit it, but Mango was sure there was a list of officers who attended that infamous convention and anyone on that list would not be promoted.

Mango and Wolfie were the last of the hell-raisers, the last of the old dogs. They were outgrowing it anyway. Both were in their thirties and raising families. Mango had left his wife, Lisa, with two sick kids so he could play on this carrier off Key West. Lisa kept telling him he would have to grow up one day. Mango knew it was true. But he would dearly miss the camaraderie, the brotherhood of warriors he had been allowed to join for a brief part of his life. The kids coming into flight school today were a bunch of neurotic engineers. They analyzed things too much, Mango thought, put too much pressure on themselves. They were like sponges soaking up every word he and Wolfie uttered, asking a million questions. Flight school would be the best years of their lives. Enjoy it, Mango would tell the students. "The most fun you can possibly have with your clothes on and they're paying you," Mango liked to say. "All you have to do while you're here is eat, sleep, fuck, and fly."

But these kids wouldn't listen. The ten chicks Mango and Wolfie had brought to Key West for their first carrier landings were coiled tighter than steel springs. Wolfie had ordered them all to speak over their plane radios in low, manly voices to try to make them feel like warriors. Before leaving their home base in Kingsville, Texas, he had passed around a Styrofoam cup filled with malted milk balls. "Take one," he ordered each student. "It's good juju and it lasts for a week." The ten nervous chicks carefully swallowed the candy as if it were a magic potion to bring them good luck.

Dunn had descended in his jet to 800 feet. He was now three miles from the carrier, just off its right side. The training squadron's students would be landing on the USS *John C. Stennis*, a brand-new, 100,000-ton, nuclear-powered supercarrier named after a dead Mississippi senator whose only claim to fame had been larding the Navy with billions of dollars of pork when he chaired the Armed Services Committee. Dunn glanced down and to his left. The ship still looked damn skinny even from this lower altitude, he thought. Dunn could feel his heart now racing. Pablo and Baby Killer followed him in their three-plane formation.

Pablo was Paul Rasmussen, a twenty-five-year-old lieutenant junior grade who got his call sign because he had learned to speak Spanish fluently working summers with Mexican-American dishwashers in a steak house. Pablo was like a character out of *Fast Times at Ridgemont High*, Wolfie thought. Short blond hair spiked in every direction. High-strung like he just stuck his toe in a light socket.

Mango had given Lieutenant Junior Grade Brian Burke the call sign "Baby Killer" to make him feel like a warrior. Burke, also twenty-five, was one of those thoughtful silent types, Mango said. A preppy who read *U.S. News & World Report* every week, Burke kept everything churning inside.

None of the ten chicks qualifying on the *Stennis* this weekend were female. The instructors let their guard down more when no women were in the detachment. The language became saltier. The Goshawk no longer was a jet but a "pointy nose pussy getter." There were three rules to live by when landing on carriers: one, when your jet hit the deck keep it at full power to take off again in case the tailhook didn't catch the arresting wire; two, follow the deck crewmen when they directed your plane after landing; and three, the landing signal officer was never wrong. The instructors gave the students one other rule when they went on liberty in Key West: "Never get caught in bed with a dead woman or a live boy."

Dunn and the nine other students all grew mustaches to show manly solidarity for the carrier qualification—or at least tried to grow them. They decided to order T-shirts commemorating the event; printed on them, a cartoon of a woman performing oral sex on a pilot as he landed his jet on the carrier. The instructors were a little edgy about that one. Don't buy anything you couldn't wear in front of your mother, they warned. The ten chicks deliberated. What the hell, they decided. The prudes in the Navy would probably disapprove of any shirt, so let's go for the crudest and keep it hidden from mom.

Dunn raced past the right side of the carrier at 250 miles per hour. So many things suddenly seemed different from the hundreds of landings he had made on the field back at Kingsville. This airfield moved constantly. On land there was always *land* beyond the airstrip. At the end of this 350-foot airstrip was a sixty-foot cliff that dropped into water.

Dunn was surprised at the smooth ride. The ground had different temperatures, which created bumpy thermal pockets for the pilot flying above. But the water temperature was the same around the ship so there were no pockets.

Dunn tried not to let the new experiences overwhelm him. He had grown comfortable in the Goshawk over land. But this was all new, like when he slept over at a friend's house the first time as a kid, listening to unfamiliar noises, lying wide awake, keyed up.

Dunn banked his Goshawk sharply to the left so it flew in front of the bow of the carrier. Pilots called this the "break," the beginning of the final landing pattern before touching down on the ship. He started to descend the plane another 200 feet so it now flew at 600 feet. Dunn willed himself to pay attention to his flying. Students became so mentally overloaded by their maiden trip to the carrier that the first couple of passes could become a blur and minds went blank. Some students forgot their plane numbers when the carrier tower radioed them, even their names.

Now flying downwind in the opposite direction that the ship sailed, Dunn quickly lowered the aircraft's landing gear, flipped out the speed brakes to slow down the jet, then checked his seat harness to make sure it had him locked tightly to the backrest. He didn't want to fly forward from the violent stop he would make.

Dunn's jet reached a point perpendicular to the rear half of the ship. It was called the "abeam point." He had hit it perfectly—600 feet altitude and one mile to the left of the ship just parallel to the LSO platform where Mango and Wolfie stood. Dunn keyed the transmission switch on his throttle. He felt like he had a sock stuffed down his throat. He could barely choke out the words. Forget the manly talk.

"Echo four abeam," Dunn said, feeling almost out of breath. Echo four was his call sign for this flight.

Dunn had accumulated a long list of personal call signs. The other students had first tagged the twenty-five-year-old lieutenant junior grade with "Muddy" because once after a raucous party in Pensacola he had raced a four-wheel-drive Jeep over a sand dune and into a muddy bog. It remained stuck there for two days until low tide and his buddies could help him dig it out. Mango alternated between calling him "Under Dunn" or "Dunn Deal." Dunn could be his own worst enemy, Mango believed, another one of those wound too tightly who thought too much. It took Wolfie an hour of shooting hoops with him before Dunn stopped calling him Lieutenant Wolfard.

Mango didn't know how well Dunn would do today. He could land on the boat if his head was screwed on tight. But he could also unravel. Mango and Wolfie made their own private bets on who would succeed or fail at carrier landings. They were right only about half the time. Dunn was a toss-up, they thought.

For Rob Dunn, fear of failure now overwhelmed any anxiety he felt about the danger of carrier landings. A student could excel at dropping bombs and dogfighting in the sky, but if he couldn't land his jet on the boat, he was useless to the Navy as a combat pilot. If Rob failed to qualify today, he would be given a second chance. If he failed again, he was out of jet school, perhaps even out of aviation entirely. Sometimes students who failed would be transferred to another school training the service's land-based pilots, but often they refused to take the jet school's rejects. A career could be destroyed in the next few seconds.

If he qualified, the grades he made landing this weekend would be used to decide what kind of jet he would fly. Score poorly and it hurt your chances of flying the highly coveted F/A-18 Hornet. Poor landers would also be steered away from the F-14 Tomcat fighter, which was difficult to fly onto the carrier, or the EA-6B Prowler, which, crammed with sophisticated electronic countermeasures gear, was too expensive to crash. An instructor had chalked the day's date in large letters on a board in the squadron's temporary ready room at Key West. For the ten chicks it would be a day as seminal as their wedding or the birth of a child, the instructors told them.

Rob Dunn couldn't stand the thought of failure. He had been clawing his way through flight school from the beginning. Short and wiry, Dunn seemed born with a wild streak. A Chicago boy, he spent his last two years in college tending bar in the evenings and sleeping in a flat he rented from the bar owner just across from Wrigley Field. He stayed on the go every waking minute, from riding a bike fifteen miles each day to attend classes at the University of Illinois at Chicago, to pouring beer for drunken Cubs fans, to catching glimpses of the game from his bedroom window. Dunn studied hard and played hard.

It was the playing hard that always seemed to land him in trouble. After one wild party at his apartment, Dunn and several buddies sneaked into Wrigley Field to run the bases waving flashlights. Neighbors telephoned the police. Dunn barely escaped. If he had been caught and arrested, the Navy would have taken away his scholarship to be an officer and packed him off to the fleet as an enlisted sailor. The scare was sobering.

But Dunn couldn't seem to stay out of bad scrapes. He had scored in the top of his class during primary flight training. But at Meridian, Mississippi, where he began jet training in the T-2 Buckeye, every student came from the top rank. Dunn's grades were average. When he finally soloed in the jet, Dunn was so excited he began flying loops and victory rolls, then roared into the landing pattern 100 miles an hour faster than he was supposed to. Nobody would notice the hotdogging, Dunn thought.

He was wrong, of course. An instructor pilot flying nearby took down the tail number of his jet and spent an hour on the ground later chewing him out. From that day on, Dunn had a reputation as a pilot who might not be trusted in the air. The instructors started tightening the chain on him, watching anything that might prove the rap.

The chain finally got yanked after a weekend Dunn spent drinking and chasing girls in Pensacola. He rolled into work Monday morning hung over and dehydrated and flew miserably the next two flights. When an instructor asked what was wrong, Dunn made the fatal mistake of telling the truth.

Within a day, he was standing at attention in his crisply pressed khaki uniform before a Performance Review Board. The PRB was the Spanish Inquisition for student pilots, a panel of instructors who reviewed the cases of problem students and decided whether they should be thrown out of flight school. Dunn managed to convince the board he was salvageable. He escaped with another tongue-lashing that left him watery-eyed.

But the next stop was the squadron commander's office. Waving Dunn's training file in front of him, the CO began to carve out his piece of Dunn's rear.

Dunn then made his fourth mistake. "Sir, have you looked at my file?" he asked defiantly. "I couldn't be as bad as they're saying I am. Just to get here you had to have good grades."

The skipper turned beet red and flung the file across the room. Its papers showered down around Dunn. "Now you listen to me!" the commander bellowed. "I'll take a pilot who follows the rules any day over a hotshot with a stellar file!"

Dunn scooped up the papers and practically crawled out of the office. He became a model student afterward.

But bad luck continued to dog him. He cracked a joint in his back one weekend during a rough game of touch football. Dunn visited the flight surgeon the next day. His back was broken, the doctor warned. He might not be able to fly in any plane with an ejection seat. The blast out would be too incapacitating.

Dunn was distraught. His dream to become a fighter pilot could be destroyed by a silly sandlot game. He flew to Pensacola for a second opinion from a Navy specialist.

If his back healed correctly, Dunn might be able to survive an ejection seat launch, the specialist concluded. Dunn waited. His back finally did heal correctly. He was allowed to continue jet training. Of course, he might still have to quit if a back problem flared up, the doctors warned. We'll be watching you.

Dunn transferred to Kingsville to finish the final phase of his jet training in the T-45 Goshawk. He was glad to be out of Meridian. The problems in Mississippi hopefully wouldn't follow him. Kingsville would be a fresh start. The flight surgeons there wouldn't be aware of all the back problems he had had unless they dug into old records.

So far, Dunn had had no more severe pains. The back ached a little when he pulled Gs, but if anyone noticed him rubbing it, they wouldn't immediately associate it with an old injury. He just had to make it through flight school. Then the flight surgeons would stop watching him like hawks. The doctors were always pickier with students, or at least that's what the students thought. Once he pinned on his wings and became a pilot, Dunn figured the Navy would have too much invested in him and the flight surgeons would be more tolerant of minor ailments.

For students in training, the flight surgeon became the enemy. Navy regulations prohibited pilots from self-medicating or visiting private physicians. Flight surgeons were the only doctors allowed to see them so the service could constantly monitor the aviators' fitness to fly. But the rule was routinely violated. Too much was at stake. A flight surgeon might find a minor ailment that could keep a pilot out of the cockpit, end his career. Students instead would secretly visit private doctors for treatment. Or they would see a private physician first for a diagnosis and, if the ailment wouldn't disqualify them from flying, then go to the flight surgeon to be treated. In the weeks before they flew to Key West for their first carrier landing, the instructors privately advised Dunn and the other students that if anything was wrong, treat it themselves so they wouldn't be disqualified from this critical test.

Dunn banked the jet gently to the left to begin lining up behind the *Stennis*. He adjusted the power and the position of the jet's nose ever so slightly so the Goshawk began descending at a rate of about 250 feet per minute. He glanced at the vertical speed indicator in his cockpit to make sure the plane was descending at the proper rate. The VSI said he was. It was critical that Dunn line up at the proper altitude and distance behind the carrier; otherwise, he would have to make too many last-second corrections to touch down on the carrier deck at the right point. During the practice landings over ground, Dunn had had a problem with reaching the final lineup point too high and with the aircraft traveling too fast.

As his jet continued to bank in a wide turn to the left, Dunn's eyes darted quickly to four points on the cockpit control panel in front of him. His brain raced to instantly absorb the readings of each gauge and send a signal to his hands to make slight adjustments to the stick and throttle he gripped. He glanced first at the angle-of-attack indicator, a tiny box with colored arrows and a circle perched on top of the control panel that registered the angle of the wind relative to his aircraft. It looked fine. Next the vertical speed indicator. Fine also. The Goshawk was falling faster—500 feet per minute—exactly as it should. Next the gyroscope that measured the angle his aircraft was banking to the right. Okay as well. Twenty-six degrees. Then finally to the radar altimeter.

Damn! he cursed himself. He was high again. The same problem he'd had at the field. At the end of his left turn to line up directly behind the *Stennis*, Dunn's altitude should have been no more than 375 feet. He was slightly higher.

There was another problem. When he leveled his wings to start his final descent flying to the carrier deck, Dunn was not only too high, he was also too close to the carrier.

Ever so slightly, Dunn pulled the throttle back with his left hand so the plane would drop because it had less power. All the throttle and stick movements had to be finesse ones at this point. This wasn't dive-bombing or dogfighting where the pilot yanked the stick like reins on a horse. He had to have a surgeon's touch now.

Dunn was now at the start of the "groove," which began about a half mile behind the carrier. The groove represented the final flight path his jet had to travel on in order to touch down on the carrier. He looked outside at the ship in front of him.

You're kidding me! he thought to himself. I'm supposed to land this six-ton jet on that skinny runway? The carrier's landing strip was angled to the left so jets could land and take off at the same time and so that an aircraft could get airborne again if it missed the arresting wire. Dunn felt like he was swinging a station wagon into a tiny parking space. His throat felt like sandpaper.

The next fifteen to eighteen seconds—the time it took from the start of the groove to touchdown—would determine his future in the Navy. Of the two days Dunn would spend with his "carrier qual," no more than four minutes of it would be consumed by the critical time in the groove, the time that the instructors would scrutinize the closest to see if he could land on the boat.

Dunn, in effect, now had to thread the giant Goshawk through a long thin pipe. He had to be nimble with the stick and throttle. In order to touch down on the carrier deck at the right spot, he had to fly the jet down the groove at a precise angle. His "glideslope" for the Goshawk was a gentle 3.25 degrees down.

To help Dunn stay on his glideslope, the carrier had the Fresnel lens perched on the left side of the deck. Down the center of the light box were the five lenses, each of which was angled up to project a light. The top four lenses would shine an amber light. The fifth and bottom lens would shine a red light. From his angle flying down the groove, Dunn saw only one light, the "meatball," at any time. The lenses were set so that if the amber light that Dunn saw was in the middle of the box, it meant his jet was on the proper glideslope. He was "on the ball." If the amber light appeared at the top part of the box, it meant Dunn was above the correct glideslope and had to drop his altitude. If the ball was low, he was below the glideslope and had to increase altitude.

But staying on the proper glideslope wasn't enough. As Dunn's jet drifted down toward the carrier deck, the aircraft's nose also had to be tilted up to catch the wind at just the right angle. Pilots called this the "angle of attack." It was important for a simple reason. What mattered to Dunn was not where the Goshawk's wheels hit the carrier deck but where the six-foot-long tailhook sticking out the back of his plane struck. The tailhook, after all, was what grabbed one of the two-inch-thick arresting cables strung across the carrier deck and brought the jet to a stop. Dunn had to worry most about landing the tailhook in the right place and that could be tricky. Like a seesaw, if he tilted the jet's nose up, the tailhook in the back dropped down and struck the carrier deck too early. If he tilted the nose down, the hook went up and struck the deck too late to catch one of the wires.

To keep the plane at the proper angle so the tailhook was positioned correctly, Dunn had to maneuver his stick and throttle the opposite from what pilots who landed on the ground normally did. Landing on a regular airstrip, an aviator held his throttle at a steady speed and gained or lost altitude by pointing his plane's nose up or down with his stick. But to land on a carrier, Dunn had to keep his plane tilted at a steady angle with his stick in generally one position, then gain or lose altitude by pushing his throttle back and forth to add or take back power. Delicately pushing or pulling the stick and throttle correctly could be nerve-racking. No pilot ever ended up doing it perfectly.

Dunn saw the amber light on the Fresnel lens. The meatball was one step too high. He was slightly above the correct glideslope.

Holy shit! Dunn thought to himself. I've arrived! Years of hard work and here I am.

He felt proud. But only for an instant. He still had to put this plane on that tiny steel deck.

"Two three six Goshawk, ball, three point oh, echo four," he radioed in a nervous and high voice to Mango at the LSO platform. "Two three six Goshawk" identified his aircraft and its tail number. "Three point oh" was the fuel left in his tanks, 3,000 pounds. "Echo four" was his radio call sign. "Ball" meant he could see the meatball.

"Roger ball," Mango radioed back in a casual and almost cocky voice. Dunn, Rasmussen, and Burke were Mango's students. Wolfie had other students such as Schroder and Sobkowski. The two LSOs were the chicks' lifeline to the carrier, their umbilical cord, the calm and confident voice who would talk their nervous souls down to the deck. It was the same relationship an athlete had with a coach. Mango and Wolfie believed they could talk any chick down.

Mango gazed into the light blue sky. He didn't really need the radio call from echo four to tell it was Dunn. He could pick out Dunn just by the way the plane flew. Mango and Wolfie had guided down tens of thousands of jets to the carrier deck. Each pilot had his own way of landing, generally made the same mistakes each time. It was like a signature or fingerprint the LSOs came to recognize.

"Here comes Dunn Deal," Mango shouted to Wolfie. Dunn always started the groove too high.

A landing signal officer's job was to guide the aircraft during the critical few minutes from the time it started the groove until it touched down on the carrier deck. In the early days of carrier aviation the LSO waved colored paddles to keep the pilot on course or order him not to land because other planes on the deck were in the way. Today, Mango and Wolfie signaled the pilot over radios and pressed a red button on a pickle switch, which flashed red lights on the ship's Fresnel lens to signal the pilot that the deck wasn't ready for a landing.

Mango and Wolfie had been "waving planes" for almost a decade. It was one of the more curious professions in the Navy. A seasoned LSO developed a razor-sharp eye for distances, angles, and altitudes to the point that standing on the carrier a mile away he knew better where a pilot was on his glideslope than the pilot himself.

An aviator learned to be a landing signal officer much like craftsmen of the Middle Ages. In the four years it took to become one, an LSO would spend only one week in formal classes. The rest of the time the skill was learned on the job through a hierarchical apprenticeship. It began at jet training. A student with an aptitude for landing on carriers usually would tell an LSO he wanted to learn the craft. LSOs often were the best in the squadron at landing on carriers. The student was first given the job of secretary, standing with the LSO on the platform and writing down the complicated scores he dictated for each pilot's landing. He then climbed up three more levels of apprenticeship until certified to guide down all types of carrier planes. By then he was a seasoned lieutenant with more than a thousand hours of flying under his belt.

Dunn didn't have his tailhook down. For his first approach he would perform a touch-and-go. A pilot flew down the groove as if he planned to land, but instead of hooking one of the arresting wires his jet bounced on the carrier deck and took off without slowing down. Touch-and-go's were good practice for bolters, the times when the hook was down but failed to catch an arresting wire and the pilot had to take off quickly to try again.

Dunn first had to deal with the problem of being too high at the start of the groove. He pulled the throttle back slightly to reduce power and drop the plane. But the groove the pilot flew in was so narrow that every correction he made had to be followed quickly by a countercorrection or he'd fall out of the groove again. No sooner had Dunn dropped the plane than in the next four seconds it had dipped below the correct glideslope. As he approached the middle of the flight path down the groove, Dunn had to add back power. But though he flicked the throttle forward only slightly, Dunn had added too much power again.

A quirk in the wind also didn't help him. A breeze did funny things blowing behind a moving carrier. As the wind spilled over the back of the carrier deck when the ship steamed forward, it swooped down, then up like a rooster tail. The effect was called a "burble," which tended to push the jet up when it reached the middle of the groove, then back down when the aircraft was close to the back end of the carrier. Pilots had to adjust their altitude even more for the burble effect. But by the time Dunn had reached the middle of the groove he was too high again. What's more, he had lowered the nose of his jet when he added power so he was flying flat to the carrier with the tail too high.

Mango could see Dunn bobbing up and down in the groove ever so slightly. It didn't surprise him. Dunn Deal always seemed to be high in the middle of the groove as well. But what worried Mango more was that Dunn was flying his jet just to the left of the white centerline that ran down the middle of the carrier deck's landing strip. Most landing mishaps occurred because the pilot didn't line up his jet directly over that center line and plowed into another aircraft. Students were taught to keep the jet positioned so the white line in effect ran between their legs in the cockpit. No deviations. But keeping on the centerline could be tricky. The landing strip and the line were angled ten degrees to the left. The carrier was steaming forward. That meant the angled line constantly moved slightly to the right.

"Watch your lineup," Mango ordered over the radio with an edge to his voice.

Dunn quickly tapped his stick slightly to the right. Students were taught to respond instantly and without question to any LSO command. The right wing tilted down. The aircraft shifted to the right.

Engine roaring, Dunn's Goshawk raced across the back of the carrier no more than fifteen feet from the ramp. The rear part of the steel landing strip had thin, rusted red streaks, scars from the tailhooks of poorly flown jets that had come perilously close to the ship's stern. Dunn willed his eyes not to guide his plane down by looking directly at the carrier deck. Pilots called it "spotting the deck." Before the advent of the Fresnel lens, pilots watched the deck as they landed. Some older aviators still did, refusing to trust the meatball. But the Fresnel lens ball was electronically calibrated to guide the pilot's tailhook to just the right point on the deck in order to catch an arresting wire. If the pilot looked at the deck instead of the meatball when he landed, his plane tended to strike the deck early because from his vantage point the deck appeared about ten feet further away than it actually was. A deck spotter could kill himself crashing into the stern.

Thump! The Goshawk's tires, filled with extra air to make them rock hard, banged down on the carrier deck. Dunn felt as if he was in a giant-screen theater with the scene rushing at him. In a flash he was surrounded by the steel gray of the ship. The carrier's tower whizzed past on the right. Fleeting snapshots of deck crewmen and parked jets. Dunn shoved the throttle forward to full power and at the same time flipped a switch on the throttle forward to retract the speed brakes. He pulled back on the stick so the Goshawk banked up. In a flash, the gray below him was replaced by the bright blue of ocean water.

As the jet climbed, Dunn banked it slightly to the right, then just as quickly back to the left so his flight path wouldn't cross that of other jets being launched from the carrier's bow.

"Wow!" Dunn shouted to himself. He couldn't believe he was in a $16 million plane bouncing off a $5 billion carrier. He felt like a kid who had just been on the best amusement park ride of his life.

Dunn snapped his head to the side to shake off the wonderment. He wiped the first shock of touching down on the carrier out of his mind. It was now time to go to work. Concentrate, he ordered himself.

On the LSO platform, Mango turned to the young apprentice serving as his secretary for these flights. He thought for a brief moment and dictated, shouting into the apprentice's ear over the roar of jet engines on the carrier deck. "High start to in the middle. Too much power in close. Flat and lineup at the ramp. Touch-and-go."

The apprentice scribbled down the critique in shorthand: "HX-IM TMPIC B.LUAR"

"Four wire," Mango continued. Even though Dunn had performed a touch-and-go, Mango calculated that if his tailhook had been down and he had tried to land, the hook would have caught the fourth wire. There were four arresting cables strung across the carrier deck's landing strip, forty feet apart from each other. An ideal landing had the jet's tailhook hitting the deck between the two and three wires and catching the third wire. Catching the second wire was considered satisfactory, but not as good as number three. Hooking the four wire meant the pilot had overshot his landing slightly and was close to missing all the wires. Catching the one wire was the worst of all; the pilot had dropped altitude too early at the last second and was in danger of striking the stern.

Mango paused for another brief moment.

"Fair pass," he finally shouted. That was Dunn's score. Every landing a pilot made aboard a carrier was graded. A "perfect" pass, something akin to an A+, which pilots rarely made, was awarded five points. An "okay" pass, the equivalent of an A, was given when a pilot had minor deviations in the correct flight path but made good corrections to get back on course. A "fair" pass, in effect a B, meant the pilot had even more deviations but managed to correct them. A C was a "no grade" pass, below average as far as the landing signal officer was concerned. If the pilot hadn't set up his approach properly for a safe landing, the LSO would order him not to land and instead fly over the carrier for another try. This was called a "wave-off" and the pilot received only a one, or the equivalent of a D. There was one other grade between "no grade" and a "fair" pass. Sometimes the tailhook banged the deck and bounced over a wire, never catching it. This was called a "bolter" and the pilot usually received a low grade when it happened.

Dunn climbed to 600 feet as he banked left for another touch-and-go on the carrier deck. The students performed two of them first as warm-ups before dropping the tailhook for an actual landing. Rolling into the groove behind the carrier, Dunn this time began too fast and too high. Then he never dropped the altitude enough to stay on the correct glideslope.

"Too much power," Mango radioed Dunn. If it had been an actual landing, Mango guessed he would have missed all the wires because he had been too high. Mango gave him a "no grade" for that pass. Two points.

"Don't let yourself get overpowered," Mango lectured Dunn over the radio.

Dunn was a little irritated with himself. Too much power, too high at the start of the groove. The problems he had been having with many of his landings at the field. Bad habits were returning.

He circled the carrier for the third time. Now he was supposed to land. Dunn pushed down the gray handle marked HOOK on the right side of the cockpit panel.

He radioed to Mango that he could see the meatball. He started the groove just fine. But by the middle, his rate of descent had slowed dramatically. Dunn's plane was supposed to fall at 500 feet per minute. By the middle he had almost stopped descending. Dunn had only seven seconds to make up the correction. As he neared the ramp at the stern of the ship, his Goshawk still flew too high.

Dunn tried to make the jet sink like a stone in the last second. But too late. He felt the thump of the carrier wheels on the steel deck, then the banging of his tailhook on the metal. He shoved the throttle forward as he had been taught to do. Even though he had his tailhook down to catch one of the wires and bring his jet to a stop, Dunn had to keep the engine at full power. If the hook didn't catch a wire, Dunn had to be prepared to instantly fly off the deck to try again. If he had reduced power after touching down, his jet wouldn't have enough speed after a bolter to take off again. The Goshawk would simply roll off the deck and fall into the ocean.

It was good that Dunn had been at full power this time. He had boltered. The hook banged the deck ahead of the fourth wire. A bolter gave him two and a half points, only slightly better than a "no grade."

Now Rob Dunn was angry. To hell with the thrill of a first carrier landing. If he didn't stop making the same mistake every time—flying too high—he'd never land this damn beast on the boat.

"You blew it!" Dunn shouted to himself in the cockpit. He flew the Goshawk around again for another pass. Lined up at the start of the groove, Dunn was determined not to be too high.

He turned out to be too determined. So he wouldn't fly the groove too high, Dunn increased his rate of descent to a whopping 700 feet per minute, 200 more than he should have. By the time he closed in on the carrier's rear ramp, Dunn was dangerously low.

"Power!" Mango ordered menacingly over the radio as Dunn neared the carrier deck.

Dunn pushed the throttle forward but it was too late. He was still too low.

"Wave off, wave off!" Mango shouted over the radio.

As Dunn's Goshawk crossed the rear of the carrier deck he pulled back the stick and pushed the throttle forward even more. The jet flew twenty feet above the carrier deck, then began to rise. Dunn received one point for the pass. Below average.

Mango looked at Wolfie. They were both thinking the same thing. This was usually the time Under Dunn either pulled himself together or unraveled.

"They should calm down now," Wolfie said hopefully. Dunn, Rasmussen, and Burke all were flying a bit ragged, he thought. It usually took about three or four passes before the light bulb turned on, the anxiety went away, and they realized that landing on a carrier wasn't impossible. They could do it.

"I need you to settle down and start flying decent passes," Mango radioed Dunn soothingly as his jet banked right behind the carrier. "Easy with the power in the middle there."

Dunn set up for his fifth pass. He started at the groove slightly low so he added power to lift up the aircraft. Dunn took Mango's advice and moved the throttle forward no more than a half an inch, but it still turned out to be an overcorrection. He was slightly high again as the jet closed in on the rear ramp.

But not too high. Dunn made more subtle corrections this time for the glideslope deviations. As he was just about to cross the carrier ramp, he had the aircraft almost in perfect position.

"Right for lineup," Mango ordered quickly.

Dunn flicked the stick slightly to the right to get the jet back over the centerline.

Thump! The tailhook banged on the steel deck. Dunn shoved the throttle forward to power up in case he boltered again. He didn't this time.

Dunn felt as if the jet had hit a brick wall. He had been traveling at 120 miles per hour when the arresting cable yanked him to a violent stop in just 200 feet. Dunn felt totally out of control. The Goshawk jerked and skidded from side to side. Dunn felt himself being thrown forward, his arms flailing in front, the seat straps cutting into his shoulders and waist, his head snapped back from the whiplash. He sat breathless for a second.

"Throttle back two three six, we gotcha," a voice from the carrier's tower radioed to him impatiently. Dunn shook his head almost dazed and powered down the plane. The arresting cable retracted a little so it made the Goshawk roll back and Dunn could lift and disengage the tailhook. But Dunn unhooked from the cable too slowly. The tower had to wave off the next jet approaching the ship because he hadn't cleared the landing strip quickly enough.

Rasmussen had landed just ahead of Dunn. Pablo had been a bundle of nerves from the beginning. His right leg had been bouncing so much during the preflight briefing, Dunn thought he was going to jackhammer it through the floor. Gremlins made Pablo even more nervous. Walking to his jet at the Key West naval base, where all the chicks had taken off that morning for their first landings aboard the *Stennis*, Rasmussen realized he had forgotten his knee board with directions for the flight. He had to run back to the hangar to retrieve it. Back at the jet, out of breath, he couldn't get the damn radios in the plane to work. Frantically he motioned a maintenance man over to check it out. Pablo had forgotten to push a button in the empty rear cockpit so the system would operate.

He tried singing to himself during the flight from Key West to the *Stennis*. "The End" by the Doors. He sang it over and over again. He was still nervous.

It was eerie, Pablo thought after his plane came to a stop aboard the carrier. The ship seemed so gray and lifeless as he landed. But as he shook the cobwebs out of his head from the violent stop, the carrier deck suddenly became alive with people stepping out of clouds of steam, swarming around his aircraft. They appeared from nowhere. Crewmen in Mickey Mouse helmets, darkly shaded goggles, different colored vests—yellow, green, brown, blue, purple, white, red. It was like being in the land of Oz with munchkins suddenly coming out of the rocks, all of them waving at him silently through the din of roaring engines he could hear outside his cockpit.

An aircraft carrier deck was the most dangerous place in the Navy. The crewmen who moved planes around it in cramped quarters risked their lives every day. The work was grimy, hot, and noisy. One slipup and jets crashed into each other. Bodies too close could be ingested by jet engine intakes. Limbs could be sawed off from whiplashing cables. Accidents were frequent. If the regular brakes failed on a plane, the pilot frantically applied the parking brakes to keep the aircraft from rolling off the deck. Sometimes he had to crash into the tower to bring the jet to a stop.

The taxi directors used complex hand signals to guide the pilots in the planes. Pablo tried to pay attention. Hand signals above the taxi director's waist were for him. Below the waist, the hand signals were meant for other deck hands. Their hands moved fast or slow depending on how quickly they wanted the pilot to move his jet.

The *Stennis*'s deck crewmen were a surly lot. They had just fifteen seconds to disengage Pablo's jet from the arresting wire and move him out of the way before the next jet landed. Rasmussen saw two taxi directors in yellow vests who seemed to be signaling him at the same time. Their signals were unclear. The taxi directors, who had to rush planes off the landing area so other aircraft could land, became impatient. In the *Top Gun* movie, the hand signals were all exaggerated, he remembered, probably so the audience would understand them. The taxi directors here made more subtle gestures as they ordered him off the landing strip to the catapult. Pablo became confused on whom to follow.

Finally, an irritated taxi director ordered him to stop the jet. He marched up to the left side of Rasmussen's cockpit, looked up and pointed his two fingers at his eyes menacingly. It was the signal for "pay attention to me, stupid!" Pablo had never been cussed at in sign language. The taxi directors had other choice signals for the inattentive, such as pulling a fist out of the other hand. It meant get your head out of your ass. The deckhands had a long list of bonehead things students did in their planes, such as taxiing along with the parking brakes on, which blew the tires. That usually earned the student the call sign "Boom Boom."

Dunn refueled his jet, then slowly taxied it up to the left catapult, which would rocket him off the carrier. A square of the deck behind his jet was automatically angled up as the deflector to catch the hot exhaust from his engine. A catapult director in a yellow vest signaled him to flip down the launch bar switch at the bottom left of his control panel to the EXTEND setting.

Deck crewmen bent over and scampered underneath the Goshawk to hook the launch bar into the shuttle on the catapult

track. The crewmen have been known to have fun with the catapult when a plane wasn't being sent off. A catapult officer's last day
on the job was usually commemorated by tying his shoes to the catapult and launching them off the ship. As a prank, one carrier even
hoisted a junk car aboard and rocketed it off the catapult.

Once hooked to the catapult, Dunn moved the throttle forward
to full power. The Goshawk shuddered and roared. Dunn turned
the stick in a circle and pushed the rudder pedals back and forth
with his feet to make sure the steering controls operated smoothly.
He made a quick scan of the cockpit panel's engine readings, the
RPM and fuel flow gauges, and central warning system light. No
problems. Dunn flipped the launch bar switch up to the RETRACT
setting. Then he grabbed the catapult hand grip in the cockpit with
his left hand, locked his elbow, released the wheel brakes with his
feet, and placed his boot heels on the cockpit floor.

Dunn felt the nose of the jet squat as the catapult pulled the
launch bar down tightly. The catapult officer, the man who would
finally order Dunn's plane rocketed off the deck, had taken over.

There was nothing left to check. Dunn turned his head right to
the officer, gave a crisp salute with his right hand—the signal that
he was ready for launch.

The catapult would propel the jet to 120 miles per hour in just
two seconds. Many things could go wrong in those two seconds
and the pilot had to be prepared to react almost without thinking. The catapult or its fittings could malfunction. The jet could be
sent rolling down the deck at a slow speed with the pilot frantically
trying to brake it before the plane fell off the deck. Dunn had to
have his eyes glued to the jet's speed gauges. If the catapult didn't
generate enough power and Dunn's plane had not reached at least
100 miles per hour by the time it crossed the ship's bow, Dunn had
to decide if he should yank the black-and-yellow-striped handle between his legs and eject. Gray detonating cords of plastic explosives were laced over his cockpit canopy to blast the Plexiglas away
so his seat would rocket him up with a crushing force of sixteen Gs.
Ejecting in front of a carrier was hazardous. The ship could run
him over as he parachuted to the sea. Dunn could try to fly the
aircraft with not enough speed. But he might have to ditch the jet
into the water, then blow the canopy and wrestle out of the seat
harnesses before the aircraft quickly filled with water and sank to
the bottom of the ocean.

Two seconds was hardly enough time for all those choices. "Make your own decision on when to eject while you're comfortable at home with a scotch in your hand and a cheap date," Mango had told them. At the catapult, just react and grab the ejection seat handle.

Most students, however, were too disoriented on their first catapult to react at all. They had to take the shot on faith, just pray ahead of time that nothing went wrong.

After saluting, Dunn turned his head to the center quickly and leaned it back on the helmet rest. His right hand lightly cupped the stick. Dunn wouldn't have control of the jet until it crossed the bow.

Time now crept into slow motion.

The catapult officer, squatted outside on his haunches, raised an arm, swept it over his shoulder, patted the deck in front toward the bow, then raised his arm again to point forward. The signal to launch.

Dunn had one second.

He took a quick breath.

BAM!

Dunn tried to look at his speed gauge in the upper left hand corner of the control panel but his eyes rolled back in his head. His knees jerked up. His stomach rose to his throat. It felt as if someone had dropped a bowling ball on his chest. He couldn't breathe.

One second.

The speed gauges shot up instantly to 120 miles per hour. His cheeks fluttered from the two-second blast.

Two seconds.

The Goshawk shot across the bow.

The pressure instantly came off as if someone had just finished rear-ending him.

"Yeaaaaah!" Dunn screamed inside the cockpit. Wolfie had warned the students not to key their microphones during the "cat shots." They didn't want every sailor and officer with a ship's radio to hear them squealing over the first shot.

For a little more than a second, Dunn was frozen in a daze. He finally grabbed the stick to begin flying the plane. Quickly he banked the nose up and instinctively to the left. "Every time a bat flies out of his cave he turns to his left," Wolfie had lectured them. Dunn didn't know whether that was true, but the story served as a good trigger in his mind to bank slightly left after the cat shot.

The jet seemed to him to fly so smoothly after its violent launch off the bow. Dunn accelerated the Goshawk to climb to the higher altitude for circling the carrier.

Brian Burke's Goshawk came to a jarring stop for its third trap aboard the *Stennis*. He had caught the fourth wire. Not as good as the third, but still a decent landing. He scored a "fair" for the pass.

"Good boy," Mango had radioed him. He had bet that Baby Killer would be the surprise at Key West. Burke was "Joe average" at the field, a plodder who said little, whose landings were predictably sound. Those kinds of guys often excelled at the carrier, Mango said. Burke's passes so far had been "fairs," a solid B.

Burke had calmed down considerably from the first touch-and-go. Now he wanted to land and take off from the carrier forever.

He didn't mind admitting that he had been afraid in the beginning. Burke had always been a thinker at the Naval Academy. It made him enemies among the upperclassmen when he didn't jump like a dog to their commands. At Annapolis, he could never stop thinking about Steven Pontell, an academy graduate who crashed in a T-2 Buckeye jet trying to qualify at the carrier in 1989. Pontell had served in Burke's company. He graduated just before Burke had arrived at Annapolis. He had stalled the jet coming in too slow for a landing. It had flipped over at the last minute, cartwheeled across the deck of the USS *Lexington*, killing Pontell and four others. Although the student accident rate was less at the carrier than over the ground, landing on the boat the first time was still dangerous—for no other reason than it was the first time a student attempted it. The instructors had shown them a grisly film of Pontell's fiery crash. Before coming to Key West, Burke filled up his free time with movies or workouts at the gym so he wouldn't dwell on the dangers. If you thought about it too much, you could scare yourself into an accident, he worried. Burke's mother didn't want him to tell her when he flew to the carrier to make his first landing. The white-knuckle worry would be too much for her.

"Dunn Deal needs a good pass here," Mango said, turning to Wolfie. Mango's secretary kept a calculator in his vest pocket. He totaled up the score after the five passes the three chicks had made.

They would attempt four touch-and-go's and ten arrested landings today and tomorrow. Sprinkled among them would inevitably be an assortment of bolters and wave-offs, graded as well. In order to qualify, the chicks had to average at least 2.4, or a C+. After the fifth pass, Dunn averaged 2.3 and Rasmussen little better.

Pablo also bounced up and down in the glideslope, flying high and chasing the center ball as he did back at the field. He was a good pilot, Mango and Wolfie thought, but he second-guessed himself so much it eroded his confidence in the cockpit. "Spicolli," Wolfie radioed to him. He had his own call sign for Pablo. "You gotta keep that ball working on the crest there. You're letting it get too far out of parameter and trying to recenter it in close. Make those corrections a lot sooner! Okay?"

For his sixth pass, Rob Dunn started the groove low, then bobbed up and down along the glideslope, catching the fourth wire when he landed. Mango gave him a "fair," three points. It raised his average to 2.41, just barely above the minimum.

Dunn became frustrated. This wasn't his typical landing. He was starting the groove low instead of high and never completely recovering. Maybe he hadn't recovered from the shock of his first landing. His flying began to unravel.

Rounding into the groove for his seventh pass, Dunn had the jet positioned for an adequate start. He was still a little high in the middle of the groove and as he closed in on the carrier ramp, but he managed to drop the plane down at the last minute. Mango only had to order him to shift to the right slightly.

Dunn thumped down on the runway and pushed the power forward. Something was wrong. I should be stopping, he thought to himself in an instant. But there was no yank from the tailhook catching the arresting cable. In fact, no noise from the tailhook banging the steel deck.

Dunn raced off the carrier deck and banked his jet up. What the hell happened, he asked himself? He looked to the right control panel.

"Ah shit!" he shouted. "You idiot!"

There was his answer. The tailhook lever. It was still up. He had forgotten to lower the tailhook.

"The hook goes down on the pass," the carrier's air boss in the tower radioed to him sarcastically. Dunn felt dopey.

He came around for his eighth pass. His start at the groove again was adequate. But immediately he added too much power. The Goshawk didn't descend in the glideslope as fast as it should.

"Easy with it!" Mango radioed to him. But by the time Dunn neared the edge of the rear ramp his jet still was too high.

"Wave off, wave off!" Mango ordered. Dunn wouldn't have been close to any of the wires.

Mango shook his head. His secretary whipped out his calculator and began punching. A 2.2 average after eight passes. Dunn Deal wasn't doing well.

"Echo four, this is paddles," Mango radioed Dunn. Paddles was the call sign LSOs always used because of the colored paddles their forefathers waved to guide down planes.

"Echo four," Dunn answered.

"Just settle down," Mango radioed in a calm, slow voice. "You're making a pretty good correction in the middle. But you're leaving the power on for way too long and driving the ball way to the top there."

"Roger," Dunn said quietly.

But the ninth pass wasn't much better. Dunn's Goshawk boltered. "My boy's nervous as a whore in church," Mango told Wolfie. They were both now sweating under the early afternoon sun.

Dunn felt crappy. Two lousy passes. Was this the beginning of the end, he wondered. Would he be able to recover? He took a deep breath. Forget the last four passes, he told himself. Pretend they never happened. Start all over.

Dunn began the ninth pass. This time, he was on the glideslope perfectly. No deviations. No bumping up and down.

But just as he neared the rear of the carrier, Mango was forced to wave him off. "Foul deck," he radioed. Burke, who had been flying in front of him, hadn't cleared his jet out of the landing area fast enough.

Dunn cursed quietly. When it rained it poured, he thought to himself.

"Way to go," Mango radioed, trying to buck up Dunn's sagging morale. "It looked like you had a pretty good pass there, Dunn Deal. Keep it up."

Dunn struggled on. But the next pass resulted in another bolter. This time he flew just a little too slow at the start of the groove, the nose of his jet pointed up too high. Dunn overcorrected again. Every reaction created a counterreaction. When he pointed his jet down to gain speed, it slowed his rate of descent and again he arrived too high at the ramp. Dunn's shoulders sagged as he pulled the jet up from the carrier deck to make another pass. His wrists ached from gripping the throttle and stick. Fatigue had begun to set in. The water bottle he had been snatching sips from as he made each sweep around to the back of the carrier was now empty.

But a glimmer of light began to flicker. On the eleventh pass, Dunn still flew high as he neared the carrier ramp, but he stopped overcorrecting. As the Goshawk crossed the ramp, Dunn settled it down to catch the third wire.

Mango gave him a "fair" pass.

Dunn Deal began to feel just a little more comfortable about his landings. He started to develop a feel for making corrections in the groove. He had to think of them in threes. When he began the groove high, he had to pull back power slightly so the jet would descend quicker. But in the next breath he had to add power to climb or he would descend too much and fall below the correct glide-slope in the middle of the groove. Then he would have to pull back on power once more or he would bump up and be high at the ramp. Power back. Power forward. Power back. Think in threes, Dunn told himself. Think in threes. His hands could never be idle in the groove.

The twelfth and thirteenth passes went smoother. The corrections became more subtle. Mango gave him a "fair" for both.

Dunn had pulled himself out of his hole. The glideslope correction for his fourteenth and final pass of the day was even smaller than before. Dunn even began to anticipate the corrections he had to make in the groove. Mango didn't have to radio them. Dunn finally felt he'd mastered the fine art of carrier landings.

Mango's secretary began punching his calculator to total up the first day's average. Dunn Deal had a 2.46, not spectacular but still passing.

Pablo, however, had dug himself into a hole. His first-day average was 2.25, not enough to qualify if the score didn't go up tomorrow.

The chicks usually flew better the second day, Mango knew. They had a chance to mull over the first day's landings, put pieces of the puzzle together. Mango could count on Baby Killer to keep up his "fair" passes. Pablo needed a good day tomorrow. Dunn Deal had to keep his head screwed on.

It was after eleven o'clock the next morning before Dunn and Rasmussen climbed into their jets for the second day of carrier landings. The hot Caribbean sun baked the steel deck. The *Stennis* sailed at an easy fifteen miles per hour. The three chicks had been scheduled to launch earlier in the morning but maintenance delays had cut back the first flights. Dunn and Rasmussen had gone to the bathroom when one plane became available shortly before 8:00 A.M. Burke had grabbed his gear and hustled off to the flight deck. Burke owes us a beer for stealing the first jet, Dunn grumbled when he got back from the head. Pablo and Deal spent the next three hours waiting and fidgeting in one of the *Stennis*'s squadron ready rooms. By the time they launched, Burke had already finished his second round of flights and passed with a respectable score of 2.63.

The three lieutenant JGs had spent their previous night aboard the *Stennis*. They inhaled a heavy dinner, then explored the ship. They wandered through the two wardrooms all with pictures on the bulkheads of the old senator the vessel was named for. They leaned out the carrier tower's fourth-story balcony above the flight deck, which was called "vultures row," so they could watch the jets land in early evening. They trooped over to the flag bridge and when no one was looking, spun around in the heavily padded seat that the carrier group's admiral would sit on if he had been aboard. They hiked back to the fantail at the stern of the ship where legend had it that the carrier's homosexuals hung out.

Like football coaches at halftime, Mango and Wolfie had huddled with them in the ready room that evening to critique the first day's flights.

"Baby Killer, you're making huge power corrections, not flying on the numbers. You're always high when you touch down."

"Pablo, you're doing the right things adding power, but you're overcorrecting. Your scan is breaking down at times. You're not watching the ball."

"Dunn Deal, you're flying too far outside the parameters. Watch out, you're getting way, way overpowered in the groove. Don't do the same thing wrong over and over again. You were picking up in the end. Settle down. You're doing fine."

Mango and Wolfie were gentle now. This was a pep talk. The three chicks could land on the carrier. They had proven that today. Tomorrow, they had to prove that they could land consistently well. Tomorrow Mango and Wolfie would give them fewer "candy calls" over the radio. The instructors now wanted to see how the students landed on their own.

Burke promised to position his plane more precisely at the abeam point and approach turn. He was amazed at the constant power corrections he had to make during the flight. He never seemed to fly smoothly on the glideslope. But he was no longer afraid. He was excited. In fact, Sunday had been the most exciting day of his life, he thought. The catapults and traps had become addictive. There was a Zen to following the meatball, Burke believed. He couldn't wait to jump back into the cockpit and try again.

Go back to basics, Rasmussen had told himself. Sunday had turned out disastrous. He had been overawed by the ship at first. It had distracted him. His eyes didn't dart about the cockpit gauges or to the outside fast enough to absorb the stream of readings his brain needed in order to keep the jet on course in the groove. It had been weird, Rasmussen thought. He had been having so much fun with the landings, he didn't realize how poorly he had been flying. The first day's score now haunted him. Pablo had had a nightmare before flying to Key West. In his dream he failed to qualify at the carrier and was kicked out of flight school. It jarred him awake, his heart pounding. He had to pull up that grade.

The three chicks had spent the rest of Sunday night slouched in the ready room's cushioned chairs watching the movie *Waterworld* over the carrier's TV network, then they trudged off to their staterooms and fell exhausted into bed.

Catapulting off the carrier deck Monday morning, Dunn felt as though an alarm clock had jarred him awake and he was expected to work immediately. On day one, the flight to the carrier from Key West had at least given him twenty precious minutes to collect his thoughts and become comfortable with the jet before landing. Now he blasted into the sky, turned left, and within seconds had to hit precise points in the flight path for the touchdown.

Dunn's head wasn't in the game yet. Flying around to the back of the carrier, he overshot the point where he should have flown the groove to the deck.

"Keep your turn in," Mango warned him over the radio.

But Dunn had edged too far to the right at the start of the groove. Mango could see that he would never be able to shift left in time to line up the Goshawk along the correct glideslope.

"Wave off," Mango said in a tired voice before Dunn even reached a half mile from the carrier.

Mango looked at Wolfie skeptically and shook his head. Pablo's first two passes had been fair. He might be climbing out of his hole. But this was another bad beginning for Under Dunn.

Dunn shook it off. He finally felt warmed up by the time he banked left to line up behind the *Stennis* for the next pass, this one a touch-and-go off the carrier. His jet was high for much of the flight to the deck. But he managed to drop enough at the last second as the Goshawk crossed the rear ramp so that when his wheels banged on the deck his tailhook would likely have caught the third wire if he had been landing. Mango gave him a "fair." He gave him a "fair" as well for the next touch-and-go. Dunn seemed to be bouncing back.

For the next pass, Dunn's Goshawk flew high again as he neared the ramp but its tailhook still managed to catch the magic third wire. Mango gave him another "fair."

But Dunn had made another mistake this time. Rounding into the groove he had forgotten to radio Mango when he saw the meatball on the Fresnel lens. Mango turned to his secretary. "Put a dollar sign by that score," he said in amused irritation. "Fine him for not calling the ball." The writer dutifully made the notation.

"Deal, this is paddles," Mango radioed Dunn. "Just relax. Very smooth out there. But don't forget to call the ball, all right?"

Dunn didn't make the mistake the next time around. "Two three six, Goshawk, ball, echo four," he radioed slowly and precisely like a pupil reciting a story in class.

Mango chuckled. "Roger ball, Muddy," he radioed back.

"He wants to make sure I heard him," Mango told Wolfie.

But this time Dunn began the groove too low. He pushed the throttle forward to add power. There was always a maddening delay of a second or more before the jet responded to the throttle's command and rose. By the time his Goshawk reached the middle of the groove it again flew too high. Dunn decelerated, then the jet dropped below the correct glideslope once more. The result of this weaving up and down through the glideslope was that Dunn ended up too low crossing the ramp. His jet's tailhook grabbed the deck's second wire.

Dunn realized instantly that he had caught the second wire. The taxi director waving at him to disengage from the cable was much further down the deck because his plane had stopped short. Dunn could kick himself. That was a sloppy landing, he knew. Mango wouldn't like it.

Mango didn't. He gave Dunn a "no grade." Just two points. At least Rasmussen and Burke made generally the same types of mistakes every landing. They knew the problems to correct. Dunn was becoming so inconsistent he didn't know what to fix, Mango worried. He had his secretary punch out the averages on the calculator again. Rasmussen stood at 2.39, still in trouble. Dunn's grade: a 2.5, passing but the "no grade" scores would begin to hurt.

Rasmussen boltered during the next pass. Mango shook his head worried, turned to his writer and dictated: "He lined up to the left, too much power in close, high at the ramp."

Dunn rolled into the groove. Again, he flew too low in the approach, plus slightly to the left. The jet slammed down on the carrier deck and jerked to a stop. Dunn looked ahead.

Uh, oh, he thought to himself. The deck hand seemed to be even further away than before when he caught the second wire.

Mango, Wolfie, and the half dozen other LSOs now crowded on the platform began cheering and laughing and pumping their fists. Not because of Dunn's landing. It had been lousy. Mango gave him another "no grade."

Mango was cheering because he had just won a bottle of liquor. The first arresting cable was the worst a pilot could catch. It meant his aircraft had flown too low and near the back end of the ship. Pilots who "aced" the landing, as it was called, were kidded unmercifully by their comrades. Before leaving for Key West, Mango and Wolfie had made four booze bets with each student. A student could put up any priced liquor or beer for each bet. If he lost, the booze went to the instructor. If he won, the instructor bought him the drink. The first bet: score better than 2.75 for the passes, Mango's average when he qualified as a student. The second bet:

70 percent of the passes graded "fair." Bet three: no catching the first wire. And bet four: whether the wire numbers the student's tail-hook caught add up to an odd or even number. An LSO's liquor cabinet usually became the best stocked in the training squadron. But several students from another group in the air had been nailing their landings, making this an increasingly expensive weekend for Mango. Fortunately for Mango, Dunn thought there was no chance he would snag a one wire and had wagered a twenty-dollar bottle of Irish whiskey.

Rasmussen came around for the start of the groove. He had two more passes left. They had to be good. Pablo was dangerously close to disqualifying. Any more shaky landings and he'd have to return to the carrier another time for a second and final try. If he missed then, he wouldn't fly jets.

The LSOs stood silent on the platform as Pablo's Goshawk—engines screaming, hook down, wings wobbling only slightly for the lineup—glided to the carrier deck. They all knew that Pablo's career might be on the line with this pass. The jet snagged the third wire.

Mango thought for a few seconds before turning to his secretary.

"A little too much power at the start," he began slowly. "High in the middle."

He paused another second. "Okay pass."

His writer smiled. "That *was* a sweet landing," he agreed. Pablo had earned an A, and a much needed four points.

Pablo scored a "fair" on his final landing. He passed with a final average of 2.53. Mango could see that Rasmussen had been making steady, if slow, progress in his landings. Learning to trap on a carrier deck was like learning to read, Mango thought. Some guys picked it up quicker than others. Pablo was a late bloomer.

Dunn was another story. The light seemed to be flickering off and on for Deal.

"Despite his best efforts, Dunn is trying to snatch defeat from the jaws of victory," Mango said worriedly to Wolfie.

This was it, Dunn knew as he began the groove for his final carrier landing. He had no idea what his average was at this point, but he did know he had to touch down safely or he wouldn't qualify.

"Echo four, Goshawk ball," Dunn radioed for the last time. He glanced quickly at his vertical speed indicator. The jet wasn't dropping as fast as it should. He was also just left of where the aircraft should have been for the final approach.

He drifted just a little too much to the right in making the correction. The Goshawk bobbed up as it reached close to the rear of the carrier, then dropped down as it crossed the ramp.

Thump, thump! The tailhook and wheels slammed onto the carrier deck. The Goshawk skidded to an abrupt stop. It had caught the second wire, one short of the ideal third wire.

Mango took a deep breath. "Make it a 'no grade' and see what it adds up to," he ordered.

His secretary scribbled on a pad, then pecked at his calculator.

A landing signal officer's grades ended up being subjective. The scores a student received boiled down to how comfortable the LSO felt with the pilot's landings. Sometimes it was good for a student to be disqualified on the first trip to the carrier. The second try would give him more practice and ultimately make him a better lander. But left unspoken among the LSOs was the fact that if Mango felt a student with potential was just having a bad day, he could massage the scores.

Mango's gut told him that Dunn could be a good pilot. He had given Mango a few nervous moments these two days. Maybe it was a case of being lazy with the stick. Mango didn't know, but he thought Dunn Deal was worth keeping.

His writer finished punching out the calculations. He looked up. "It's 2.45," the secretary said.

Mango didn't have to manipulate the grade. Dunn had passed, not with flying colors, but he'd passed.

Mango rubbed his chin and squinted into the sunlight. "Yeah, that's about what it's worth," he told Wolfie. Dunn had turned in a 2.45 performance these two days.

Wolfie grabbed a microphone off the platform's control panel.

"Yeah Deal," he radioed with a broad smile. "It's eleven-eleven."

"You are the man!" Dunn radioed back, excited.

The higher-ups prohibited the LSOs from telling the students over the radio that they had passed. There had been too many cases of excited students celebrating in the plane when they learned the news. One exuberant youngster buzzed the carrier and sent his jet into barrel rolls. He was booted out of the program within two weeks. But if the LSO kept the student in the dark, he would fly back to Key West distraught that he might have failed and be just as accident-prone. One such distracted soul had wheeled his jet into the terminal parking lot for cars after he landed.

The LSOs therefore used codes to let the students know they qualified. Back at Kingsville the week before, Dunn had been telling a bawdy college story in the squadron ready room. It had been late one night, the story went, and Dunn and his buddies were tired of entertaining three girls in their apartment. The girls wouldn't take the hints and leave.

Finally, Dunn popped out of his chair and announced: "It's eleven-eleven." He explained that every night at 11:11 the four of them stripped naked and danced around the apartment wearing only their socks and sunglasses.

The girls squealed and rushed into the bathroom. "We looked at each other dumbfounded and I said, 'Well, I guess this means we have to take off our clothes,'" Dunn recounted. They then stripped down to socks and sunglasses.

In a few minutes, the three girls marched out of the bathroom wearing nothing but silly grins.

Whether the story was true or not didn't matter. By the next day, it had spread like wildfire through the squadron and Dunn had become famous for the eleven-eleven tale.

Dunn screamed after Wolfie's message came over the radio. He tried to dance an Irish jig in the cramped cockpit. He felt like Superman. A taxi director walked up to the side of his jet cockpit and angrily signaled him to pay attention. Dunn still had to park the jet on the deck and climb out for another student to take it for later afternoon flights.

Walking down a passageway deep inside the carrier, Dunn bumped into Charles Nesby, a burly Navy captain who commanded the air wing over the students' training squadron.

"Hey, what's this eleven-eleven crap?" Nesby growled.

"Oh, nothing, sir," Dunn answered at attention.

Dunn spotted Mango further down the corridor. He ran up to him, the hoses and communications cords of the flight gear that he was still zipped up in clattering.

Dunn wrapped his arms around Mango, gave him a hug, then walked off.

Mango stood there embarrassed for a moment, not knowing

what to say. Pilots didn't do much hugging. This was the first he'd ever had from a chick.

He thought for a moment.

"Yeah," he finally said with a shrug. "I guess that's what makes the job fun."

CHAPTER FOURTEEN

Celebration

ROB DUNN swayed back and forth on wobbly legs in front
of Fat Tues Day as a bouncer up the steps to the bar stared at him
with a scowl. It was several minutes past eleven o'clock on Monday
night and Dunn was about to flame out. He wore shorts and a sport-
shirt, brown sandals with socks. Someone had plopped an Austra-
lian bush hat on his head, backward. He didn't know who.

Glassy-eyed, Dunn tried to focus on the crowd of people strol-
ling by him. Duval Street had just begun to awaken for the night. A
half-full bottle of beer dangled at the neck from two of his fingers.
Dunn had no idea how he'd bought it or, for that matter, most of
the other drinks he'd gulped down since landing back at Key West.
He remembered vaguely being hungry, buying a slice of pizza from
an open-air stand, dropping it on the sidewalk, dusting off the dirt
and stuffing it in his mouth. Dinner for the night.

Duval Street was the island's main drag, crammed with seedy bars, expensive boutiques, T-shirt shops, whitewashed rickety wooden hotels with wide porches and slender columns, restored homes in New England seafarer architecture or Spanish colonial. Tropical trees and antique light posts sprouted from its sidewalks. At night, long lines of cars trolled down the street, their drivers in hunt of tanned, long-legged beauties with big breasts and miniskirts practically up to their crotches.

The Key West island, one of dozens that stretched along the Florida Keys southwest of the mainland, had been a haven for Spanish conquistadors, Caribbean pirates, Cuban cigarmakers, spies during the world wars, the U.S. Navy, and colonies of writers and artists such as Ernest Hemingway, Winslow Homer, and Tennessee Williams. By the mid-1990s, the naval base at Key West had gone to seed. No more German U-Boats or Soviet submarines or threats from Fidel Castro to keep it busy. Key West had become an amalgam of sun-starved Northern tourists, natives with brown leathery skin, fashion models and movie stars looking for hideaway vacations, struggling old writers, gays and lesbians, groupies looking for one-night stands, Naval aviators who flew in its warm clear skies by day and partied at night.

All ten chicks had qualified at the *Stennis*. Now they had two things on their minds. Getting drunk and looking for women.

They began late Monday afternoon by taking over the Key West Bar and Grill, a driftwood pub with rough wooden floors, fans whirring overhead, two dingy pool tables, and several regulars with scraggly beards who were perched on stools and who tried to ignore the ruckus that had invaded them. On a small chalkboard hanging behind the rows of liquor bottles, someone had written "First Time in a Gay Bar?" The students cheered and hooted and laughed and slapped each other on the back. Instructors wandered in to congratulate them. They drained gallons of beer, told war stories about every landing they had made the past two days, posed in dozens of different combinations for photographs. Dunn ordered up shots of whiskey for the chicks and began toasting anything on his mind.

"This will never stop," Nibblelink said, leaning on the bar, his hand gently cupping a glass mug of beer. "There's never an end to how scared you'll feel, how excited you'll be. There will always be these pinnacles of terror and joy. It never stops."

The instructors taunted Dunn about the one wire he'd caught. Dunn didn't give a damn. He cared little about what he had scored as long as he qualified.

"Yesterday was a carnival," Rasmussen shouted over the rabble in the bar. "I had a good time yesterday but I wasn't focused. It was one big blur. Today was business. Today, I felt I could do no wrong." When Mango radioed to him in code that he had passed, Rasmussen had thought, I'm not a Naval aviator yet. But I am a carrier aviator. It choked him up. He came close to tears.

Burke sat on a stool pensive, nursing a cup of beer. He couldn't get drunk tonight. The squadron had scheduled him for an early morning flight the next day, air combat maneuvers. He had to stay sober and go to bed early. Dogfighting could be murder with a hangover.

Burke felt overwhelmed by what he had just accomplished. As soon as he had returned to his quarters on base, he telephoned his parents. He wasn't afraid to tell others that his mother and father were his best friends. The constant moves during flight training had left him little time to make close friends his age or have serious girlfriends at each duty station. Nancy and Regis Burke had been his confidants, the ones he still leaned on for support.

Regis grabbed the phone when Burke called.

"Dad, I'm back," Burke told him. "I'm safe. I'm done. I qualified. Tell mom I'm okay."

Burke had decided to tell his mother ahead of time that he was flying to Key West to begin his carrier landings. She had left a message on his answering machine back at his apartment. He could hear the deep worry in her voice as she wished him well.

Burke took another small sip of beer. He thought about walking into McGarvey's Bar in Annapolis one night as a sophomore at the Academy and telling the owner, a former Air Force pilot, that he wanted to be an aviator. The boisterous owner had promised him a bottle of champagne the day he walked back with gold wings pinned to his chest. Burke was sure the owner had long forgotten him. But he planned to return to McGarvey's anyway.

Burke also couldn't stop thinking about Steven Pontell. He imagined Pontell's mother getting that first phone call after the carrier qual. Pontell didn't die in war. There would be no movie made about him. He would join hundreds of other aviators who died in obscurity in carrier accidents, remembered only by other pilots like Burke.

The students staggered out of the Key West Bar and Grill at 8:30 P.M. They wandered over to Rick's Bar, an upstairs joint with a dance floor and disco music, then to Sloppy Joe's Bar, once a favorite watering hole for Hemingway. On its wall hung a sailfish the author had caught.

By ten o'clock, they had gathered at Fat Tues Day, a daiquiri bar that boasted eighteen different flavors oozing out of wall dispensers. Calypso music banged out from overhead speakers. A photo booth stood in the center of its open-air patio. They toasted and cheered. Several hoisted Sobkowski on their shoulders until bouncers rushed up to calm them down.

By 11:00 the organized celebration, if you could call it that, began to dissipate. The sticky sweet daiquiries began taking their toll. Some students slumped in corners in drunken stupors. Others began wandering down Duval Street. Others looked for fast women. Pimps roamed the sidewalk hunting customers.

Mango wandered up to Fat Tues Day in his cups. He wore a brightly colored Hawaiian shirt hanging out over his shorts, dark sneakers, and white socks. He looked like he just stepped off the set of *American Graffiti*.

Mango and Wolfie had been patrolling Duval Street all night policing up chicks, trying to keep them together in one bar.

It now became an impossible mission.

"It's a fucking shame," Mango now said, his words slurring, his eyelids drooping. "A generation of warriors is being lost . . . You want to go to war with a killer, someone willing to drop bombs and risk his life and get drunk . . . All this PC shit is producing a bunch of pussies as pilots . . . You wanna know what a warrior is . . . A warrior is someone who can party hard, get rip-roarin' drunk the night before, climb into a jet the next morning with a hangover, and go kill people . . . That's what it's all about.

"Ah, fuck it, I'm getting out," Mango finally said.

Dunn dropped the beer bottle hanging from his fingers. The brown glass shattered on the sidewalk. White foam sprayed out. Fat Tues Day's bouncer began shouting at him, so Dunn hailed a cab, piled into the back seat, and ordered it to drive him to his base quarters before he passed out.

Mango and Wolfie gave up trying to herd chicks. They found a bar on Duval Street, ducked into it, and huddled at a corner table to talk quietly by themselves. There wouldn't be many more nights like this.

Bill Sigler steered the Goshawk jet down the taxiway toward the hangar at the Kingsville Naval Air Station. He could see far ahead a cluster of a dozen students milling around the hangar's front entrance with buckets of water and fire extinguishers. It was shortly after noontime and the clear skies over South Texas were oppressively hot and muggy. The air was thick like a sauna.

The dogfight had not been one of his best. Two jets piloted by students versus one an instructor flew. Ziggy had accidentally fired his simulated machine gun against his fellow student. Even with just three planes in the fight, the skies could become confusing and a pilot had to be careful not to mistake a friendly plane for the enemy.

But Ziggy didn't dwell on the flight now. It was far more important to him that this was the last flight he needed to complete air combat maneuver training. Ziggy had now finished jet school. He would pin on his wings.

Just four days after announcing that no more retreads would be allowed to fly jets, the Navy reversed itself in Ziggy's case. His training squadron commander had been on the phone for hours lobbying Washington on his behalf. The Navy was under tremendous pressure from Congress to close the door on retreads, but Ziggy's flight scores, which the squadron had faxed to the Pentagon, were too compelling. It would be tragic to waste a fine pilot because of a kneejerk decision, the squadron's senior officers argued.

The decision finally was bucked up to Admiral Jay Johnson, the vice chief of Naval operations who would take over the service after Boorda committed suicide. Johnson approved a waiver. Ziggy could fly jets—by himself. His Naval career had been saved.

The victory though was bittersweet. The pressure had become almost unbearable as he waited for the Navy brass to decide his fate. Ziggy was the only retread granted a waiver. He had come within a hairbreadth of seeing his dream snuffed out. He felt incredibly lucky, but, he thought, lucky like a squirrel who has his tail snipped off by a speeding car as he scampers across the road. The squirrel is grateful to be alive, but he's still missing his tail. The entire experience for Ziggy had been sobering. He had survived. But his faith in the Navy had been severely shaken.

The students and instructors had gathered at the hangar for Ziggy's "wetting down." Later, gold wings designating him a Naval pilot would be pinned on his dress white uniform in a formal ceremony. This afternoon, Ziggy would receive the leather identification patch that stuck to his flight suit with the wings engraved above his name. They were called the "leather wings." By tradition the students and instructors then washed him down with buckets of water and fire extinguishers.

Sigler didn't think he would be overwhelmed with emotion by this day. He had already had one winging as a Naval flight officer. He had been proud of that accomplishment. He almost decked a Kingsville student once when he said, "Oh, you're here to get your real wings." He had worked hard for the NFO wings, which he now would replace with the pilot's wings.*

But taxiing down the line of parked Goshawks to the hangar, Ziggy felt his chest heaving, his throat knotting up. He felt overpowered by the rush of emotion, the realization that years of struggling to become a jet pilot had finally reached a happy ending. This was his last flight in the Goshawk. He would now become a fighter pilot. An air warrior.

Tears welled up in his eyes.

After he parked his jet, Ziggy climbed out and walked around the aircraft several times, pretending to inspect it.

He really did it to regain his composure. Ziggy didn't want his comrades to see him like this.

---

* The two wings have only a minor difference in design. For a Navy pilot's wings, one anchor sits upright behind the crest separating the two wings. For a Naval flight officer, two anchors are crossed behind the crest.

4

FLIGHT OF
THE HORNET

CHAPTER FIFTEEN

Gucci Warfare

CRACKER'S F/A-18 Hornet cruised north of the El Centro Naval Air Facility over the Salton Sea, a huge dead lake turned to a greenish muck by years of chemical runoff. It was early September and Southern California was a desert oven, 107 degrees always by noontime. The Hornet's air-conditioning system strained to keep the cockpit temperature bearable. Even so, Cracker drank gallons of water before each flight to keep from dehydrating.

The Hornet was such a complex machine it reacted to brutally oppressive weather almost like the human body. In Arctic temperatures it took longer for the Hornet's circulation system—the miles of tubes carrying fuel and hydraulic fluids for powering the plane—to warm up just as a sprinter would for a race. The plane's joints could become stiff, the muscles of its motors sore. Breathing the hot stale air in El Sweato's desert, the F/A-18's engines could have what amounted to a heat stroke baking on the blistering hot tarmac. Radars would fry if turned on too soon while on the ground. Cracker considered himself lucky to have this plane up in the air in these conditions. The heat had grounded about half the squadron's ten jets each day the past week.

Crossing the Salton Sea, Cracker quickly spotted the long ridge of the Chocolate Mountains to his right. He shifted his stick slightly to the right. The three other Hornets in his formation did the same to fly over the mountain range.

The power Cracker had under each fingertip never ceased to amaze him during these flights. With his right hand wrapped around the stick and his left perched on the throttle of the sleek gray warplane, Cracker felt like a concert pianist. You want to be a Hornet pilot, the instructors at the Naval Academy had told him, buy a Gameboy and begin limbering up your fingers. The fighter–attack jet's throttle and stick were ergonomically designed so every motion of the human hands as they gripped the two instruments could conveniently serve scores of other functions besides steering or accelerating the plane.

No finger movement was wasted. Cracker's left hand rested comfortably on what were actually split throttles, two levers side by side. His palm could shift one or the other forward to power either of the Hornet's two turbofan engines. His left pinkie finger could flip the throttle's radar switch. His ring finger could engage an auto-pilot button. The left middle finger controlled a button for moving radar antennas. His left forefinger moved around a throttle designator controller like a computer mouse to lock his weapons systems on to targets. His left thumb could wiggle up and down to punch a slew of buttons: one activated speed brakes to slow his jet, another changed frequencies on his radio, another dispensed chaff and flares, another activated his air-to-air missiles.

His right hand gripping the stick played just as many notes. Its pinkie finger could push a button controlling the jet's nose wheel or unlocking the plane's air-to-air missile system. The right ring and middle fingers moved the stick to fly the jet. The right fore-finger squeezed the red trigger to fire the jet's machine gun. The right thumb could punch any of four buttons: the pickle button to drop bombs, the sensor control switch for detecting the enemy, the trim switch, and the weapons select switch (push forward for a

Sparrow missile, right for an AMRAAM missile, back for machine guns, down for a Sidewinder missile).

Cracker could fight the whole war with his hands never leaving the throttle and stick.

Cracker. It was ludicrous how he got that call sign, he thought. His real name was Richard Whiteley. Twenty-six-year-old Lieutenant Junior Grade Rich Whiteley . . . Sounds like Rich Whiteboy . . . Okay, Georgia cracker, they decided in his last training squadron. Ridiculous. He was the furthest thing from a Southern redneck. Up until the time he entered the Naval Academy, Rich Whiteley had been British.

He had been born in West Germany to English parents. The family had moved first to Brazil, then to Ridgefield, Connecticut, when Rich was ten years old. Brian Whiteley was determined that the children not cut their ties to Queen and country. They remained British citizens. His mother and father never lost their thick accents. Rich kept getting the crap beaten out of him in grade school whenever he spoke. He quickly shed the accent when he was with friends. But to this day there was still a slight lilt in his voice from the mother country. And even at home he would often revert to his parents' English. It would later freak out Academy roommates he brought to the house for holidays.

Whiteley had wanted to fly since he was eight. Maybe he caught the bug on the many flights back to England to visit relatives. He didn't know. But the notion of soaring through the clouds, doing loops, being completely free captivated him. In his room hung a poster with a line from "High Flight," a poem by John Gillespie Magee Jr.: "I have slipped the surly bonds of Earth and danced the skies on laughter-silvered wings."

Rich could have returned to England and joined the Royal Air Force. He certainly looked like one of Her Majesty's pilots. Thin, erect, angular features to his face, grayish blue eyes, blond hair brushed to the side. But Rich had never lived in his native country. Though he still held a British passport, he considered himself as American as any kid in his school.

He couldn't fly for the U.S. Navy, however, unless he became an American citizen. Brian Whiteley at first was reluctant to change allegiance. But finally he relented. In Rich's senior year of high school, the family dressed up in its Sunday best and drove to the federal courthouse in Hartford where they swore an oath to the United States. Rich Whiteley's half-brother, Iain, Brian's son by another marriage, was a Royal Army officer in the King's Own Scottish Borders. Brian Whiteley now boasted that he had two sons in the military—on opposite sides of the Atlantic.

For Cracker, flying the Hornet was the fulfillment of a goal. He divided his life between goals and dreams. A goal was making it through F/A-18 training, something he would probably achieve if he didn't botch this flight and others in the next three months. A dream was climbing into a car at 2:00 A.M. with two buddies, reaching Cape Canaveral, Florida, by sunrise, watching a space shuttle launch, shielding his eyes from the blinding light of the fiery rocket exhaust, feeling his chest rumble from the roar of the boosters, and imagining himself an astronaut one day.

The formation of Hornets headed thirty miles northeast to control point Chevy, a mountain peak just west of Graham Pass. If Cracker couldn't spot control point Chevy looking outside his cockpit—which was a distinct possibility since there were dozens of mountain peaks below him—he could still pick it out on the small video screen between his legs that displayed his plane flying over a road map. It was called the "multipurpose color display." Receiving constant position updates from satellites in space, the Hornet's inertial navigation system always knew where the jet was on earth. The color map on the screen, "a floating Rand-McNally" as Cracker liked to call it, moved as the plane flew.

Cracker could see Chevy on his electronic map about fifteen miles ahead of the symbol for his plane.

Jumpin' Joe began ordering the jets in the formation to break away and climb to higher altitudes, which they would use when circling over Chevy. Jumpin' Joe Gelardi (he got the call sign because skydiving was a hobby) was the senior lieutenant in the four-plane formation and the flight leader for this hop. The pilots flying the other planes were Marine Captain Javier Ball and Navy Lieutenant Curtis Carroll. Gelardi had flown F-14 Tomcats for six years but was ecstatic when the Navy allowed him to go back to school to learn to fly the Hornet. There was no future for a pilot in the Tomcat, which was being phased out. But so far, relatively few of its pilots had been allowed to switch to F/A-18s. Gelardi's last job had been as an instructor at a jet training squadron at Kingsville, Texas, a dead-end assignment Tomcat aviators tried to avoid unless they planned to leave the Navy for the airlines. Jumpin' considered himself lucky. At thirty-one, he was being given a second chance as an air warrior.

"Three detach," Jumpin' ordered ten miles from Chevy. That was shorthand for Cracker's radio call sign, Roman 53.

Cracker banked the Hornet up, climbed to 8,000 feet.

Actually, Cracker didn't fly the jet to that altitude. He asked a computer to do it for him. The Hornet had two sophisticated high-speed computers on board that controlled the plane's flight and weapons systems. A pilot didn't really move the jet's ailerons, flaps, or rudders when he pushed the stick around in a Hornet. The stick only activated a computer. The computer then took the pilot's stick movements and decided how to move the ailerons, flaps, or rudders to best fly the plane in the direction the pilot wanted to go. The pilots liked to joke that they were only a voting member of the jet's flight control system. The plane practically flew itself. If Cracker banked left, the computer made sure the ailerons and trims shifted in a way that gave him the smoothest turn possible. The computer would even feed a slight resistance back to the pilot's stick to give him the sensation of wind resistance when he moved the instrument.

Even so, the computers flew the jet so smoothly the pilot could easily lose all sensation of flight. Traveling at 150 miles per hour in the Hornet felt as silky as 350 miles per hour. The slight purr the pilot heard from his jet engines behind the cockpit sounded no different at either speed. That could create a dangerous situation. The Hornet might lose speed and drop out of the sky before a pilot's body would sense it. Cracker had only about thirty hours' flight time in the jet. He would need at least 100 before he would have a good feel for all the subtle sensations the Hornet gave in flight. Then he would begin "wearing the plane," as its pilots liked to say.

Cracker began fiddling with the "Gucci stuff" in the cockpit. He punched the STORES button on his left video screen. Light green lines and symbols immediately flashed on the screen showing him the position of the four bombs hooked to the underneath of his wings. The F/A-18 Hornet was a testament to the Pentagon's obsession with American technological superiority to vanquish any foe.

The jet was the most complex flying machine the Navy had ever built.

The Hornet was supposed to be a cheap and lightweight attack fighter so the Pentagon could manufacture more of them to overwhelm an enemy with numbers. It ended up being anything but that. First produced in 1981, the plane finally cost about $28 million a copy and weighed up to twenty-five tons after all the bombs and fancy new equipment were loaded onto it. Because the Navy couldn't stop itself from adding new bells and whistles to each batch of jets rolling off the assembly line, servicing the aircraft became a spare parts nightmare. No two squadrons' planes were built quite the same. The jet also ended up never flying as far or carrying as much ordnance as the Navy first claimed.

But pilots loved to fly the plane. Not surprising. Aviators always love the planes they fly. But it was well they did in this case. The Hornet had become the Navy's warplane of the future, in fact its only one for the future. After the Pentagon canceled the Navy's stealth jet program because the service had mismanaged it, the admirals in the mid-1990s settled on just building a new version of the F/A-18, called the F/A-18 E/F or "Super Hornet." This new jet was expected to end up performing only marginally better than the older models but be far more expensive.

The service's aging A-6 Intruder attack bomber had been retired. The F-14 Tomcat air combat fighter would be phased out by the early 2000s. Both jets had two airmen in the cockpit. The Hornet, with just one pilot in the cockpit, would take over both of their missions. The "F/A" stood for fighter and attack. Instead of having one jet designed to bomb targets on the ground and another jet to dogfight enemy planes in the sky, the admirals were betting that one souped-up warplane, the E/F version of the Super Hornet, could do both—at a cost that could reach $53 million for each new aircraft. Some pilots feared the Hornet was becoming too complicated for one human to fly. The admirals, however, were counting on new aviators like Cracker to master a job it once took four airmen to perform. Indeed, the future of Naval aviation rested on his Nintendo generation and how well it could manipulate the expensive Gucci controls.

When he first climbed into its cockpit four months ago, Cracker had been stunned by the complexity of the Hornet. The volume of information its computers could throw at him during a flight was mind-boggling. The forty-year-old A-4 Skyhawks he had flown in Meridian, Mississippi, during jet training were jalopies compared to this Lexus. After Meridian, the Navy had pinned a shiny new set of gold wings on his chest because he could fly a jet and land it on a carrier. Now he had to learn in just eight months how to be what amounted to a systems manager in aerial warfare.

It was no small chore. Cracker wasn't a computer geek. The Hornet had what aviators called a "glass cockpit." Instead of a maze of gauges and dials, most of the F/A-18 cockpit panel in front of Cracker was taken up by three square video screens, one at the top left, the other the top right, and the third in the middle between his legs. Surrounding each screen was a line of small square buttons that looked like gray Chiclets. Punching them, Cracker could call up dozens of computer "pages"—video screens full of numbers and symbols that gave him every minute detail of the plane's flight.

Between the top two screens and just above the middle screen were the up-front controls, a panel that consisted of a keypad for entering numbers, fourteen option buttons, and a column of a half dozen thin display windows with instructions for the pilot that looked like an electronic stock ticker. As he would a computer keyboard, Cracker used the panel to program the plane's autopilot, plot his course with the inertial navigation system, guide his radars, select the weapons he wanted to fire. Above the up-front controls was the cockpit's heads-up display, the concave glass that projected symbols that Cracker used for navigating and targeting the enemy.

Flying the Hornet became an exercise in flipping computer pages. A fuel page pinpointed how every gallon of gas was distributed in his plane's six tanks. A flight control system page could alert him to any problems with the flaps and rudders. A "situational awareness" page gave him a top-down view of the direction his plane was flying. A radar page, with a thin green line sweeping back and forth, showed him mountains on the ground or other planes in the sky. An engine page read out fuel temperature, thrust, and revolutions per minute in his engines. A communications page alerted him to problems with the radios. Weapons pages described the armaments on board. A built-in test page spotted problems in the computers and radars. A checklist page reminded him of the toggles and switches he had to flip on takeoff or landing.

It took two people in the old A-6 Intruder to fly the plane and read all the dials and gauges for dropping bombs, two people in the F-14 Tomcat to fight aerial combat. But in theory, a Hornet pilot could be both a good bomber and a good dogfighter if he could shuffle the computer pages on the screens before him like a Las Vegas card dealer. In one instant, the screens could give him a bomber's cockpit. In the next, they could flash up a dogfighter's cockpit.

That was the theory at least. Cracker at first had found it confusing and nerve-racking manipulating screen pages. The computers could become so mesmerizing, an inexperienced pilot would forget to fly the plane and crash into a mountain. In just six months of training, however, Cracker was expected to make theory a reality.

After winning his wings and being assigned F/A-18s as reward for performing well in jet training school, Rich Whiteley had been sent to VFA-106 in May. One of three fleet readiness squadrons the Navy operated to train Hornet pilots, VFA-106 was headquartered at Cecil Field, a backwoods air base just west of Jacksonville, Florida.* Whiteley was now in what amounted to graduate school for pilot training. Reserve squadrons such as the 106th tried to be more collegial. Many of its trainees were pilots like Jumpin' who came from other jets or they were Hornet veterans merely back for refresher classes.

Even the new aviators like Cracker were called "replacement pilots" instead of "students." It seemed less disparaging. In the training command instructor pilots kept their distance. Students were treated as second-class officers. Now they were considered part of the varsity team, winged aviators with whom instructors were expected to socialize more (although a replacement pilot who questioned an instructor's authority could find himself treated like the lowest of students in a heartbeat).

Gone also were downs that could automatically flunk an officer out of the program. Instead, a "signal of difficulty," or SOD, would be given if a trainee had problems with a flight and faculty boards would be convened to devise remedial courses. Attrition at this point was low; the Navy had too much invested in the young aviators to begin washing a lot of them out as they neared the end of their training.

But finally being treated like a pilot had its price, Cracker soon

---

* The Navy decommissioned Cecil Field in 1999.

discovered. The course load was crushing. The workdays stretched longer. Cracker had to learn quickly. No more spoon-feeding in the plane. Stacks of manuals were dumped in his lap and he was simply told to memorize them. After just four flights with an instructor in the back seat of the jet, Cracker was expected to fly solo.* Then the instructors threw a new procedure at him for each flight. Formation flying, night flying, flying in packs of four, flying in foul weather, radar operation. There was no time to catch his breath in the cockpit, no time to practice what he had just learned from a previous hop. There was always something new to learn about the Hornet.

A young pilot had to think creatively in the cockpit. If he tried to memorize just one way to fly a mission, he usually performed poorly in the Hornet. His mind had to be flexible, to jump quickly from one task to another. Because there could be so many different ways to operate the complex jet, instructors became technique advisers as well. Trainees tried to fly with as many seasoned teachers as they could in order to sample the different styles as they would a buffet.

Cracker and his classmates were deployed at El Centro for two weeks to learn the fine points of tactical flying and attacking targets on the ground. They had been bombing at night when the desert was pitch black and from as high as 30,000 feet using radars to pinpoint the target. They had practiced dropping ordnance from different angles of attack and rocketing the plane 10,000 feet up into the sky, then back down for a stomach-churning dive-bomb maneuver called the "Hornet high pop." To infiltrate under enemy radar coverage, they had flown 420 miles per hour just 200 feet off the ground. That close to the earth, they practiced special eye movements to glance inside the cockpit for as little as one second so their brains could absorb readings from the video screens. A pilot could not afford to take any more time away from watching ahead. At that speed and altitude, the Hornet could crash into the ground in one and a half seconds.

The close air support hop they were now on was the culmination of the detachment, the most difficult exercise they had to perform.

---

* In the fleet, the F/A-18 was a single-seat jet. But some of the planes were built with two-seat cockpits so instructors could sit in the back to train pilots.

Everything they had learned the past two weeks—the bombing, computer navigation, low-level flying—would be put to use. Close air support was military jargon for bombing enemy ground forces who were attacking your troops. Fighter pilots had never considered it a glamorous mission. A jet jockey became an air ace painting enemy jets on the side of his fuselage, not tanks.

But the days of dogfighting Soviet MiG-29s only were expected to be long gone. The new air warriors like Cracker were being manufactured at a time when the Navy was undergoing tremendous changes in its tactics. In addition to Russian MiGs, the enemy was now Chinese, North Korean, or Iranian jets. In addition to preparing for war with Russian ships in northern waters, the U.S. Navy was parking carriers off coasts to intimidate Third World nations, or fighting in cramped waters such as the Persian Gulf, or conducting peacekeeping operations off Bosnia. Projecting power or fighting conflicts at the shoreline—littoral warfare, as the strategists called it—would be important Navy battles in the post–Cold War world and the F/A-18's more important job would be providing air cover for Marines hugging the beaches.

"Falcon, this is viper!" Cracker suddenly heard a voice screaming frantically from his radio earpiece. "We're taking some pretty heavy artillery fire on the ground. I hope those F/A-18s you're bringing in have some good bombers!"

"They're the best we've got out here," Falcon radioed back. "They'll be there in four minutes."

Cracker could hear muffled explosions in his earpiece. He chuckled to himself.

They were sound effects from Joe Bags, one of the instructors circling the target area in a beat-up T-34C turboprop plane. The instructors used the T-34Cs as spotter planes for the practice strikes because they were far cheaper to fly than the gas-guzzling Hornets.

There were no troops under attack. Joe Bags was pretending to be a forward air controller on the ground with the call sign Viper. He was inventing the dialogue for dramatic effect. The instructors tried to make the hops realistic, distract the students with harried radio chatter that they would hear in real combat.

Lieutenant Joe "Joe Bags" Evans loved these "dets." That was the shorthand the pilots used for the two-week detachment trips they took to practice bombing or aerial combat or carrier landings at places far from Cecil Field. Many pilots hated to be away from home. Not Joe Bags. He tried to make at least one det every month. He was a thirty-two-year-old divorcé. His marriage couldn't survive the stress of a Navy life's constant moves and separations. Joe Bags now had organized his life around these dets. All his credit card and house bills were paid automatically from his bank account. There was nothing better in life than being away from Cecil and his desk piled high with paperwork. On det there were no administrative headaches, no sailors to baby-sit or get out of jail, no Navy brass looking over your shoulder, no nagging wives and girlfriends. On det all he did was fly and drink beer at the officers' club and pal around with his buddies in the squadron. Joe Bags reveled in the camaraderie, the joking, the bonding. His best friends had been made on det. The best flying a pilot ever did was on det.

For the son of a Montana rancher, the Navy had been an adventure of the spirit. There would be no more exciting time in Joe Bags's life than flying in the storms of the North Sea. Waves crashing over the top of the ship. High winds driving the temperature so low the plane handlers could stay outside for no more than thirty minutes. The carrier deck becoming so icy slick the jets would skid sideways and wing walkers would rush alongside with chains to anchor the planes down so they wouldn't slide off and into the ocean. Ocean swells pushing the back of the carrier deck up and down as much as forty feet. The landing signal officers timing the cycles of the waves to guide the jets to touchdown just as the ship's stern reached its peak. Catapult officers launching planes when the carrier's bow crashed down into the water, hoping that the two seconds it took the aircraft to reach the end of the bow would be enough time for the swells to push the front of the ship up so the pilots would fly into the sky instead of the ocean. Some of the most breathtaking scenery Joe Bags would ever see had been on those NATO exercises in the Arctic. It was scary. It was exhilarating.

Too bad it all would end one day, he realized. Nobody made admiral flying on dets constantly. The Navy expected you to be an officer first, a leader second, a pilot third. It was sad, he thought. You were trained to be an aviator but you only had fleeting moments to enjoy it. Joe Bags had been offered a job as an admiral's aide in Norfolk. Everyone had told him it would be good for his career, not the kind of offer an officer turned down. Maybe he'd take it. Who knows, he thought, maybe he'd learn to enjoy it. Maybe a break from flying wouldn't eat him up inside.

Joe Bags looked out the side of his cockpit and spotted a pickup truck near the two crude airstrips that formed an X on the western side of the Chocolate Mountain range. The airstrip and several dozen charred remains of tanks nearby were the target area the four Hornets would use for dropping their bombs. Joe Bags pointed the nose of his turboprop down and began buzzing the pickup until the idiot behind the wheel finally drove away. Scavengers looking for bomb parts were a constant problem at the military ranges. One this week had even set up a tent there and camped overnight. If an instructor flying low in a T-34C hadn't spotted him and chased him off, he could have been killed. From higher up where the Hornets bombed, the tent looked like a tank hulk.

Cracker turned down the radio frequency that was yapping with Joe Bags's useless chatter. He had more important things to concentrate on at the moment. The mission consisted of four bombing runs. Each run involved traveling from a control point like the one he was now circling over to another landmark miles away called the "initial point" and from there flying to the target, which he had to attack within plus or minus ten seconds of the time the forward air controller wanted it hit. (Actually, Cracker calculated that he had only a four-second margin of error for reaching the target. The bomb took about six seconds to fall from his jet to the ground.)

The first two runs were preplanned. The pilots had received the control points, initial points, and targets before they took off. For the final two runs, the pilots didn't receive the checkpoint information until they were in the air. That way it simulated a spur-of-the-moment attack.

Cracker was uneasy about this flight. Its timing had to be exquisitely precise. There were constant adjustments to be made in the jet's airspeed along the way, dozens of calculations to perform in his head, no room for errors in programming the navigation computers. Garbage in, garbage out. One degree off in a latitude or longitude entry could send the aircraft sixty miles off course. Arrive at the target too early and in real combat your jet could be hit by friendly artillery raining on the enemy before your attack.

The four pilots had set out on this exercise earlier in the morning. Cracker had had everything memorized for the first two pre-planned runs. He had flown them in his head a dozen times. He could mentally picture the checkpoints he would see on the ground. But the pilots had to abort that morning flight after several of the planes developed engine problems. The instructors had quickly ordered up new checkpoints, new routes for the bombing runs, and then given the young aviators only an hour to study them before rushing into new planes and taking off again at noon. Cracker hadn't had time to think through the first two runs. He didn't have the "God's eye view" that he always wanted for how a mission would be flown.

The four Hornets flew around control point Chevy at staggered altitudes like a mobile. Gelardi circled at 6,000 feet, Ball at 7,000 feet, Whiteley at 8,000, and Carroll at 9,000. Whiteley now had a lot of calculations to perform, which would keep his head pointed down inside the cockpit. Before he did them, however, he punched in the barometric altitude hold command on the up-front control panel. He shoved the throttle forward to the speed he wanted, tilted the stick to the left, then switched to autopilot. The computer would now keep his plane cruising at 8,000 feet from the ground in a wide circle. Cracker at first found it unnerving watching the stick between his legs move on its own when the computer became the pilot. It was also a bit embarrassing. The computer could fly the plane smoother than a human.

Cracker began punching buttons to arrange the computer pages on his screens. The left screen displayed the same bomb-targeting information he saw in his glass heads-up display. The right screen displayed symbols showing the bombs he could drop. For this exercise they were using 500-pound inert bombs, which rather than explode gave off puffs of smoke so spotters could mark where they hit. The center screen between his legs had the moving map. Eventually the map would also have dotted lines over it to show the flight path he should take to the target.

Cracker turned up the radio volume. He could hear Jumpin' Joe talking to the forward air controller and he knew it was important.

"Falcon, Roman five one checking in as fragged," Jumpin' radioed to the controller. The radio call sign numbers were arranged by the order the pilots would be dropping their bombs. Jumpin's was Roman 51. Ball's was Roman 52, Whiteley's was Roman 53, and

Carroll's was Roman 54. The call sign numbers could become confusing. The four pilots rotated the bombing order for each run. On the next run, Ball would be Roman 51 and be the lead plane in the attack, Whiteley would come in second as Roman 52, Carroll would be third as Roman 53, and Jumpin' would move to the end of the line as Roman 54. If a pilot became confused about which call sign he was during a run, he could easily bomb the target at the same time another jet was attacking it, risking a midair collision as the two planes dove to the same bull's-eye.

"Okay Roman five one, stand by for a target area brief," the forward air controller radioed back. The controller in this case was Lieutenant Tim Taylor, an instructor who was riding in a second T-34C circling the target. Along with Joe Bags, "Tinkle" Taylor would be calling in the air strikes and spotting where the bombs struck. Taylor had been stuck with his nickname during a carrier tour when a squadron commander who had been potty-training a toddler back home noted that Taylor's initials, T.T., sounded familiar. Taylor had ever since been trying to rid himself of the humiliating call sign. He kept marking his handle on briefing charts as "Tool Man" after the *Home Improvement* television character bearing his name. No luck. His buddies kept erasing it and writing in Tinkle.

"Okay, for the target area we have an airfield," Tinkle radioed Jumpin' a minute later. "Two 8,000-foot runways. One north–south. The other on a two-three-eight run." There was also a row of fifteen to twenty tanks just south of the center of the X that the two runways made, he added. "I need some close air support on some key locations here so we can shut down this airfield."

Jumpin' already knew what the runways looked like from flying over them earlier so he didn't bother to write this information down on the pad strapped to his knee. The runways were actually little more than wide dirt paths carved into the dusty brown ground. From his altitude they appeared like two crossed toothpicks, a lighter shade of beige than the brown around them.

The runways would be easy to see in the bombing runs. The more difficult targets to spot would be the several dozen tank hulks south of the airfield, which from his altitude looked no different from the hundreds of patches of dark brown scrub brush surrounding the strips.

"Any friendlies?" Jumpin' asked. That was one bit of information he didn't have.

"You've got some friendlies east of the field," he answered. "The last section of Tomcats didn't do any good. We need more air support."

Tinkle wanted the four young pilots to have to thread their bombs into a tiny target area. That was how they would have to strike in this day and age if the bombs were real, he knew. Tinkle had had to drop two laser-guided bombs on a Serbian storage facility thirty miles south of Sarajevo during the brief NATO air campaign in 1995. The pilots spent days planning each mission to be sure their bombs hit just military targets. A stray ordnance falling on civilians could cause an international incident.

"Roman, your first mission will be twelve tac one," Tinkle continued. "Stand by for TOT."

Twelve tac one was the first of the two prearranged air strikes Jumpin' and the three other pilots would fly. For this one they were to travel from the Chevy control point to the initial point labeled Golf on their maps, a mountain peak twelve miles west of Chevy and nine miles north of the airfields on the other side of the Chocolate Mountains. Jumpin' now waited for his TOT, his time on target when the bomb should hit.

Joe Bags interrupted with more radio chatter pretending to be another squadron coming in for a run at the target.

"Falcon, this is Dogbreath, flight of four AV-8s with eight mark eight-sevens, what do you need?" he babbled.

Tinkle told him to hit the north end of the runway, then turned his attention to giving Jumpin' his time for striking the field.

"Roman five one, TOT plus three nine," Tinkle finally told Jumpin'.

Cracker had been monitoring the radio chatter from his cockpit. He scribbled "39" on the notepad strapped to his left knee. On the pad, he had already made a photocopy of charts and way point times he would need for each flight with blanks left for the 39 and the other numbers he had to fill in. Lucky he was left-handed, he thought. He was trusting the computer to not fly the plane into a mountain while he had his head buried in the cockpit with these calculations. But just in case, his right hand was always free to grab the stick while he wrote.

He now had to perform the arithmetic in his head. The Hornet cockpit had practically every new navigation gizmo you could think of. But it didn't have a simple calculator. The math wouldn't have been terribly complicated if Cracker had been computing it on the ground. But in a cramped, steamy hot jet, with radios chirping in his ears, computer screens glaring at him, and his seat tilted 45 degrees to the left as the autopilot flew him in a circle, the computations could take three times as long to complete. He felt as if he was juggling balls and doing math in his head at the same time. In fact, as practice some pilots on the ground would juggle three tennis balls and recite emergency aircraft procedures at the same time.

Cracker squeezed his eyes shut for an instant and willed himself to concentrate. Sweat poured down from his forehead and stung his eyes from the salt in it. On dive-bombings, the G force caused the sweat to pop out of every pore of his face spraying a film on the inside of his sun visor so he could hardly see.

Okay, Cracker thought to himself, sucking in a deep breath from the oxygen mask. A time on target of plus three nine meant Jumpin' had to have his bomb striking at exactly thirty-nine minutes after noontime. Each Hornet was supposed to hit the airfield one minute after the other. Ball was ahead of Cracker. That meant Ball's bomb had to strike at forty minutes after the hour and Cracker's bomb had to be there at forty-one minutes past.

Now he had to quickly calculate when he should leave control point Chevy so he would be at the target at exactly 12:41 P.M. The start time was called his "push time." He worked backward, first figuring out how long it would take to fly from his initial point at Golf to the airfield. Golf was nine miles from the airfield. Cracker quickly scanned down a long thin chart on the photocopied sheet clipped to his kneepad. At a constant speed of 480 miles per hour—that was the speed Cracker intended to keep during the run—the chart told him it would take one minute and eight seconds to fly nine miles from Golf to the airfield.

He subtracted one minute and eight seconds from his time on target, 12:41 P.M. That made it 12 noon and 39 minutes and 52 seconds.

Now he had to determine the distance from Golf to control point Chevy that he was presently circling. Cracker checked a second chart on his kneepad. The distance from Golf to Chevy was twelve miles. He glanced quickly at the first chart. Twelve miles flying at 480 miles per hour took one minute and thirty seconds. He subtracted one minute thirty seconds from 12 noon, 39 minutes, and 52 seconds. That came out to 12:38 P.M. and 22 seconds.

But Cracker had to subtract one more number before he came up with the exact time he pushed from the control point. As he neared the airfield he would have to bank the jet sharply to one side for a quick look at the target before diving to it. That sharp bank took an extra ten seconds. He had to subtract another ten seconds from his number.

Cracker's push time from control point Chevy was 12:38 P.M. and 12 seconds. No sooner, no later. Only a ten-second margin of error was allowed for the flight.

Cracker scribbled the number on his kneepad. Next to it he wrote "PUSH" in large letters and circled it so he wouldn't forget when he was supposed to leave Chevy.

The manual work was done. Cracker now had to quickly set up the Hornet's computer system for the run. Before he had taken off, Cracker had already entered into the computer the exact longitude and latitude coordinates for the target as well as for each control point and initial point in the exercise. The computer could process as many as twenty-four points on the map at one time. Cracker had also assigned a number for each set of coordinates. Point eighteen was the target. Points seven through nineteen were for control points or initial points.

Cracker's fingers began dancing over the computer buttons. On the up-front control panel—the keypad between the two top video screens—he punched the button marked DATA, then the button for SEQUENCE. The ticker tape commands on the tiny window screens around the keypad began flashing with orders for different numbers Cracker had to enter into the system. For sequence, he punched 8 first, the number for control point Chevy where he would start the run. Next he punched 16, the number for his initial point at Golf. Then he punched 18, the number for the airfield, and told the computer this was the target. The final number in the sequence that he punched in was 19, which designated a second control point just over the Little Mule Mountains to the east, where all the jets would rendezvous after dropping their bombs.

Cracker entered the ground speed he wanted the plane to fly: 480 miles per hour. He punched in his time on target: 12:41 P.M. Then he began poking at the gray Chiclets surrounding the center video screen between his legs. This was the screen with the glowing color map that moved as his plane moved. Finally, he pushed the Chiclet marked for AUTO.

The computer began churning out solutions for him. Green dotted lines connected to two circles and one diamond suddenly appeared over the moving map along with a tiny plane symbol. The two circles represented initial point Golf and control point Chevy. The diamond represented the target. The plane symbol showed him where the Hornet was flying over the map at any time. The dotted lines showed him the flight path he had to take over the map to reach the Golf initial point, then the target. All he had to do was fly the jet so the tiny plane symbol on his screen stayed over the dotted line and he'd be on the correct flight path.

Cracker let out a deep breath. His cheeks ballooned around the oxygen mask plastered to his face. He felt somewhat calmer now. The computer now knew the time he wanted to bomb the target, the flight path he wanted to take to target, and how fast he wanted to fly the flight. The autopilot was set. The computer now had already calculated what Cracker had figured out in his head—the time the Hornet needed to push away from control point Chevy to reach the airfield at 12:41 P.M.

In fact, if Cracker had been a complete "computer cripple," as he liked to call it, he could have relied on the machine to decide when his push time was. Sometime around 12:38 P.M. and 45 seconds he would feel the jet speed up to 480 miles per hour and he would see the stick between his legs begin to turn on its own to position the jet to leave control point Chevy on its own. It gave him some comfort that if he became frazzled and couldn't think, the computer at least would start the run for him on time and point him in the right direction.

Of course, no pilot would stake his life on the computer flying him to the target by itself. There were horror stories galore about the Hornet's computers sending aviators off to the netherworld because of bugs in the system or mistakes aviators made entering in data. Some older pilots wouldn't even rely on the computer at all. They still did all the navigating by dead reckoning because the fancy gadgets had led them astray so many times before. Cracker would always back up the computer with his own math. Still, the glowing green screens in front of him now felt like a warm cuddly security blanket.

"Roman five one's pushing," Cracker heard Jumpin' announce over the radio.

Cracker looked down at the digital timer in his cockpit. Twelve o'clock, 36 minutes, and 12 seconds. Jumpin' had left control point Chevy exactly on time by his calculation, two minutes ahead of when Cracker was supposed to push.

"Emerald," Tinkle radioed back. Emerald was a prearranged code word that meant "continue your flight." If Tinkle had wanted him to cancel the run he would have said "diamond" over the radio. If he had wanted Jumpin' to change course, he would have said "jade."

"Yeeeeeeeeeh-hah!" Cracker heard over the back radio the four pilots used to chat without the instructors listening in.

Cracker chuckled to himself.

Carroll had let out the rebel yell to announce the official start of the run. That was typical of him, Cracker thought with a smile.

Curtis Carroll was busy as well playing the piano inside his cockpit, setting up the computers for his run one minute behind Cracker's. Tall and lanky, Carroll had spent most of the twenty-nine years of his life stumbling from one accident to another. Three scars connected together in a long line under his chin from three different mishaps: a swing that crashed into his jaw when he was seven, a bicycle accident when he was thirteen, a dirt bike spinning out of control at sixteen. By the time he was eighteen, Carroll had been in the hospital seven times. Racing motorcycles he had broken ribs, his left collarbone, and left ankle. He'd lost count of the number of fingers that had been broken. Returning home after a New Year's liberty in 1994, he was sideswiped by a car as he ran across a four-lane road at a truck stop in North Carolina. He ended up in the median ditch with a broken leg and crushed right toe that would keep him out of the cockpit for eight weeks. His wife, Donna, was terrified his next accident would be in an F/A-18. His classmates had nicknamed him Roadkill.

Tinkle ordered the pilots to aim their bombs at the point where the two runways intersected. Jumpin's bomb hit the intersection dead center and struck just one second after the time it was supposed to be on the target. That was well within the ten-second window the instructors demanded.

Javier Ball, who was following Jumpin' in the second jet, took a quick look at the tiny airspeed box in the upper left corner of his heads-up display. Inside the box, the numbers indicating his jet's speed changed constantly as the aircraft sped up or slowed down. Just underneath that box a "carrot" symbol that looked like a tiny coolie hat drifted right and left. The carrot was Ball's prompt from the aircraft computer that told him he needed to fly faster or slower if he wanted to reach his target on time. If the carrot was perfectly centered under the airspeed box it meant the Hornet was traveling at the correct speed to meet his time on target. If the carrot drifted to the left, it meant he was flying too slow and needed to push the throttle forward to pick up speed. If the carrot slid to the right, he was too fast and had to pull back power if he didn't want to arrive at the target ahead of time.

As Ball flew west toward initial point Golf, the carrot under the speed box sat far to the left. He would be late reaching Golf unless he put the engine into afterburner to gun the jet up to 580 miles per hour.

Ball didn't want to do that. The engine inhaled gas when the pilot raced it in afterburner. That wouldn't leave Ball with enough fuel to finish the bombing run. Javier hadn't been too flustered by the aborted run earlier that morning and then all the changes to the mission afterward. A lot of guys liked to have everything planned to a T before they launched on a mission, everything filled out on their kneepad, every chart perfectly drawn. Javier liked to stay flexible. All those careful charts went out the window once you launched into the heat of battle, he knew. He liked to take only a few sheets of paper with him on the flights, which had only the numbers he really needed to know.

Javier decided to improvise now. Instead of flying directly to Golf and then making a sharp right turn to the airfield, he cut the corner early and flew diagonally south to the airfield to make up the lost time.

Thirty-year-old Javier Jeronimo Ball was one of the few Mexican-Americans flying Hornets for the Marine Corps, as far as he could tell. He came from proud lineage. His great-grandfather's name had actually been Balli, but an immigration officer Anglicized it when he entered the United States and the family settled in El Paso, Texas. One of his father's uncles had ridden with Pancho Villa. His mother's family could trace their roots back to the famed

Indian warrior Geronimo (Javier was given the middle name but spelled with a J).

But Javier's had been a life filled with tragedy. His father, Francisco, had been a decorated combat officer in Vietnam but the war had changed him. At home he was a Great Santini, whose discipline his son came to despise, and a hard drinker who committed suicide when Javier was just thirteen. Cecilia Ball clawed and scraped to keep creditors from repossessing the property she and her five sons had after the death. Javier got a part-time job and turned over half his paycheck to his mother each week.

Javier eventually graduated from Southwest Texas State University in 1989 and joined the Marine Corps to fly Hornets. But he flunked the visual test for flight school because the day before a stick had poked him in the eye during an infantry training exercise and blurred his vision. He became a logistics officer and shipped off to the Persian Gulf War in 1991.

But several days before the ground war began, Javier received a message from the Red Cross. His brother, Juan, who had led a troubled life, committed suicide as well. Javier hopped the first flight back, but didn't arrive in time for Juan's funeral. He returned to Saudi Arabia, but the war had already ended by the time he rejoined his unit.

After the war, Javier married Beverly Nelson, whom he had met in North Carolina before Desert Storm. He arranged for another flight physical, this time passed it, and packed off to flight school. They began building a family. A daughter, Victoria, was born in 1992, then Geronimo came in 1994.

But tragedy would strike once more. Beverly was involved in a violent automobile accident in Kingsville, Texas, where Javier was taking jet training. Beverly and Victoria escaped with minor injuries but eight-month-old Geronimo suffered brain damage and slipped into a coma. He died a week later.

Javier took thirty days' leave to deal with the crisis. But by the end, he was going nuts not flying and just wandering around the house looking at framed pictures of Geronimo. He ordered himself to climb back into the cockpit.

Javier had long steeled himself to the dark cloud that followed his life. Geronimo's death was "just another fucking thing" like his dad's and brother's suicides that he would have to deal with, Javier would say. You cry about it. Not a day went by that he didn't think of the little boy. But you still continued living. Nothing you could do would change what had happened.

The way Javier handled the death caused problems with Beverly. Javier accepted it. Beverly couldn't. The accident devastated her. Javier insisted on putting it behind them. No shrines for Geronimo. No moping about the house forever clutching pictures of him. Javier had been raised Roman Catholic but he was not religious. In fact, he had no use for religion. He believed in himself. That would get him through this crisis.

His first jet flight after the death was abysmal. Then Javier disqualified his first time trying to land on the carrier. A human being could compartmentalize only so much, he realized. But gradually, Javier's air skills returned. He won a coveted slot as an F/A-18 pilot. He made Beverly find a job to get her mind refocused. And in June 1996, they had another girl, whom they named Alissa.

Streaking south over the Chocolate Mountains at 3,000 feet, Ball took a quick glance at his altimeter to make sure he was high enough to clear the range and not slam into one of its peaks. He was only minutes from the target now. There was no time to shift throttle power to change his arrival time. He had to keep his speed constant at 480 miles per hour. He'd already bought the time on target he would have for this run by the flight adjustments he had made back at the initial point. Now he had to concentrate on getting ready to bomb.

For an instant he forgot what point he should be hitting.

"Falcon, confirm target for Roman five two," Ball said hurriedly over the radio.

"Intersection of the two runways," Tinkle repeated curtly.

That's right, Ball thought to himself.

"Five two's approaching," Ball radioed back. He was about five miles from the airfield and just about to swoop to the side, then down for his dive-bombing.

"Emerald," Tinkle radioed back, the code word to continue.

Ball made a final check of his weapons page to make sure he would be dropping the right bomb. He was. He quickly flipped the MASTER ARM switch in the top left corner of the cockpit up to the on position and punched the A/G button above it so that the computer would know he was firing from the air to the ground and flash the correct symbols on his heads-up display. Then he took a quick look at his radar altimeters to make sure they would warn him in case his jet dipped below 500 feet during the dive-bombing. All set. With his left forefinger, he began wiggling the throttle designator control, the tiny computer mouse on his throttle that would pinpoint the target on the ground.

Four miles from the airfield, Javier cut a sharp turn to the right so he could see the field clearly by looking out the left side of his cockpit. In the next second he rolled the jet to the left, then upside down so its nose dipped down quickly toward the airfield. It was critical that Javier be diving down now at a 15 degree angle when he flipped the Hornet back right side up. He was.

Javier squinted at the glass heads-up display in front of him. He could make out the crossed runways looking through it. In the older training jets he had flown, Javier had had to estimate in his head when to drop the bomb based on the jet's altitude, its dive angle, and airspeed, as well as the winds outside blowing against the plane and the Gs it was sustaining in the dive. Now he was auto-bombing. The computer did most of the headwork, sensing altitude, angle, speed, wind, and Gs, then calculating, based on all these factors, when the bomb should be dropped so it hit the target.

Quickly Javier pushed around the throttle designator control mouse with his left forefinger so it moved a tiny diamond on his HUD screen directly over the point where the two runways crossed. That designated the target. The computer would aim the bomb for where the diamond was. A vertical thin line also flashed up on the HUD screen at the same time along with a tiny plane symbol. The vertical line, which was now slightly to the left of the diamond, represented the steering line Javier's jet should follow in the dive toward the target. The line was slightly to the left of the diamond because the computer's sophisticated sensors had already taken into account a crosswind blowing over the airfield that would throw off Javier's bomb if he dove directly toward the bull's-eye. As long as he steered the jet so he kept the little plane symbol over the steering line as he dove down, the bomb would hit the bull's-eye by the computer's calculation.

"Five two's in," Javier radioed Tinkle when he had the airfield in his bomb sight.

"Cleared hot, cleared hot!" Tinkle gave him the final command to drop his bomb.

Javier took one last quick look at his HUD.

Diamond on the bull's-eye.

Plane traveling down the steering line.

Everything perfect.

With his right thumb he punched the red button on top of the stick—the pickle button.

But the bomb didn't drop immediately. In the Hornet's case, pushing the pickle button merely gave the computer permission to drop the bomb. The computer then decided at what point in the dive the bomb should be released so it scored a direct hit.

Javier waited.

One second. Two seconds. A tiny cue marker appeared across the steering line that now kept moving down the line like the New Year's Eve ball in Times Square, letting Javier know the computer was waiting for the perfect moment to drop.

Suddenly a light on the HUD screen flashed. The signal that the computer had dropped the bomb at about 1,200 feet.

Javier immediately yanked the stick back with all his might to pull the jet out of the dive. His G suit inflated as the force of five Gs from the pull-up crushed at his stomach and chest. He had no time to look back to see where the bomb struck. In the next second, he jinked the jet sharply to the left, then to the right to simulate avoiding antiaircraft fire on the ground. Next he simulated releasing chaff and flares to deceive surface-to-air missiles. He then pushed the WEAPONS SELECT switch on the stick down with his right thumb so the Hornet would now fire Sidewinder missiles. New symbols and markers flashed on the radar on his right video screen. The computer had instantly switched his weapons system to air-to-air combat in case he had to dogfight his way out of the target area.

Javier's bomb had struck nine seconds ahead of time, just within the ten-second window, but it had landed about eighty feet short of the bull's-eye. Auto-bombing, Javier knew, didn't eliminate all errors. If he didn't keep the diamond exactly on the target he wanted to hit, the bomb would be thrown off. If he jerked the plane around during the dive, the computer often couldn't adjust to the changes quickly enough to reset its sights. If his eyes weren't in the correct position to view the glass heads-up display on top of his control panel, it could throw off the aim. (He fixed that by adjusting his seat up or down so his eyeballs were at the right angle from the HUD.)

Javier finally banked the jet left and headed for the next control point west of the Chocolate Mountains. He had only minutes before Tinkle would be radioing with instructions for the next bombing run.

"Nightmare you're cleared hot, cleared hot."

"Nightmare break right, break right, missiles right there."

"Nightmare two's popping."

"Three abort! Hard left!"

"Coming hard."

"Three, those are friendlies!"

"Falcon, from down here on the ground. What are you doing up there? You almost took us out!"

The hysterics from Joe Bags never stopped. The four young pilots couldn't hear themselves think.

For the second run they had to hit a taxiway at the south end of one of the runways. That became far more difficult for them to spot than the center of the X.

The run seemed jinxed. Cracker struck on time but his bomb sailed an embarrassing 300 feet away from the target, kicking up dust on the bare desert. The release for Jumpin's bomb malfunctioned so the ordnance never dropped. Javier, who thought he was lined up to hit a perfect bull's-eye, was ordered to abort the run because Tinkle and Joe Bags didn't spot him quickly enough coming into the dive. The instructors had to make sure they saw each plane clearly in the dive to prevent accidents.

Only Roadkill had a strike even close to the target. His bomb landed fifty feet away. But it struck twelve seconds late, outside the limit the instructors set.

Roadkill banked the jet steeply to the right and pointed it back to Chevy, the control point for their third rendezvous. He climbed to 7,000 feet and relaxed for a second.

But something felt funny, he thought. Something felt cool.

Strange, he thought. The cockpit normally was an oven in this weather.

Something on his butt.

He looked down on his left side.

A white glob of ice had formed around one of the cords that connected to his oxygen tube. The frozen cord had been touching his thigh. It was leaking liquid oxygen, which the plane converted to a gas for him to breathe.

Roadkill cursed to himself. At this low altitude he could take the oxygen mask off and breathe the air that the air-conditioner pumped in. But he knew he still had to go home. He couldn't fly with liquid oxygen spewing into the cockpit. It might combust.

Roadkill never worried that his accident-prone luck would catch up with him in the cockpit. He didn't take reckless risks in the air and he wasn't about to this time. He was fatalistic about flying the Hornet, though. Piloting combat jets was dangerous work, more dangerous than outsiders ever realized. So take every precaution, then if something happens it happens. Nothing you can do about it.

Donna had a hard time accepting that attitude. She worried constantly. Every time there was a jet accident, Donna would become moody with worry. They would talk endlessly about it. You take a chance walking out of your house, climbing into a car, and driving on the highway, Curtis would tell her again and again. "This is what I love doing." Donna would eventually come around. But then there would be another aircraft accident that would trigger the gnawing fear once again.

Roadkill radioed Jumpin', their flight leader, who passed along the report to Tinkle that his cord was leaking liquid oxygen and he would have to fly home.

"Hey Roadkill, Jumpin', you need any help getting back?" Gelardi radioed Carroll.

"Naw, I'm good," Roadkill said, dejected. He pointed the jet south toward El Centro.

The remaining jets circled again over control point Chevy. It was Cracker's turn to lead the third strike in.

"Is everyone ready?" he asked the other pilots on the back radio the instructors weren't tuned to.

No answer.

"I understand everyone is ready," Cracker repeated.

The others were still busy with their computers.

"Okay, that must mean everybody is ready," Cracker said. If they weren't, they would have been shouting "no, no, no" over the back radio.

Cracker flipped to Tinkle's frequency.

"Falcon, Roman five one, mission one tac three checking in," Cracker radioed.

"Roger," Falcon answered. "Stand by to copy nine line."

Cracker flipped to a blank chart on his kneepad and got ready to write. This run wouldn't be preplanned as the other two had. The information on the target and how to fly to it would all be new to simulate a spur-of-the-moment attack. It was called a "nine line" because there were nine pieces of critical information the pilots needed for the bombing run, which they filled out on nine lines on their knee charts.

"Roman, this is Falcon. Nine line," Tinkle finally radioed.

Cracker looked down at his kneepad to begin writing. His seat again tilted to the left as the autopilot kept the Hornet in a wide circle over the control point.

Tinkle started reading off the instructions: "November, two two seven, Romeo, eight point five."

Tinkle paused for a second to let Cracker scribble and digest what he had dictated. "November" was the initial point for the run, a small hill just south of Surveyors Pass. "Two two seven" was 227 degrees on the compass, the heading his jet should take from initial point November to the final target. "Romeo" meant that as he approached the target, he was to make his hard bank to the right so that he could spot it with his eyes for the dive-bombing.

Tinkle continued: "Six hundred, airfield, north three one eight four nine, west one one five two four two eight."

He paused again to let Cracker write. "Six hundred" meant the target was 600 feet above sea level. It was the same airfield they had bombed before. The string of numbers after that were the latitude and longitude coordinates for the spot where Tinkle wanted the bombs to land.

Cracker scratched down the last of the numbers and took a breath.

Taylor continued: "No mark, friendlies two kilometers east, egress south to Pontiac."

No one on the ground would set off a smoke grenade to mark the target for the pilots. Friendly forces were dug in two kilometers to the east. After bombing, the jets should fly southeast and circle over a gravel pit that was control point Pontiac.

Cracker finished and waited for the final piece of information, the time Tinkle wanted the jets to bomb the target.

"TOT, zero four," Tinkle finally said. He wanted them to strike at 1:04 P.M.

Cracker looked at his watch. That was about seven minutes away. If he didn't think the pilots could make all their computations and fly their jets to the target in seven minutes he was supposed to radio back, "negative zero four" and wait for Tinkle to come up with a new deadline. But in this case there was enough time, Cracker calculated.

"Roger, zero four," he radioed back.

Cracker began scribbling down the computations to come up with his push time.

"Everyone concurs with lead's push time at thirteen oh two," Jumpin' said over the radio several minutes later. Jumpin' was always faster with the math. He had come up with Cracker's time to leave Chevy before Cracker had and had radioed it as a backup to what Cracker would calculate.

"Thanks," Cracker radioed back gratefully. He began punching the sequencing buttons on the up-front control panel to set up the autopilot for the bombing run.

"Bingo, bingo," a soft, sexy female voice chirped in his radio earpiece. It was "Bitching Betty."

Instead of irritating beeps or buzzers, the Hornet's computers warned the pilot with a calm feminine voice that he needed to pay attention to a reading in the cockpit or that something wrong was about to happen. Aviators called the warning alarm Bitching Betty and in this case she was telling Cracker the jet's tanks had only 6,000 pounds of fuel left. So conserve gas.

"Roman, this is Falcon," Tinkle interrupted. "Amplifying target information."

"Go," Cracker radioed back.

"Dogbreath helped out with some of the artillery pieces," Tinkle said. "I've got one tank and an artillery piece still east of the intersection one quarter up the runway length. The large group of tanks, linked from east to west. We need you to attack the furthest west tanks in that column."

"Roger," Cracker answered, copying down a few notes. Jesus Christ, he thought to himself. Tinkle wanted him to put the bomb on a dime practically. From 3,000 feet up he was supposed to pick out from all the scrub brush below one tank and one artillery piece that were on the west side of the runways. And they'd have only a few seconds rolling into the target to spot the tank and gun.

"Roman, this is Falcon," Tinkle interrupted again. "Need you to expend the rest of your ordnance on this run." They were running out of time. Another group of planes needed to use the bombing range, so they would make just three runs instead of four.

"Roger," Cracker answered and relayed the information to Ball and Gelardi.

Even though he wasn't exactly sure which tank he was supposed to be hitting, Cracker had the best drop of the day. A bull's-eye and only one second early.

Jumpin' followed Cracker in the run this time. His computer told him he had to speed up quickly if he wanted to make the target on time. But if Jumpin' gunned the engine he would burn up too much fuel and risk not having enough to comfortably get back to El Centro. The easiest way to pick up speed without stepping on the gas was to fall like in a roller coaster from the 7,000 feet he was at, to about 3,000 feet.

Jumpin' pushed the stick forward. The nose dipped down and the jet accelerated. Just as if he was riding down a roller coaster he suddenly felt light in his seat. The quick dive had the effect of creating zero gravity in his cockpit.

Jumpin' didn't notice his black canvas navigation bag, no bigger than a lunch box, begin to float up from his right side, then drift back over his right shoulder. The bag was stuffed with manuals and checklist books he might need during the flight.

Jumpin' was completely absorbed with the run-in now, scanning his heads-up display to make sure the carrot was centered under the speed box so he would be on time, shifting his stick ever so slightly to make sure the plane symbol in his center screen stayed on the dotted line, the flight path the computer had programmed to get him to the airfield. He made a quick check of his right video screen and quickly punched buttons so the computer would drop two bombs this time instead of one.

"Roman five two approaching," Jumpin' radioed hurriedly to Tinkle when he was about four miles from the airfield.

"Emerald, five two," Tinkle radioed back. Continue.

Jumpin' could make out the airfield ahead, but not the tank he was supposed to hit on its western side. He assumed he'd finally be able to spot the vehicle in the few seconds after he rolled in for the dive.

Jumpin' banked the jet quickly to the right to begin the roll-in, then tipped it upside down and to the left. The nose pointed down and as he began the dive he turned the plane right side up. The heads-up display flashed the diamond he now placed over a tank. The vertical steering line glided to the side of the bull's-eye. Shifting the stick with his right hand, Jumpin' lined up the plane symbol along the steering line.

"Five two's in," he quickly radioed to Tinkle.

"Cleared hot, two, clear hot!" Tinkle radioed back just as fast.

Jumpin' pressed the pickle button with his right thumb.

One second.

The HUD light flashed. The computer had dropped the bomb.

Jumpin' took a quick look at the altitude indicator in the upper right corner of the HUD. One thousand four hundred feet.

He began to pull his stick back to come out of his dive.

But it wouldn't move.

Jumpin' tried to yank the stick back with all his might.

It budged only slightly. It seemed stuck. He couldn't pull the plane out of the dive.

"Flight controls, flight controls," Bitching Betty intoned sweetly. The computer also sensed that something was wrong.

A fiery hot flash coursed through Jumpin's body.

Time compressed.

The next three seconds seemed like three hours to him.

His eyes darted to the video screens. They had nothing to enlighten him on why he couldn't pull the stick back.

The jet was barreling toward the ground at 500 miles per hour.

The altitude reading dropped and dropped. One thousand four hundred feet, 1,300 feet, 1,200 feet.

Instinctively, Jumpin's left hand inched its way quickly off the throttle, then over his lap to near the black-and-yellow-striped canvas ring in front of his crotch. The ejection seat handle.

One thought kept shooting through his mind like a bullet: I'm going to have to eject. I'm going to have to eject.

If the nose didn't pull up before about 400 feet he'd have to eject or there'd be no time to escape before the Hornet crashed.

He tugged harder and harder on the stick.

Jumpin' braced his shoulders back on the seat, tilted his chin up so his spine would be as straight as possible for the ejection.

But the last thing he wanted to do was punch out of the Hornet.

Ejections could be brutal on the body. Nylon straps wrapped around his thighs and ankles would jerk his legs back. But his arms could easily flail and be broken as his seat rocketed up. Neck and back injuries were common from the sixteen Gs when the seat rockets blasted. The faster a pilot flew the more hazardous an ejection became because the onrushing wind slammed into him like a truck. At 500 miles an hour this low to the ground Navy studies showed that Jumpin' stood a two-out-of-three chance of crushing several bones. One ejecting pilot was decapitated at this speed because his chin strap had been loose and the blast of wind caught under his helmet, ripping it off. Jumpin' in a flash wished he had had time to stow away his kneepad; it could become dangerous shrapnel in an ejection.

Eleven hundred feet.

I can't believe I'm going to have to do this, Jumpin' thought.

One thousand feet.

Jumpin' pulled with every ounce of strength he had.

Gradually, the stick moved back.

The nose of the Hornet began to jerk up like a speedboat being bounced up by crashing waves.

Jumpin' immediately moved his left hand away from the ejection handle and back to the throttle. That had been the closest he had ever come in his flying career to reaching for the handle.

The muscles in his clammy body relaxed. His throat felt dry as sandpaper. The numbers on his altimeter box had stopped ticking down. The jet had leveled out of the dive and now in fact was gaining altitude, albeit at a far slower rate than if something hadn't been snagging his stick.

The jet climbed laboriously to 6,000 feet just south of the airfield. Jumpin' checked all the caution lights on the cockpit controls. Nothing flashing. He quickly punched up the computer page on his left screen that would show if anything was wrong with the complex machines in the jet that moved the elevators on his rear wings to push the plane's nose up or down. If there was any malfunction, little green Xs would appear in a column of boxes. The computers monitored every widget in the aircraft.

There were no Xs. Something else was binding the flight controls.

"Tinkle, Jumpin'," Gelardi radioed somberly.

"Go ahead, Jumpin'," Tinkle answered.

"Yeah, I'm up at ten miles south of the area now, climbing out," Jumpin' began almost matter-of-factly. Rule one on combat flying. Never sound panicky on the radio. "Apparently I've got some binding flight controls. That's what it feels like. I don't have any Xs or anything. The aft stick authority is severely degraded. I had some problems pulling off the target there."

The radios went silent. The excess chatter stopped. Everyone knew immediately that this was serious. Jumpin' had escaped a smoking hole in the ground. But he still had to land the jet.

"Okay, copy that," Taylor acknowledged. "We're going to stay on this frequency with you, Jumpin'."

"Okay," Jumpin' said curtly.

"Stand by for further instructions," Taylor said.

They decided to switch to a little-used radio frequency so Jumpin' and the instructors could talk without interruptions from ground controllers.

Jumpin' began taking an inventory of his cockpit. If the machines the stick commanded weren't malfunctioning, it must be something in the cockpit that had jammed it. He began on his left side and started looking up and down for anything missing, anything not where it should be.

Nothing in the left part of the cockpit or the center.

Jumpin' turned to his right. He looked down.

And froze.

"Ahhhh, shit!" he shouted to himself.

The small ledge where he had placed his black navigation bag was empty. No bag. Most of the jets had a strip of Velcro on that ledge. The navigation bag's Velcro could be stuck to it. But the strip had been missing in this plane. So Jumpin' had wedged the bag into the ledge thinking it would stay put. But apparently when Jumpin' dove the plane from the control point, creating the effect of zero gravity in his cockpit, the bag had come loose and floated up. Jumpin' was flying in one of the two-seat Hornets that could have an instructor in the back.

The back seat was empty this time. Jumpin' guessed that the bag floated up and back over his right shoulder, then plopped back in the rear seat area when he came out of the earlier dive. The bag had to now be wedged between the front edge of the seat and the stick in the rear cockpit. That was jamming the rear stick, which in turn jammed the front stick because the two operated in tandem.

Jumpin' felt somewhat relieved. He now knew what the problem was. He didn't have to worry that the flight control machines or computers had gone haywire and that even now they might throw his plane into a wild spin. He knew what he was dealing with.

But he also felt a little silly. His loose bag had caused all these problems.

"Uh, I believe I have an idea what the problem is now," Jumpin' radioed Tinkle and Joe Bags sheepishly. "My nav bag is nowhere to be found and I have stick authority problems, so I'll bet you it's in the back seat somehow."

The cramped cockpit made it impossible for Jumpin' to un-buckle his harnesses and climb into the rear seat to dislodge the bag. He would have to land the jet with the bag remaining where it was.

"I'm going to do a controllability check out here," Jumpin' ra-dioed to Tinkle and Joe Bags.

Jumpin' pulled the stick back as much as he could. The jet climbed slowly to 15,000 feet. He then banked it into a gentle right turn so it flew west toward the Salton Sea. A controllability check, in effect, was a practice landing high in the sky to see if there would be any problems when he actually tried to land. Better to find them out now at 15,000 feet when there was plenty of airspace to recover than when he was about to touch down when there was no room for error. (If he couldn't recover now from the check, at least the jet was over water so no one would be injured on the ground if it crashed.)

Jumpin' unstrapped his kneepad and stowed it away. If he had to eject again, at least that wouldn't be in the way. He lowered the landing gear. It came down easily. He put the wing flaps halfway down, which would give him slightly more lift so the aircraft would glide better for a smoother landing. No problems there. With his right thumb he manipulated the trim switch on top of the stick.

The trim worked fine. Finally he slowed the jet and dipped the nose to simulate going in for a landing.

After descending about a thousand feet, Jumpin' pulled the stick back and shoved the throttle forward to full afterburner. In the real landing, there was always the chance that something might happen just before touchdown and he would have to pull the jet up again, fly around the airfield, and make another try. Jumpin' had to know if he could lift the plane up for a second try or if his first landing attempt would be his only one.

The stick would move back. Not much. But enough to wave off a landing if he had to.

"Tinkle, this is Jumpin'," Gelardi radioed finally.

"Go ahead," Tinkle replied anxiously.

"I had no problems here with the dirty up," Jumpin' said. "Dirty up" was aviation slang for lowering the landing gear for a touchdown.

Jumpin' banked the jet left and headed south toward El Centro. He looked down at the map display between his legs and plotted a route back that avoided populated areas on the ground—just in case.

Jumpin' switched frequencies and radioed the air station's control tower that he wasn't fully in control of his jet. He would fly a straight-in, emergency landing. Nothing fancy.

The tower's air traffic controllers went bananas, or at least that's what Jumpin' thought. To clear a path for him, they ordered planes off all the runways and the jets nearing the airfield to fly into holding patterns. An alarm went out and firemen rushed to crash trucks.

Commander Bill Gortney, the forty-one-year-old leader of VFA-106, stood hunched over a radio set in the squadron's makeshift ready room back at the El Centro air station. The set had been monitoring the frequency Jumpin' used to talk to the tower. Gortney had just happened to be standing near it several minutes earlier chatting with other pilots in the squadron. As soon as Gortney had heard the words "controllability problems" and "emergency landing," he knew one of his jets and its pilot were in trouble.

Gortney could feel his stomach tightening. Accidents were a squadron skipper's worst nightmare outside of combat casualties. A seasoned pilot who had been flying combat jets since 1978,

Gortney looked like Robert Conrad. He was short and wiry, always busy like a terrier, who kept his fingers in every aspect of the squadron's operation. His call sign was Shortney.

In this case, Gortney planned to make sure that Gelardi would only have to land once. He grabbed the radio set's microphone.

"Jumpin', this is Shortney," he said, keying the mike. "Take a trap on this one."

Stretched across the runway at El Centro was a thick steel arresting cable just like the ones on aircraft carriers. A pilot who wasn't sure he could bring his jet to a stop on his own could lower the tailhook and have the cable snag him. In Jumpin's case, a trapped landing meant he wouldn't have to worry about pulling the jet back up for a second attempt.

Jumpin' could see the El Centro airfield straight ahead. He lowered the flaps halfway and began a slow, gradual descent. Takeoffs and landings always could be dangerous. As the jet slowed in its approach to land on the runway, it neared stall speed when it was the least controllable. The pilot had to carefully watch the pitch of the plane so its nose didn't move up or down too much and send him crashing into concrete. The last thing any pilot needed in a landing was a balky stick. Still, Jumpin's nerves had calmed. He now felt in command of the problem.

Three yellow crash trucks raced up to the end of the runway and stopped, engines running, waiting for Jumpin' to land.

Jumpin' flipped up the landing checklist on his right computer screen and began turning switches for the final touchdown.

Steady, steady, he said to himself. This had to be a smooth landing. No big corrections as he prepared to touch down. Not with this stick.

He pushed the nose down slowly. The Hornet glided smoothly to the beginning of the runway. Engines roaring, the wheels touched down just ahead of the arresting cable.

Jumpin' felt himself being jerked forward. But the trap didn't seem nearly as violent as on a carrier.

He powered down the engines. The three crash trucks raced up to him and came to a screeching stop.

Jumpin' gave the drivers a thumbs-up from the cockpit. He was fine. He began taxiing to the hangars, the crash trucks driving alongside him as escorts.

Jumpin' pulled the F/A-18 into a parking space in front of the squadron hangar. He shut down the engines and popped open the cockpit canopy.

Maintenance crewmen began clambering up the jet's ladder to begin checking what happened to the controls in the back seat.

Jumpin' wrenched himself around.

"Don't touch anything until I get back there!" he shouted at them.

The crewmen backed off until he could wrestle out of the seat harnesses.

Jumpin' hoisted himself out of the front cockpit, bounded onto the wing, and stepped back to the rear seat. He bent over and looked inside.

Sure enough, his navigation bag was wedged between the stick and the rim of the seat.

Jumpin' shook his head and smiled. What a tragedy it would have been to lose an expensive jet over such a minor mishap.

The paperwork to check back in the plane with the maintenance crew took longer this time because Jumpin' had to write an explanation for his emergency landing. Twenty minutes later he finally walked back into the squadron ready room cradling his helmet under his arm. His parachute harness and survival vest hung unzipped and loose from his shoulders. Sweat stains dotted his flight suit, his crew cut hair was matted and sticky.

Jumpin' made a beeline for the refrigerator in the ready room. He grabbed a Gatorade bottle and began sucking down the cool green liquid.

Gortney walked up to him. The skipper had already telephoned the maintenance shop and ordered that, along with fresh strips of Velcro, hooks be installed in all the cockpits to latch down the nav bags.

"Three teenies for making me wait so long before you radioed that you had landed safely," he growled, then let a slight smile creep across his face.

The pilots gave each other teenies, a nickname for points assessed, for little mistakes made during the two-week detachment. They were recorded on a blackboard in the ready room. At the end of the detachment the teenies were totaled up for each pilot to determine how much he would pay for a squadron party. If a pilot complained about a teenie, he could receive even more for being a whiner.*

Jumpin' chuckled and marked them on the board by his name.

The Mirage

* Even I got a teenie, for nodding off during a debriefing after a flight. It was the pills I took for air sickness, I protested. The twenty-five-milligram tablet of promethazine I gulped down before each flight to fight the nausea made me sleepy. The amphetamine I swallowed to counteract it lasted only the hour during the flight. The promethazine was good for six hours, which meant that for five hours after the flight I was zonked with only the downer in my system, I tried to explain. The pilots just laughed and told me to quit griping or they'd slap me with another teenie.

CHAPTER SIXTEEN

The Mirage

COMMANDER Jack "Hooter" Holt sat on a bar stool and leaned on a high cocktail table sipping beer from a plastic cup. It was Friday night at the Mirage Club. Corona night. Mini-bottles of the specialty beer sold for ninety-six cents apiece. The Mirage Club's bar was spacious. Its walls were all painted black like a disco club. The usual assortment of squadron plaques hung on them along with NFL posters and lighted plastic beer signs. Large inflatable plastic footballs, helmets, and a Budweiser can with goal posts sticking out of it hung from the ceiling. Perched overhead from blackened posts supporting the roof were television sets flashing MTV rock videos. A raised disc jockey booth stood at the back wall opposite the long bar at the other end, between them a checkered red and white dance floor. Two worn pool tables sat to the side along with a dart board.

The Mirage was an "all-hands" club on the El Centro Naval Air Facility, open to both enlisted sailors and officers. Every other night that week it had been nearly empty, only a few sailors drifting in occasionally for a beer, along with Hooter and the other pilots from the VFA-106 squadron who were temporarily deployed here on the detachment.

But the club had been advertising Corona night for several weeks. Finally the bar was filled with sailors who worked on the base, pilots from the 106 squadron and others temporarily deployed at the station, along with clusters of young women from the towns off base or just across the border in Mexico. The heads of the women swiveled around constantly like submarine periscopes hunting for male prospects. The lure of ninety-six-cent Coronas, even though the bottle held only ninety-six cents' worth of beer, was enough to pack the bar. The DJ played hard rock and rap CDs at ear-splitting decibels. A nearby beer distributor had trucked in the Coronas along with two Corona girls—one blond, the other auburn-haired, all gelled in frizzes, with thick makeup, navy blue minidresses painted on, and breasts stacked up with cleavages as long as the San Andreas fault. They roamed the crowd like bookends, painting rub-off tattoos on anyone willing to produce a bare arm.

Corona night was also billed as "Attitude Adjustment Night." It had all the ambiance of a prison Christmas party. Outside the Mirage, base policemen in bright orange vests waving flashlights stood guard at every entrance into the parking lot ordering cars to spots. At the club's entranceway, a security detail demanded identification cards from every person walking in, even middle-aged captains and commanders, to confirm they were old enough to drink. Those who were had plastic hospital bands snapped around their wrists. Inside the bar, a dozen grim guards wearing red T-shirts and radio earpieces patrolled the crowd like Iranian mullahs ready to pounce on anyone having too good a time and getting out of line.

The revenge of Tailhook. The Navy had never been adept at reacting subtly to complex social problems in its ranks. The service had dealt with the raucous lifestyle aviators led five years ago by taking scrub brushes and lye soap to it. The Miramar officers' club on the West Coast in California, where Top Gun pilots once raised hell in the Tom Cruise days, was now a ghost town at happy hour. The

Oceana O' Club on the East Coast in Virginia had long cleared out the four o'clock strippers who performed each Wednesday and Friday.

The Mirage Club now had more rules than a seventeenth-century Salem Congregational Church. A sailor walking in with anything sexually suggestive printed on his T-shirt would be immediately ordered out. All week, Hooter and the other squadron pilots had been sneaking in aviator dice games at the bar, like schoolkids pitching pennies behind the library. The complicated dice games, played with a cup, were Horse and Ship's Captain and Crew. The pilots had been betting dollars on the rounds until the ayatollahs decreed this was gambling and must stop. They could only play for beers. The pilots played for money anyway, hiding the dollars under the table. Two nights earlier, several of the squadron's aviators, just back from night flights, famished and thirsty, had to eat pizzas they had ordered from Domino's on the lawn outside the club, because it barred food being brought in to eat with the beer. As they sat on the grass munching slices, a group of young women walked into the Mirage to celebrate one of the girls turning twenty-one that day. The birthday girl was turned away. Her high heels didn't have straps on the back; club rules prohibited women inside with shoes minus straps on their heels. One of the pilots lent her his flight boots. She clump-clumped back into the club.

"None of the clubs are like they used to be," Hooter said, taking slow sips from his cup. The clubs used to be packed every night with aviators unwinding from a day of hard flying along with flirting girls and loud music and cheap beer. "Now they're all closed or gone to seed. An institution of military life is vanishing," Hooter continued. "Tailhook was a seminal event in U.S. Naval history. The pendulum has swung too far toward Puritanism. It's having a helluva effect on the poor women in the squadrons. They feel isolated." The PC Nazis have taken over.

Hooter was about to assume command of a Hornet squadron and had been taking refresher training with VFA-106. He had enlisted in 1974, part of the post-Vietnam generation whose social permissiveness was now outlawed. At forty years old, he had a young face with short brown hair well groomed that fell across his forehead. He looked more like a Wall Street stockbroker than a Navy fighter pilot. The new aviators in the squadron thought he walked on water. He was a good stick and a natural leader, as far as they were concerned. But Hooter's last boss had dinged him on a fitness report. "You're one of the dinosaurs," the boss had told him. "Your time has passed." Hooter's crime: becoming too festive at a New Year's Eve party.

Structural changes killed the clubs as well. Attendance began dropping at most officers' clubs, so to save money the Navy combined the officers' facilities with the enlisted clubs to form "all-hands" clubs that both officers and sailors could patronize. But an officer had to be constantly on guard in an all-hands club, never knowing if the person in civilian clothes with whom he was striking up a conversation was really enlisted. Officers were forbidden from fraternizing with sailors. Even worse if the person in civilian clothes was an enlisted female. That would end a career. An officer could never really be relaxed in an all-hands club. Some aviators would never walk into one unless they wore their flight suits with the rank clearly displayed to warn off enlisted women.

"The clubs used to be the center of an aviator's social universe," Hooter said. "But now they've got the Shore Patrol stationed outside the clubs waiting to give pilots Breathalyzer tests so they can charge them with drunk driving. The pilots don't feel welcome anymore, so they go to bars off base where they feel freer to have fun and don't have to worry about offending anybody."

The Corona girls were busy pasting tattoos on arms. Some of the squadron pilots placed a private bet among themselves. First guy to get the two Corona girls to walk into his room at the base's Bachelor Officers Quarters that night would win $100. He didn't have to do anything with them. Just talk them into walking across the threshold.

The Corona girls seemed to have heard every pickup line in the book, however. The pilots got nowhere.

They tried another maneuver: talking the Corona girls into driving back to the base the next day to pose with the squadron in front of an F/A-18. They'd negotiate what the Corona girls would or wouldn't wear when they got there.

The Corona girls said they would think about it and gave the pilots their phone number.

The aviators tried it the next day, but got only an answering machine that said in a breathless voice: "You can leave an oral or verbal message." The pilots didn't know if this was sexual innuendo, or the Corona girls not quite grasping the grammatical concept of a redundant sentence.

Hooter loved the Navy, its traditions, its culture, the bonding of the squadron, the banter, the jokes, the sharing of danger. That was why the changes made him so sad. "The Navy no longer lets you make one mistake," Hooter tried to explain carefully. "In the old days, a young pilot could screw up and his squadron CO would take care of him. It wasn't a cover-up. It was just looking out for your own. Today a young pilot makes a mistake and the first instinct of the service is to protect the system. The pilot is out.

"That's why pilots are leaving in droves. There's no camaraderie to fall back on, no sense of identification with the group, no feeling that the command will look out for you if the going gets tough. All he has is his plane and the flying and nothing else. Hell, the airlines offer about as much camaraderie as the squadrons now. So why shouldn't he pick the airlines where there's less stress on his life.

"We need people with stars on their shoulders to finally say enough is enough. The Navy is heading in the right direction. Tailhook is over. I'm laying down my stars. The junior officers are desperately looking for leadership, someone to hang their hats on, someone to follow. A John Wayne."

CHAPTER SEVENTEEN

Practice Bleeding

IT was like swimming in a bowl of milk. Phillip Clay looked outside the cockpit of his F/A-18 and saw nothing but a thick white haze. He had been launched at what the pilots called "pinkie time," the minutes just before sunset when the horizon lit up in reds and oranges. And now the carrier's air traffic controllers had ordered him to fly through the blanket of clouds over the ship. The strobe lights from his wings reflected off the dense fog, creating a flickering flash of light that made him feel as if he was almost dancing drunk on a disco floor. Worse still, he had lost his sense of where he was in the world.

The clinical name for his condition was vertigo. A disoriented state of mind. For a pilot it often occurred while flying at night or through clouds. His sense of position became confused because his eyes could not distinguish earth from sky outside the cockpit and the nerve endings in his ears did not give him a seat-of-the-pants feel for which way his body tilted.

The condition could prove deadly. Pilots flying through clouds sometimes felt they were soaring up into the sky when their planes actually were plummeting toward the ground. In Phil Clay's milk bowl, he could swear that his jet was making a hard bank to the left. He leaned to the right like a bike rider to compensate. In reality, however, the Hornet was flying straight and level.

Clay began to feel clammy. Sweat soaked the collar of his flight suit. He felt hot fear. This was his first bout with vertigo. The pulsating strobe light began making him dizzy. I know I'm in a left turn, I know it, he kept thinking to himself, becoming angrier that the controllers had placed him in this soup. The displays on his cockpit computer screens told him otherwise. He began to stare at the electronic indicators that showed his jet was not tilted and realized quickly it was dangerous to fixate on one aspect of his flight when he must monitor so many other computer readings from the Hornet. Fixation could get him killed.

"Screw it," Clay finally said to himself. He quickly punched buttons on the up-front controls and his throttle in order to kick in the autopilot. With vertigo, he was better off letting the computer take over the plane or he would end up flying himself into the ocean.

Clay had more important things to worry about. He was soaring over the ink black Atlantic Ocean, minutes away from landing on a carrier at night. Then he would take off for the first time at night. They were a Navy jet pilot's most difficult and dangerous maneuvers. Veteran aviators called it "practice bleeding." A night of cat shots and traps could leave them wobbly-kneed and exhausted. Pilots loved the thrill of catapulting and trapping off the steel postage stamp during the day. They dreaded it at night. Crossed chicken bones would be hung over the front of the ready room television set that broadcast flight deck operations—a maudlin joke to hex the horrors of the night.

But it was a skill Clay had to master. America's armed forces on the ground, in the air, and at sea waged war at night, when darkness masked the attackers and a sleepy enemy was less prepared to defend. Clay had to be able to fight with his Hornet at night, then land in a black void where the sea and horizon blended and only faint dots of light like far-off stars gave any hint of where the ship was.

Qualifying in night carrier landings and catapults was the final rite of passage for Clay before shipping out to the fleet as a carrier pilot, the scary culmination of three years of training. He had just completed air combat maneuver training, much of which consisted of mock aerial dogfights over Key West, Florida, where his Hornet was pitted against special adversary squadrons flying complicated maneuvers that foreign air forces used. He learned to fight from afar with secret radar tactics for long-range missiles, then battle up close with eight or more jets in the fray. It was the best flying he had ever experienced in the Navy, dizzyingly complex aerial chess games with the pieces moving hundreds of miles per hour. But he still had to learn how to go back and forth to work on the night shift.

For the past month, Clay had been practicing night carrier landings on the ground strip at Cecil Field, Florida, with the VFA-106 squadron. Now he had to do it for real. The world's other navies that had aircraft carriers allowed their pilots to fly for years before they attempted night landings. The U.S. Navy threw its aviators into the dark waters when they were rookies. In operational squadrons, nugget fliers had to land and take off at night far more times than the veterans so they would gain experience quickly. The more seasoned a pilot became at night, the more he realized its dangers and the less eager he was to catapult and trap then. Senior officers would try to schedule their night launches only when a bright moon was out or stars lit up the sky so a horizon and the ship could more easily be picked out. These were called "commanders' nights."

Tonight was no such night. The chill of late October had set in along the Atlantic seaboard and the week had been rainy. Lightning flashed from thick threatening clouds on the horizon when Clay had catapulted off. A layer just below 5,000 feet now blanketed the skies over the carrier while the wind began to whip up to almost forty miles per hour. No, this was a "varsity night," the pilot's euphemism for dangerous flying conditions.

Clay had been ordered to wait his turn to land by hovering in what was called a "marshal stack," twenty-three miles northwest of the carrier. Jets in the stack flew in circles one on top of the other at 1,000-foot intervals. To get his mind off the vertigo, Clay busied himself with chores in the cockpit as the autopilot flew the Hornet. He flipped on the moving map display on the center video screen between his legs and began selecting way points to direct him to land in case there might be some problem getting the jet aboard and he had to divert to an airfield. He checked his fuel state and punched in the level at which he wanted Bitching Betty to warn him that he was low. He fine-tuned the brightness of the left and right video screens. The left one repeated the picture of the heads-up display he saw in the concave glass perched atop his cockpit rim. The right screen displayed a sweeping radar that blinked with the location of the carrier. Next he checked the settings for the instrument landing system and automatic carrier landing system on his up-front control panel, then the altimeter settings to warn him if the jet flew too near the ocean.

"Roman three oh one, established angles five," Clay radioed the *JFK*. He was now set in his circle in the marshal stack. "Angles five" was shorthand for his altitude of 5,000 feet. The fear of vertigo had finally seeped out of his body.

The USS *John F. Kennedy* steamed slowly north about fifty miles off the North Carolina coast. The *Kennedy* was an aging carrier, whose keel was laid in 1964. It had once been a symbol of a young and vigorous president eager to project American power on the seven seas. In its wardrooms and galleys, photos still hung from the Kennedy family albums. "Any man who may be asked in this century what he did to make his life worthwhile," Kennedy, a former PT boat captain, once mused, "I think can respond with a good deal of pride and satisfaction: 'I served in the United States Navy.'" By the mid-1990s, the *JFK* was a step away from mothballs, spared only because defense budgets were so tight and carriers fewer, the Navy couldn't afford to retire it. The carrier would be decommissioned in 2007.

Phillip Zane Clay was not even born when Camelot ended. He had grown up in a poor neighborhood on the south side of Oklahoma City, the son of an oil roughneck who eventually set up his own business selling supplies to the thousands of rigs that dotted the state. Lee Clay managed to keep food on the table, but the business never thrived and he spent long hours on the road practically every day of his life to eke out a living. He still worked hard though he was eighty-one years old.

Phil resented the fact that his father toiled such long hours, that he was never there for ball games. Lee Clay was a God-fearing Christian man. He was a kind soul who treasured his family. He was a Good Samaritan who would always stop along the road to help a motorist with a flat.

Phil used to hate it. Nice guys finished at the bottom, which was where the family always seemed to be financially. Lee Clay's integrity and hard work had gotten him nowhere, as far as his son was concerned.

Oklahoma's public schools had been dismally poor, Phil's teachers mostly indifferent. He cared little about studying. His peers experimented with drugs. Tall and muscular, he became a high school football star. His only goal in life: play football in college, then maybe the pros. Phil enrolled in Southwestern Oklahoma State University seventy miles east of Oklahoma City and played defensive back. When he wasn't on the field, he drank beer, chased women, and skipped classes. He was becoming a big, dumb jock with a mean streak.

But Phillip Clay had been saved. With his father away so much, Phil's four older sisters had helped their mother raise him during childhood. They bossed him around, picked on him when he was little. They also came to have a stake in him succeeding. Phil was the only one of the five children to go to college. The sisters had all chipped in money so he could attend and now he was throwing away the opportunity. He would end up a football has-been working in a convenience store like many of his hometown friends.

Finally one of the sisters, Paula, jacked him up. A petite woman with a soft voice, Paula had an inner strength that could be overpowering. A paramedic, she would later join hundreds of rescue workers climbing through the rubble of the 1995 Oklahoma City bombing looking for survivors.

"Do you really think you're going to play football all your life!" Paula dressed him down one day, her eyes flaring. "No, you're going to end up in a dead-end, just like your friends. Or, you can educate yourself and open up doors for yourself."

The lecture finally began to sink in. Clay transferred to Oklahoma University and dropped football. He majored in aeronautical engineering, enrolled in ROTC. He was not the brightest student. But he made up for it through long hours of study and graduated with a B average. He became a gentle, caring person, with warm blue eyes and a soft manner. He now displayed the work ethic he had so resented as a child. Through flight school most of the other student pilots were brainier—or so he thought—but he drove himself hard in the cockpit and proved a natural with the stick. He performed at the top of his class. He became a favorite of his instructors, a class leader among fellow students.

He also fell in love, with Julie Talkington, a former rodeo cowgirl whom he had met at an air show the year before. Julie had become his oasis, his steady rock of support as he worked his way through the ultracompetitive flight schools. She filled a void he felt when he left his family in Oklahoma City to fly for the Navy. Now Phillip Clay wanted a family with Julie. He wanted children and Thanksgivings and Christmases with them. He wanted a home full of love and lives he could watch grow.

He wanted to be like his father, he finally realized. He had come to admire Lee Clay. They were now very close.

"Roman three oh one, left for final bearing zero three three," the radio voice from the *JFK*'s air operations center ordered Clay. The center referred to each plane by the number painted on its tail, which for Clay's aircraft was 301.

"Roman three oh one," Clay acknowledged curtly and banked the jet to the left so it flew to the right of the carrier and downwind toward the ship's rear.

The *JFK* may have been long in the tooth but its air operations center buried just below the flight deck was crammed with the latest technology to track aircraft and guide them to the carrier deck. Several rooms full of mainframe computers powered the center's Precision Approach Landing System. In another darkened room lit only by colored buttons and green-glowing video screens, air traffic controllers sat chattering quietly into radio headsets before five large, circular radar screens. At other control booths, sailors stared at computer videos that guided jets down their flight path toward the carrier deck. Officers sat on padded chairs behind them nervously watching the choreography.

Three air traffic controllers would guide down Clay's Hornet, one handing him off to the other as the flight progressed. A marshal controller first monitored the flight in the stack to make sure the jets stayed separated. When Clay was ready to begin his landing, he would be handed off to an approach controller, who would guide his plane out of the holding area and give him turn directions until he reached four to six miles from the back end of the carrier. Then the final controller took over.

The final controller had three ways he could guide the Hornet down. If the weather was so bad or the pilot so incapacitated that he couldn't land the jet himself, the center's computers could link up with the Hornet's computers and fly the aircraft down by autopilot. Clay would sit back and let the computers move the stick and throttles to land for what the operations center called a "mode one" approach. This was the last-resort option. No pilot wanted to trust his life to a machine landing him.

The aviators preferred the "mode two" approach. The automatic landing system computers on the carrier would constantly feed the Hornet information on the path the plane should take to make a perfect approach to the carrier. This was nicknamed the "needles." The carrier's computer would flash the needles flight path onto the Hornet's heads-up display, which the pilot would then use to stay on the correct landing path.

If the computers were cranky and not talking to each other, there was always the "mode three" approach. An air traffic controller simply talked the jet down. The final controller sat in front of a wide TV screen that had two thin dotted lines running parallel and diagonally down. The lines represented the correct flight path the plane should take as it approached the carrier deck. Radars constantly tracked Clay's jet, which would appear as a plane symbol on the screen. If he flew the correct flight path, the plane symbol stayed between the two lines. If the symbol drifted outside the lines, it meant Clay was off course and the controller would radio him directions to push the jet up or down, or turn left or right.

The final controller guided the plane down until it reached three quarters of a mile from the ship. At that point, the pilot could see the ball from the Fresnel lens on the carrier deck, and the landing signal officer on the ship's left platform radioed final approach instructions before the touchdown.

It all sounded complicated and it was. Tensions always ran high in the air operations center at night. Antacid pills were popped constantly. Voices snapped at one another. Moods were always testy. Dozens of planes buzzed around the carrier in the black sky like blind bats with only radars and computers and technicians watching flickering dots on screens to keep them from crashing into one another or the ocean. The slightest misstep—a controller's attention diverted, a handoff mishandled, one jet confused for another—could result in a catastrophic accident.

Clay had flown to about eight miles behind the ship when the approach controller radioed him to make a U-turn so he could begin lining up for his approach in. The several minutes it took him to reach the back of the carrier had given him time to collect his thoughts after his nightmare with vertigo. He kept the HUD image displayed on his left video screen. On the right screen he punched the gray Chiclets to bring up the flight control system page and began running tests to make sure the computer-controlled trim worked properly. It did. Then he lowered the landing gear along with the wing flaps. His jet was now 1,200 feet off the ocean. He reset his radar altimeter so the alarm would sound if he dropped below 400 feet. Then he decelerated to 150 miles per hour.

The lower speed was a relief. He needed to slow the world down so he didn't become overwhelmed with the tasks he now had to perform in the cockpit. He was like a motorist in rush hour traffic at night trying to read street signs and a map at the same time; the natural tendency was to ease off the accelerator to do it all.

"Roman three oh one, hook up this pass," the approach controller radioed. To qualify at night, Clay had to perform two touch-and-go passes, where his jet wheels just bounced on the carrier deck and he took off immediately, then six arrested landings. Tonight he would perform one of the touch-and-go's, then four traps. That would leave him exhausted enough for the first round. The second touch-and-go and the final two traps would be flown the second night.

"Three oh one," Clay acknowledged in shorthand. He punched up the attitude direction indicator page on his right video screen to display how his jet was tilted in the sky.

The Hornet's fancy computers and flight controls made it far easier to land than other planes in the Navy inventory. A training squadron of F-14 Tomcat fighters, with their 1970s-era technology, was also practicing night landings aboard the _JFK_. The new Tomcat pilots were constantly boltering or being waved off. Their huge jet was extremely difficult to land, especially in bad weather. Its cockpit was antique compared to the Hornet's, no advanced heads-up display or sophisticated computers to fly the plane and guide it down. The old jet was also difficult to control in the final moments before touchdown, its pilot having to constantly bob his head up to look outside and down into the cockpit to read gauges and dials. The Tomcat instructors expected many of the new pilots not to qualify at night. It often took several tries.

With all the state-of-the-art equipment in their plane, Clay and the other new Hornet aviators were expected to qualify on the first try. There were no excuses if they couldn't land the F/A-18.

It was almost 8:00 P.M. and the sun had disappeared from the horizon. The night now was black. Clay flew below the clouds but he could see nothing outside the cockpit except for a tiny white speck ahead of him that he guessed was lights from the aircraft carrier. As he neared six miles from the ship, a different radio voice—this one from the air operations center's final controller—took over his flight and began spitting out directions.

"Roman three oh one, lock on six miles, say needles," the voice commanded. "Say needles" was shorthand for asking Clay to radio when the Hornet's computer was locked on to the carrier's automatic landing system.

Clay punched buttons on the up-front control panel to activate the automatic carrier landing system in his plane. The system, "needles," tried to link up the carrier's landing computer with the one in his plane to guide him precisely down the flight path, or glideslope, to land on the carrier. A tiny circle with a short line protruding from its top flashed on his heads-up display. Pilots called it the "spermie." To stay on the correct flight path for landing, Clay had to maneuver the jet with the stick and throttle so that the plane symbol in his HUD eventually overlapped the spermie. Then as long as he kept the plane symbol over the spermie, his Hornet was flying the correct path to the carrier deck.

But it was easier said than done. The computers were having trouble talking to each other tonight. The spermie kept disappearing from his HUD screen. That was one reason pilots were never willing to depend only on the computers to fly the jet down. The automatic carrier landing system had a tendency to lock and unlock.

Irritated, Clay radioed the operations center. "Three oh one dropped lock." The spermie kept fading from his screen.

"Roman three oh one, fly bull's-eye," the operations center radioed back.

"Bull's-eye" was the nickname for the carrier's instrument landing system, a backup system not as accurate as the spermie. The bull's-eye reading on Clay's heads-up display consisted of two crossed lines that floated in different directions. To fly the correct flight path to the carrier, Clay had to steer the jet so the HUD's plane symbol eventually remained centered in the bull's-eye's crosshairs.

Even that became difficult. Clay felt as if he was driving on a dark bumpy road full of potholes. The wind outside manhandled his jet. He tried to move the stick lightly to make only slight corrections so the plane symbol remained on the crosshairs, but the symbol kept jumping about on the screen.

The jet lurched to the right. Even strapped tightly to his seat, Clay felt like he was being slung around the cockpit. He punched the Chiclets on his right video screen to bring up his inertial navigation system page. That would tell him how fast the winds were whipping up outside. The screen flashed thirty-four miles per hour. Within the parameters to land, but not great. The pilots ideally wanted twenty-four- to thirty-mile-an-hour winds blowing at them in order to provide the jet with the lift it needed to touch down on the carrier deck at a slow speed.

Five miles from the ship, Clay again tried to lock up the automatic carrier landing system. The spermie appeared briefly, then faded. He kept his eyes focused instead on the crosshairs of the bull's-eye, struggling to keep the plane symbol at its center. There was nothing to see outside, the white speck from the carrier lights was only slightly larger. Clay concentrated instead on the heads-up display, flying off the plane symbol and bull's-eye as he would in a video arcade.

At three miles, Clay could make out different lights on the ship. Tiny white dots formed a long rectangular box with a centerline of lights running down the middle of the box. The box, which at the moment looked no bigger than the nub of a pencil, was his landing strip. Another thin line of lights dropped down from the rear of the box. These were the drop lights that trailed off the back of the carrier to help the pilot line up his approach.

Clay tipped the nose of the jet down and began the gradual descent toward the carrier. Now he had to keep drifting down the glideslope at a 3 degree angle. He also had to make sure the jet remained tilted slightly up as it drifted down. He did so by keeping the plane symbol in his HUD also centered on a bracket to the side, which looked like an E.

Most of his course corrections he made with just tiny shifts of the stick in his right hand and the two throttle levers his left hand gripped. The closer he got to the carrier, the more delicate the throttle and stick movements became. He had to have a surgeon's touch at this point. The slightest slip could send him far off course, or even prove deadly.

The week before, Clay and the half dozen other Hornet aviators making their first night landings aboard the *JFK* had been put in a classroom and shown a videotape of gruesome crashes that had killed other pilots at night. It was surreal. Rock music blared in the video's background as planes careened on carrier decks in fiery explosions and hapless pilots ejected, futilely trying to escape death. Some of the other young aviators in the classroom had whistled and laughed and whooped it up at the horrifying crashes, as if they were watching hard tackles in a football game. False bravado, Clay realized. He didn't know why the trainers always made them look at those damn videos. They were supposed to be a slap in the face, to wake up the students to the fact that what they did was dangerous and they had to be careful. But Clay hated dwelling on death. He hated going to funerals. He hated watching stock car races because of the crashes. Julie had once seen the videos and like the other pilots' wives or girlfriends had found nothing in them to cheer about as the males always did. It had gotten so that Julie did not want Clay to tell her what he planned to do in an upcoming flight. She wanted to hear about the flight only after he had flown it and was safe on the ground.

At a mile and a half from the ship, the lit box appeared only slightly larger. Clay paid little attention to it. He was still concentrating on the images in his heads-up display.

"Roman three oh one, slightly above, slightly left," the final controller radioed. The controller could see on his screen that Clay's jet had strayed up and to the left of the parallel lines that represented the perfect glideslope to the carrier deck.

Clay pulled the throttles back slightly. The high winds tended to shove the plane down too far when he let off the power. He didn't want to overcorrect and find himself bouncing up and down through the glideslope.

"Roman three oh one, slightly above, on course, call the ball," the final controller radioed back. Clay was now three quarters of a mile from the carrier. His jet was still just above the correct glideslope but lined up properly to hit the center of the flight deck. At three quarters of a mile, the final controller turned the pilot over to a landing signal officer for the last twenty seconds of the flight.

Clay looked out the cockpit quickly. He could see the cross of the Fresnel lens on the left side of the carrier deck. More importantly, he could see the amber light shifting up and down on the center column of lights—the meatball the pilots used to guide their planes down during the final seconds. Clay was surprised how bright and clear the ball and the other landing lights appeared in the black night. The glowing drop lights down the ship's stern in fact made it easier for him to line up the Hornet and keep it from drifting right or left.

"Three oh one, Hornet ball, six point six," Clay radioed the carrier hurriedly. The message informed the carrier that he could see the meatball and his jet had 6,600 pounds of fuel left.

By now, Clay's eyes had left the heads-up display and were taking quick looks outside the cockpit. For the last twenty seconds of the flight, his eyes darted constantly from the meatball on the Fresnel lens (fly the plane so the amber light stays in the middle of the column) to the lineup lights (keep the plane pointed to the center-line down the runway) to the angle of attack indicators in the cockpit (make sure the jet is positioned as it drifts down so the tailhook in the back will catch an arresting cable). Meatball, lineup, angle of attack. Meatball, lineup, angle of attack.

Clay's eyes never stopped moving. They couldn't spend too long on one reading or the others would be neglected and the flight could be thrown dangerously off balance. If he fixated on the meatball too long he would even fail to catch its slight movements up and down that warned him his plane was too high or low. Instead, he had to sweep his eyes back and forth on the Fresnel lens's horizontal line of green lights so he could see better the up or down movements of the meatball.

"Roger ball, you're lined up a liiiiittle right," Lovey radioed him in a silky calm voice as if he was having him thread a needle.

"Lovey" was the call sign for Alex Howell, the landing signal officer on the carrier platform. Howell was Clay's instructor in the squadron. His call sign had a somewhat ridiculous lineage. He was named after the nickname the character Thurston Howell III gave his wife on the television show *Gilligan's Island.*

Lovey Howell was as young as Clay, but far more experienced in the cockpit. At twenty-seven, he was already a Top Gun pilot who had flown combat missions over Bosnia. The son of an Air Force jet jockey, Howell grew up in Europe. He spoke fluent French. By the time he had graduated from high school, he had logged more than a thousand hours flying aerobatics in biplanes. At nineteen, he joined the Navy and went directly to pilot school under a special aviation cadet program that allowed students to fly while they earned their college degree.

Lovey was a no-nonsense instructor. Short and muscular, he would wear a scowl that sent chills through students. But he had the flying skills to back up the game face. On the LSO platform Lovey was sparing and almost gentle in his radio communications to the young pilots learning to fly Hornets. At this point, they should be able to land a jet with less hand holding from the LSO, he believed. In wartime, the landing signal officer's radio transmissions could be intercepted by the enemy so the less said the better. Besides, at night a busy pilot couldn't afford to be distracted by excess chatter in the final seconds before touchdown.

"Power, come left," Lovey radioed Clay. His jet was less than a quarter mile away and Lovey could make out its shape from the fuselage and wing lights and see that Clay was flying slightly below the glideslope and still to the right.

From his vantage point, Clay could now finally see the dark gray carrier lit up from its deck lights. He tilted the wings slightly to the left to line up properly, then pulled the throttles back less than a millimeter as Lovey had commanded.

The back of the carrier deck now rushed to him as if it was popping out of a black curtain. Clay kept up his scan: meatball, lineup, angle of attack; meatball, lineup, angle of attack.

Stick and throttle movements became ever so slight. He had never concentrated on anything so intensely in his life. The margin of error now was unforgiving. To remain on the correct glideslope he had to fly this multiton hunk of steel through what amounted to a three-foot-square window. At touchdown, the margin of error was only nine inches.

"Power!" Lovey ordered with more edge to his voice. As Clay's Hornet neared the ramp of the carrier it was still slightly low.

Clay nudged the throttle forward ever so slightly.

The F/A-18 roared across the rear ramp. Thin white plumes of condensation trailed its wing tips like long ribbons.

Clay's eyes never left the meatball.

Suddenly he felt a jarring thump in his seat. The Hornet's wheels had touched down on the steel deck of the carrier.

In an instant he wished his hook had been down and this had been an arrested landing. It had been one of his better touchdowns. Lovey would give him a "fair" grade for it. But this was the first of two required touch-and-go's. Clay's hook was up.

He shoved the throttles forward. The exhaust outlets for the Hornet's two engines lit up like giant Roman candles. Through his peripheral vision he caught quick blurry glimpses whizzing by of lights, and gray jets, and a giant white 67 painted on the *Kennedy*'s tall carrier island, which had a half dozen radar dishes spinning atop it.

Clay pulled back the stick. The jet lifted up effortlessly off the carrier deck and into the dark sky ahead of the carrier.

"Roman three oh one, I'll turn you downwind momentarily," the approach controller radioed him several seconds later. Downwind in this case meant flying the opposite direction the carrier steamed so he could eventually line up on the ship's stern again.

"Roman three oh one," Clay acknowledged and waited as his jet kept climbing and flying north.

"Roman three oh one, left two eight zero," the controller radioed thirty seconds later.

"Three oh one," Clay answered and banked the jet left to the marshal area northwest of the carrier.

"Roman three oh one, take angels three," the controller radioed two minutes later. He wanted Clay flying at 3,000 feet in the marshal stack. Clay felt far calmer now. The lightning far off had stopped. The layer of clouds 2,000 feet above him blocked any star- or moonlight so he felt as if he was in his bedroom at night with the covers over his head. But at least he hadn't been ordered to circle in the muck above. He began setting up the computer screens for his second pass.

At 8:10 P.M., the carrier ordered Clay to turn further left to begin his flight downwind to the back of the ship. The trip took less than five minutes.

"Roman three oh one, left to the final bearing zero three three," the approach controller radioed. Clay tilted the jet in a hard bank to the left to make the U-turn behind the carrier about eight miles out. He pushed down the handle on his right console to lower the tailhook. He wanted the carrier's cables to stop his plane this time.

"Roman three oh one, take angels one point two," the controller commanded next.

"Roman three oh one," Clay radioed and complied by dropping his jet to 1,200 feet. He would keep this altitude until three miles from the ship when he would point the nose down again to fly the glideslope to the carrier deck.

Seven miles from the ship.

"Roman three oh one, lock on six out, say needles."

Clay had been turned over to a final controller in the operations center. But this time the voice on the radio was a feminine one with just a hint of a Southern accent. The JFK had joined the Navy's fleet of coed carriers. Clay didn't feel awkward with women aboard the ship. He didn't really have enough experience with sea duty to feel anything. Everything was still too new. Yesterday was the first time in his life that he had ever had the chance to set foot on a carrier. That was when he was allowed to climb out of his Hornet after it landed during the day. During his first landings in a T-45 Goshawk, Clay had never gotten out of the plane. The students made their landings and takeoffs during the day, then flew immediately back to shore.

Clay thought nothing of squeezing by women in the passageways of the JFK. He didn't consider himself a part of the old Navy that was still hung up on sex. He had no opinion one way or the other about women flying combat jets. If he had to think about it, he probably would prefer men going to war than women. Certainly, he never wanted Julie dying in combat. But he'd met women in the Navy who wanted to fight, women who were good with a stick. Why not let them fight, he thought. You couldn't walk around the ship naked and you had to watch the dirty jokes. The Navy didn't have to tell him that. Lee Clay had raised his son to always be a gentleman around women. But the Navy brass was now so paranoid, guys in his squadron were afraid to say hello to women pilots when they passed them in the hallway or to pay them a compliment. He did it anyway.

Clay punched buttons on his up-front controls to lock up his computer with the automatic carrier landing system aboard the *JFK*. The spermie blinked on in his heads-up display. And this time it stayed on.

"Three oh one, on and up," Clay radioed the final controller. He had needles.

"Roman three oh one, concur, fly mode two," she answered. Mode two meant the ship's landing system would stay locked to his plane's computer to constantly feed it signals on the flight path to the carrier deck.

Clay now maneuvered the Hornet so the spermie remained just above the plane symbol in his HUD.

Three miles from the carrier he pointed the nose down so it would fly along the correct glideslope to the carrier. The spermie in his HUD drifted down so it now overlapped the plane symbol. As long as he kept the spermie on top of the symbol his plane was flying on the proper glideslope.

"Roman three oh one, slightly right, three miles, on glideslope," the final controller prompted. His jet was pointed down at the correct angle but positioned just to the right of the line it should be taking to the ship.

Clay moved the stick a fraction to the left. The winds had picked up again and buffeted the plane.

"Roman three oh one, going on three quarters of a mile, call the ball," the controller radioed for the last time. It was a quarter past eight.

Clay could see clearly the amber light on the center column of the Fresnel lens.

"Roman three oh one, Hornet ball, four point nine," Clay radioed, a little less hurried than before. He had 4,900 pounds of fuel left in his tanks.

"Roger ball, thirty-five knots," Lovey radioed back the wind speed in a voice as relaxed as a classical music disc jockey's.

Clay was slightly below the glideslope. He pushed the throttle forward to position the jet so the meatball moved just above the center. Right where he wanted it. With the winds gusting so much it was better to be slightly high in the last few seconds before landing.

His eyes flitting from the meatball to the lit centerline to the angle of attack readings in his cockpit, Clay tapped the throttles back ever so gently with his left hand. He held the stick with three fingers from his right hand like a violinist would a bow.

Fifteen seconds before touchdown. The hand movements had to be almost imperceptible.

Clay had inched the throttle back too far. The meatball began dropping too fast. With these winds, the slightest overcorrection for too much throttle could send the ball sinking like a stone.

Lovey could only see the Hornet's exterior lights—two red ones on the left wing tip, two green on the right, an amber light on the center fuselage. From years of watching planes land on carriers, he recognized the problem instantly.

"Power," he ordered over the radio.

Clay inched the throttles forward again.

Too much. The Hornet bobbed up and just above the glide-slope.

"You're a little high," Lovey said with a bit more urgency.

Clay worked the meatball back down to the center as the stern of the ship rushed toward him.

The Hornet raced over the ramp of the ship.

Clay's eyes never left the meatball. No more corrections to be made with the throttle and stick. A half second until touchdown.

Clay heard a banging sound, metal on metal. The tailhook struck the steel deck sending up a shower of sparks like Fourth of July fireworks.

He felt almost in the same instant the hard bump under his seat from the wheels slamming onto the deck. Clay shoved the throttle forward to power up the Hornet's engines in case the tailhook didn't catch and he had to fly off. The engines roared and the exhausts glowed bright orange in the night.

But the tailhook grabbed the third wire. Clay's head jerked forward, then snapped back on the headrest. His body tried to fly out of the seat. The straps and seat belt yanked painfully at his stomach and shoulders. The stop was far more violent than in the training jets. The Hornet weighed more and was traveling faster when it caught the arresting cable.

Clay pulled the throttle back to power down the engines. His heart raced. He could feel the veins in his head throbbing. He caught his breath and twisted his neck to make sure his head was still attached to it. He looked front right and spotted the outline of the taxi director in the dark signaling him with white flashlight wands to pull back. Clay lifted the balls of his feet off the top of the rudder pedals, which braked the plane. He could feel the arresting cable tug his aircraft backward. He reached over with his right hand and flipped up the arresting hook handle in his cockpit. Outside, the tailhook raised and disengaged from the wire. With his finger Clay flipped a switch down on the left front console to lower the wing flaps to the half level. With his right hand he pulled a lower knob up, then turned it counterclockwise to crack the wings at midpoint and raise the last six and one half feet of them up. The smaller wingspread made the Hornet thinner so it could fit into tighter spots on the cramped deck.

Signaling with his white wands, the taxi director ordered Clay to turn right and roll away from the landing area. Another jet was about a minute away from landing behind him. He pushed the right rudder pedal in with his foot, moved the throttle forward for power, and began driving with his feet.

While his feet steered, his right hand reset the radar altimeter to forty feet, the altitude he definitely did not want the Hornet to fall below when it was later catapulted off the front of the ship. When he shot off the bow, Clay wouldn't be able to see if his jet was dropping to the water. The radar altimeter alarm, which sounded if he sunk below forty feet, would be critical.

Once his jet had rolled outside the thick white line that marked the boundary of the landing area, the taxi director ordered Clay to stop. The *Kennedy*'s deck crewmen, who had been impatient with the taxiing planes during the day, slowed the pace at night. Rushing pilots now could cause accidents.

Clay had a brief moment to take in the scenery outside the cockpit. It seemed as if a black curtain surrounded the ship, but he was surprised at how well lit the carrier deck appeared from the few lamps burning and the crewmen's light wands. There was something both majestic and primal about an aircraft carrier deck at night. No other spot on earth could possibly be as dark or barren as the lonely and cold Atlantic waters the *JFK* steamed through—outer space beyond the stars. In this black hole, on greasy steel ground that puffs of white steam danced over like tumbleweeds, fearsome gray Hornet and Tomcat jets lumbered about like prehistoric pterodactyls, the throaty roar of their engines engulfing everything around them in bloodcurdling sound, their exhausts belching hot, sickly-sweet kerosene fumes. Tiny red lights blinked on the wings of these beasts. From their heads, the faint green glow of cockpit lights illuminated the lonely shapes of the air warriors in their warm cocoons. Military strategists might question whether such huge floating monsters with their growling metal birds were now needed in an age of microchip weapons and precision warfare. But no one could stand on an aircraft carrier deck at night and not feel a profound reverence for the mighty raw power that modern man had created.

The taxi directors motioned Clay to thick black fuel hoses on the starboard side of the ship. His Hornet gulped 800 pounds of gas with each pass at the carrier, so its tanks needed a fill-up. Clay moved the jet slowly. At night, a pilot could experience vertigo even on the flight deck. With the dark providing no visual references, Clay could mistakenly think his plane was moving when it was actually the taxi director walking with his lighted wands. Some aviators parked in their aircraft near the edge of the deck had almost ejected from their cockpits when a taxi director with lit wands walked away from them and it appeared to the pilots that their planes were sliding backward into the ocean. For that reason, taxi directors were under strict orders to stand still when they motioned planes to move at night.

The Hornet's tanks were filled with several thousand pounds of fuel. Two crewmen wearing purple vests uncoupled the hose from the jet's fuselage and dragged it away. Clay was grateful for the fifteen minutes it took to refuel. The respite gave him a chance to collect his wits. He unsnapped the oxygen mask plastered to his face. Its rim had left a red line around his cheeks and nose. He reached into a pocket of his survival vest and pulled out a stick of Spearmint, unwrapped its tinfoil, and stuffed the gum into his mouth. Julie had given him a bag of Halloween candy but Clay had been afraid he would be too busy in the cockpit to snack. The gum would have to do for now.

He powered up the throttle and pumped the rudder pedals with his feet to slowly turn the aircraft left toward the number one catapult. He blocked out of his mind the anxiety he felt being hurled into the black void ahead of him.

The cockpit became hectic once again. Clay punched up the

takeoff checklist on his left video screen and the flight control systems page on his right screen. With one eye on the taxi director ushering him outside with the wands, he quickly began the checks. He ran through the columns of readings on his FCS page looking for any Xs by boxes that might indicate malfunctions with wing ailerons or flaps. None. He made sure that the cockpit canopy was snapped shut, that the belt fittings for his legs and torso still locked him to the seat. He looked down to his right to see that the yellow-and-black-striped handle was pointed down in the armed position so the seat would rocket out if he had to eject. He rechecked the radar altimeter alarm to make sure it would sound at forty feet, then set the computer-controlled trim for takeoff. That done, he changed the computer page on his left video screen to show the same heads-up display he saw in his glass HUD atop the cockpit controls.

The taxi director motioned him past the giant rectangular jet blast deflector, which tilted up off the deck when the Hornet moved in front of it. The white wands made a second signal ordering Clay to spread out his wings. He again pulled the lower right handle in his cockpit up, then turned it clockwise to the notch marked SPREAD. The ends of both wings slowly lowered down. Clay glanced right, then left to make sure both wings folded out and the circular, beer-can-like red tabs at their joints popped down to ensure they were locked in place.

A deckhand carrying a box with lighted numbers walked up to the left side of the jet. He hoisted the box over his head so Clay could see it. The weight board flashed "35,000," the number of pounds the catapult officer estimated Clay's Hornet now weighed with its tanks full. Clay agreed. He signaled to the sailor with his flashlight to set the catapult power to blast off an aircraft with that weight.

The taxi director signaled him to slowly inch forward to the catapult shuttle that would launch the plane. Clay lowered the launch bar by flipping a toggle down on the lower left control panel. Crewmen scurried under the belly of his jet to connect the launch bar, just in front of the two nose wheels, to the catapult shuttle. Another bar behind the nose wheels was hooked to keep the jet in place when Clay revved up the engines.

With the two bars attached, the catapult officer waved his wand to order Clay to pull up on the launch bar. Clay flipped the launch bar toggle up with his left hand. The Hornet's nose dipped. The catapult was cocked and ready to fire.

Clay placed his feet flat on the cockpit floor. He didn't want them accidentally pushing the rudder pedals forward on the launch to activate the brakes and slow him down. He moved his left hand quickly up to the throttles, pushed them forward to put the engines at full power, then locked his elbow. He didn't want the force of the cat shot pulling his left hand back so he would accidentally yank the throttles and lose power.

With his right hand he quickly moved the stick forward and back, then right and left. The ailerons and rudder all fluttered as they should. He punched up the flight control system page on his right video screen for one more check of his engines. All fine.

He poked more Chiclets on the right screen to bring up the page displaying the attitude direction indicator to tell him if his jet was pointed up properly after the launch.

Clay took one last look out the cockpit ahead of him.

God, it's black out there, he thought to himself.

He peered back into the glowing glass HUD. He would try to watch two readings during the first three seconds of his launch: first the airspeed box at the upper left hand corner of the HUD, and next a horizontal line on the screen with a W in the center of the line. In the two seconds it took the Hornet to be shot off the bow, the jet had to reach a speed of at least 160 miles per hour to go airborne. If it didn't, Clay would likely have to eject. The numbers in the airspeed indicator would race up in a blur, but Clay had to be sure that he saw at least three numbers in the box, which would indicate the plane had shot past 100 miles per hour. The horizontal line with the W was called the "waterline," which indicated how far the jet's nose was angled up. For the launch, Clay had to make sure the waterline was about 10 degrees above a parallel line below it, which indicated the level of the horizon. If the waterline and horizon line got too close, he would fly into the ocean. If they were too far apart, it meant the Hornet's nose was tilted up too much and the plane might stall.

Clay let go of the stick and grabbed a thin black bar on the top right of his cockpit called the "towel rack." The Hornet's computers flew the jet during the first few seconds of launch, automatically setting the ailerons and trim tabs so the aircraft would point up after it crossed the bow. The pilot wasn't supposed to touch the stick and take control of the aircraft until the Hornet had flown about 100 feet from the carrier. If he grabbed the stick too soon, it would throw off the precision heading the computer had set and send the plane bumping up and down. During the first couple of launches in daytime, Clay had refused to grab the towel rack and had kept his right hand in his lap near the stick, just in case all the assurances the instructors had given him that the Hornet would in fact fly itself turned out to be false. (Many veteran Hornet pilots in the fleet kept their hand in their lap as well.) For now he was convinced he was better off holding on to the rack so his hand didn't accidentally bump the stick and interfere with the computers' work.

Clay's chest was thumping. He took two quick breaths to calm himself.

He was ready.

Pushing a switch on the left throttle, he turned on all the Hornet's exterior lights—the signal to the catapult officer to launch the jet.

He braced himself and leaned his head back on the padded headrest.

He jammed the throttle forward to afterburners. That much power wasn't needed, but Clay didn't want to take any chances with his first night shot.

The Hornet roared. The white-hot exhaust plumes lit up the front of the carrier.

BAM!

With total darkness outside, his eyes saw nothing to tell him he was moving. He only felt his chest and stomach being crushed to the seat.

One second. His head shaking, the G force blurring his vision, he tried to focus on the airspeed box in the HUD. He could make out three numbers.

Two seconds. The Hornet shot across the bow. The launch bar disengaged from the catapult.

Clay felt like he was first slamming into a wall, then dropping off a cliff into darkness in the same instant.

But the pressure on his chest eased just as quickly and he felt the soft cushion of flight.

The jet had dipped slightly but the computers performed as they were supposed to, pointing the nose up and pushing the waterline in his HUD to 10 degrees above the horizon line.

Clay finally grabbed the stick. With the afterburners on, the Hornet quickly passed 200 miles per hour.

"Roman three oh one airborne," he radioed to the *JFK*.

Clay felt elated. Finally after all this work, he had landed on and taken off from an aircraft carrier at night. The excitement new pilots felt with their first night traps soon wore off, however, when it dawned on them that they now had to do it for a living. In the next several passes at the carrier, Clay learned quickly how difficult that job would be. The second trap went smoothly. On his third landing attempt, however, he flew the approach too low and Lovey had to wave him off. Cussing at himself in the cockpit, Clay brought the jet around to the rear of the ship for another try, but this time flew too high in the final approach and boltered.

"I think you're fixating a little bit," Lovey finally radioed him after the second miss. "Keep working. Keep your jet energized all the way down. Make sure you're scanning the ball all the way toward touchdown."

Clay was furious with himself. Then doubt crept in for an instant. Why am I up here tonight? he asked himself. A varsity night and I'm just a virgin with this stuff. The thought was fleeting, however. He returned to being mad. Clay had become his worst critic, more so in fact than the instructors. He would lie awake at night in his bed for hours reflying flights, churning over in his mind what he would do differently, imagining what buttons to push, what switches to flip the next time he was in the air. He would dream up emergencies in the cockpit, freak accidents he might face. What would be his response? Would he measure up to the challenge?

Rounding the carrier for a third try at landing, Clay's heart raced because he faced a more pressing problem. Fuel. His jet was low. If he didn't trap this time, he would have only enough left to fly to his divert field on land, the Marine Corps air base at Cherry Point along the North Carolina coast. The rules couldn't be bent. If he missed this trap, then rounded once more and missed again, he might not have enough fuel at that point either to make another landing attempt on the carrier or to reach Cherry Point.

Flying to Cherry Point would be a pain in the ass, he knew. By the time he arrived, it would be too late to refuel and head back out to the carrier tonight. He would have to stay overnight on land, then return to the carrier the next morning with his tail between his legs and catch up with the other students, who would be ahead of him in landings.

This was not how Clay had envisioned his first set of night landings. Like every new pilot in the squadron, he desperately wanted to be the "top hook," the student with the best landing grades. With a wave-off and bolter, he'd probably blown that chance. Now he faced a "night in the barrel," pilot slang for bad flying. On those nights, aviators often climbed out of cockpits with their faces white as ghosts, their legs shaking. Some even cried. Clay's night in the barrel was in danger of being capped by the ultimate humiliation—running out of fuel and heading for the beach.

Clay tried to shake the mistakes from his mind. He started his next approach a little high, overcorrected a bit too much and settled low, but finally managed to bring the jet back up to the proper glideslope.

He trapped. Lovey scored it a "fair" pass.

Relieved, Clay taxied the jet to the purple shirts holding the hose so his Hornet could take a long drink of gas.

He catapulted off the carrier once more and this time made his best landing of the night. Lovey scored it an "okay."

As Clay taxied the Hornet left and out of the landing area, the carrier's control tower directed him back to the catapult.

Clay quickly checked the flight card on his knee. He had performed one touch-and-go and four traps, all he was supposed to accomplish the first night. It was almost 10:00 P.M. He was exhausted. His throat felt raw with thirst and his stomach growled from hunger. That was enough bleeding for one night, he thought.

Clay keyed the radio and diplomatically reminded the tower of how many traps he'd made. Thankfully the message came back to park the jet and climb out.

Clay would have to perform more day and night landings tomorrow to finally qualify. Then he would be off to an operational F/A-18 squadron in the fleet. The reputation he had made in the training squadron would now be important. The Navy's community of Hornet pilots was not large. Like a small town, most in it knew each other personally or had at least heard gossip on how well or poorly each pilot flew.

The fleet squadrons informally operated what amounted to an NFL draft to have newly minted pilots assigned to their units. A Hornet squadron's commander would designate several of his lieutenants as scouts to check out the strengths and weaknesses of VFA-106's graduates. The scouts would make back-channel phone calls to old friends among the training squadron instructors for gouge on the rookies. Who was the best lander? Who was safe in the plane? Who still had problems? Who was strong in air combat? Who was the most accurate bomber? Like a football coach looking for a speedy defensive back or a hefty guard to shore up a weak line, squadron skippers might try to recruit a good bomber if the unit scores had been low, or a good dogfighter if it had been beaten too much in carrier competitions. The squadron COs rotated who had the first draft pick with each graduating class.

For the next three years, Clay would be working harder than he ever had in his life, learning how a carrier operated at sea, being stuck with hundreds of extra duties nugget pilots were always assigned, quietly trying to fit into the squadron's band of brothers and sisters but at the same time struggling to move ahead of his peers. The competition would never end. He would always be under a microscope. It would take at least one six-month sea tour before he would feel comfortable in the squadron, at least a year more before he could consider himself a skilled pilot. He would also spend as much time learning to be a good officer. He would have to manage maintenance crews and training schedules in the squadron. The Navy ultimately valued organizational skills more than good stick work. Eventually he would be promoted out of the cockpit to head up departments aboard ships or on land. Then, if he was lucky and the long sea tours away from home hadn't broken up his family, he would take command of a squadron, and finally a vessel.

As he powered down the jet and climbed out of the cockpit, Clay fought back sentimental feelings. He realized he had started from the bottom. None of his old friends from Oklahoma City could say as he now could that he had landed a high-performance jet on a carrier at night. And a varsity night at that. He did not want to feel proud of what he had accomplished, however. That might breed complacency. Everyone in the squadron had accomplished more than they ever imagined. Get too complacent and the competition would quickly eat him up, he realized.

Clay did not want to be satisfied with just becoming an air warrior. He would only be satisfied when he was the best.

EPILOGUE

AFTER his cruise in the Adriatic, **John "Tuba" Gadzinski** was promoted to lieutenant commander. He resigned from the Navy, however, ten days after pinning on his new rank. Tuba had grown tired of spending so much time away from his family on sea duty. Even when the carrier was in port and he was not flying, Tuba spent long hours at work with administrative duties. He began flying with Southwest Airlines and planned to try out for the Virginia Beach Symphony Orchestra. **Kristin "Rosie" Dryfuse** and Chad Jungbluth married in May 1995 after the *Ike* returned to its home port in Norfolk, Virginia. In the two years since, they had been together only seven months because both had been flying in different squadrons with different sea tours. Kristin was flying in the F-14 Tomcat aboard the USS *Theodore Roosevelt* in the Mediterranean Sea. Chad was flying the CH-46 Sea Knight helicopter in the Persian Gulf.

Charles McKinney's dream of flying jets and then joining the astronaut program suffered a setback. The Navy had other plans for him. McKinney was ordered to fly helicopters. He would be piloting the CH-53E Sea Stallion out of Sigonella, Sicily, hauling cargo throughout Southern Europe and the Middle East. Though disappointed at first, McKinney discovered he enjoyed flying helicopters. But he was still determined to become an astronaut one day. It was possible even for a helicopter pilot. His game plan: work hard at Sigonella, impress his superiors, and, as soon as he could, send in his paperwork to transfer to jets. The application might be rejected the first couple of times, but he would keep submitting it. He was still young. There was plenty of time. He refused to give up hope. Not yet.

Mary Margaret Kenyon did realize her dream. She was promoted to captain and flew the UH-1N Huey helicopter out of Camp Pendleton, California. Her first choice had been choppers. Jets might be the glamour aircraft, but helicopters were the workhorses the Marines depended on to haul them to battle, drop in supplies, and cover them with protective fire. She would be closer to her guys on the ground. Mary Margaret could not be happier. This was the first job she had ever had that had not become boring. She had gone from being a scared rabbit to a confident aviator, a self-assured officer. Life had begun all over for a thirty-two-year-old divorcée.

They never realized when they started what a mental challenge it would be flying military aircraft. But the study group survived the competition. **Stacie Fain** piloted the Coast Guard's HH-60J Jayhawk helicopter out of Elizabeth City, North Carolina. She flew search missions along the Atlantic coast rescuing boaters in distress. **Brian Hamling** and **Bill Perkins** got the plane they wanted, the P-3 Orion sub hunter. Hamling flew out of Brunswick, Maine, patrolling the waters of the North Atlantic. Perkins served as a navigator on patrol in the Pacific waters off Hawaii. He remained addicted to science fiction books. **Dan Smellick** scored high enough to be selected for F/A-18s, his first choice.

Doug Spencer thought he had planned for everything, stretching himself before his flight physical so he would meet the height requirement, studying obsessively for tests so no question would trip him up. But there was one thing he couldn't plan for—that his stomach would stay settled when he was flying. Spencer could not stop throwing up in his early plane rides in the T-34C. He visited doctors, tried pills for a while, endured a spinning machine that helped some pilots get over airsickness. Nothing worked. Finally, he decided to accept a medical discharge from the flight training program. It was the toughest decision he ever made in his life. He became a public affairs officer fielding questions from reporters at the Navy Information Office in Atlanta.

Bret Hines and **Danny Johnson** never stopped competing to be the first in their classes. That was the only way to ensure that they would eventually be jet pilots. The two Marines kept their pact to get the best grades and were assigned to fly F/A-18s.

Athletics finally brought **Bill Mallory** some bad luck. He broke his finger and ankle playing flanker for the Pensacola rugby team. He couldn't fly with those injuries, which along with other scheduling delays meant he spent almost a year completing primary flight training in the T-34C. But he finally was graduated with scores high enough to be assigned jets. He also settled down and got married.

Ross Niswanger mellowed. At first he was depressed when the news came after the Stacy Bates accident that Naval flight officers would no longer be allowed to retrain as pilots. But he came to enjoy the work of an NFO. He was assigned to the S-3 Viking sub hunter, which flies off carriers. Niswanger began as a sailor in subs; now he would attack them. As the jet's tactical coordinator, he organized the search and strike packages against submarines that threaten the carrier battle group. Just as in the Tom Clancy novels, he liked to say. Maybe one day the door would crack open again and he would be able to transfer to the pilot program. He was taking private flying lessons just in case. You can't have everything you always want, but if you worked hard the Navy would still reward you. Niswanger still believed that.

Alwin Wessner paced nervously in his gym shorts, T-shirt, and running shoes. It was the day after the dunkers and he had already passed the running and sit-ups tests. But they weren't the problem. The critical test, the one he had flunked twice and now had to pass to stay in the flight program, was push-ups. He bent down and dropped to his knees. One of the physical training instructors stood over him with a clipboard.

Wessner began pumping his elbows, making sure to keep his

back straight as he went down, then up. He had to perform at least forty-two push-ups and they had to be perfect. No arching the back. Wessner pumped up and down quickly taking short breaths. Thirty-one, thirty-two, thirty-three.

"That's it, keep it going," the PT instructor said to encourage him. Other instructors nearby stopped what they were doing and glanced in his direction. Everyone on the training staff knew this was Wessner's last chance. They were all pulling for him.

Thirty-four, thirty-five, thirty-six. His elbows locked, his back in the up position, Wessner paused to take a rest. The instructors stared at him intently. Six to go if he was going to pass.

Wessner took a deep breath and continued the push-ups. Thirty-seven, thirty-eight, thirty-nine, forty, forty-one, forty-two. The instructors smiled. "Now it's all for pride," one of them said.

Wessner knocked out ten more. He collapsed on the canvas mat at fifty-two. When he picked himself up, the instructors shook his hand.

He would always have to spend more hours studying than the other students, but Wessner proved a natural with the stick. He scored high on his flights in the T-34C and began training to be an F/A-18 pilot.

Jenna Hausvik was assigned the EP-3 electronic warfare plane and was glad of it. After primary flight training, she had no deep urge to fly jets and live the rest of her life on aircraft carriers. She liked the idea of flying from a land base and having a crew in the plane for backup as well as companionship. She went to a squadron based in Rota, Spain. **Magnus Leslie** finally began to relax in the training, enjoying the flying more than worrying about his grades. He was assigned to pilot the C-2 Greyhound, which hauls people and cargo to the carrier. Amy joined him in Norfolk where they have set up house. **David Perrin** already had the Tom Cruise haircut and the aviator sunglasses when he flew prop planes at Whiting Field. Now he has jets. He flew a T-45 Goshawk on a cross-country trip to Las Vegas. Won $160 in the casino during the stopover. "It's a blast," he says. He is trained to fly the Hornet.

Life has been good to **Mike Sobkowski** as well. He finished in the top of his class and got his first choice of aircraft to fly. He trained to be a Hornet pilot. But before beginning F/A-18 instruction with the Marines, Sobkowski was assigned a temporary job at the Kingsville air station teaching new jet students the basics of the aircraft he was leaving, the T-45. What could have been better? Five months of extra flying in the Goshawk—fun flying where you're the instructor instead of the nervous student—while he awaited orders to the Marine air base at El Toro, California. **Rob Schroder** graduated from Kingsville with good grades as well, but with bad luck of the draw. He had desperately wanted to fly Hornets, but the service ordered him to the Marines' AV-8B Harrier. Schroder was disappointed, but guessed that, as other pilots before him, he too would eventually fall in love with the aircraft he had been assigned to fly.

Jonathan Wise scored an excellent on his next solo bombing flight at El Centro. After the wedding, he also successfully completed his day carrier landings in the T-45. Then the best news came. The Navy assigned him to Hornets. He joined his wife, **Maria Grauerholz**, at the Lemoore Naval Air Station, near Fresno, California, where the two trained in the F/A-18. They bought a house near the air station. But Jonathan and Maria didn't know how long they would be together in it after they finished Hornet training and began their adventure as air warriors.

With the highest grades in the training squadron, **Bill Sigler** had no trouble receiving his first choice. He flew as an F/A-18 pilot.

Pablo, Baby Killer, and Dunn Deal got their first choices as well. After qualifying at the carrier, **Paul Rasmussen**, **Brian Burke**, and **Rob Dunn** completed their training in the T-45 and were assigned to F/A-18 Hornets.

Rich Whiteley, **Curtis Carroll**, **Joe Gelardi**, and **Javier Ball** flew their second close air support hop the next day. It went horribly. Few of the jets attacked on time and none of the bombs landed on the targets. Altitude was the reason. During the first day's hop, the pilots flew the bombing routes at 4,000 feet. The second day, the routes had to be flown at 200 feet, the altitude at which a Hornet pilot would more likely fly in combat to escape detection by radar. But flying that low, it was far more difficult to stay on course and pick out the target while at the same time not crashing into the ground. The pilots felt miserable. But the instructors assured them that beginners weren't expected to fly it much better.

The four young pilots did graduate from the Hornet training—

but not at the same time. Whiteley and Gelardi easily passed the air combat maneuver phase of the instruction at Key West, then qualified in night carrier landings aboard the USS *John C. Stennis.* Whiteley flew combat air patrols over southern Iraq. Gelardi was part of the air wing staff aboard the USS *Dwight D. Eisenhower.* Carroll was able to escape any more accidents that might keep him out of the cockpit, but not the flu. A nasty strain of the virus kept him curled up in bed at the Key West BOQ while his classmates completed the air combat maneuver flights. Carroll returned to Key West two months later to make up the flights and was assigned to the VFA-87 Golden Warriors aboard the USS *John F. Kennedy.* His call sign was still Roadkill.

Javier Ball climbed the gangplank to the *Stennis* confident he would have no difficulty landing the Hornet at night. He had won a trophy during the class's detachment at El Centro because he had flown so well on the bombing hops. The air combat maneuver training at Key West also had been a breeze. But because of scheduling delays aboard the *Stennis,* he sat in the squadron ready room for four days waiting for his first flight. He felt like a football player all pumped up for the big game but stuck in the locker room. Ball lost his edge. After he finally catapulted off in the Hornet he kept boltering so many times trying to land that he ran out of fuel and had to fly back to Cecil Field. He remained at the field overnight, barely sleeping because he was nervous that he would now fail to qualify. The next morning, he flew back bleary-eyed to the carrier, where his air work unraveled even more. The instructors disqualified him and ordered him to return to the carrier three months later for a second try. It might well be his last if he didn't qualify the second time. Pilots usually only had two chances to pass night landings. Flunk again, and he probably wouldn't fly Hornets.

Javier was furious with himself. Sure, the weather had not been perfect, but he blamed no one but himself. He was depressed for days. He felt like someone had kicked him in the nuts. It was an "ego buster," the instructors said, but not fatal. He would pass the second time, they assured him.

Three months later, Javier flew back to the *Stennis* for his second try. His practice landings at Cecil Field had been near perfect. His day landings aboard the carrier went off without a hitch as well. But the Hornet he drew for the first set of night landings turned out to be a Christine, a horror craft with an electronics problem. After Javier catapulted off the *Stennis* into the darkening sky, the screen illuminating the heads-up display in his cockpit went blank.

Javier nervously radioed the carrier's control tower for advice on how to fix it. The tower radioed back instructions and Javier hurriedly flipped switches as his jet flew around to the back of the ship. But nothing he did made the HUD turn back on.

Javier landed on the carrier deck, one of the cables thankfully grabbing his tailhook the first time. Maintenance men rushed over to his jet after he had parked it to the side and began replacing faulty video screens on the cockpit control panel.

A half hour later, Javier catapulted off the carrier a second time. But now all three video screens in his cockpit went blank. That frightened him. The sky had no moon and he could barely make out the horizon. He frantically began the troubleshooting procedures pilots constantly rehearsed to make the display screens light up. Finally, the HUD screen illuminated as his jet was about to touch down on the carrier. But Javier boltered. The HUD screen again turned blank as he flew away from the *Stennis*.

Javier tried two more times to land, but his tailhook failed to catch an arresting wire. The HUD continued to flicker on and off. Finally, the landing signal officers waved him off the pattern and ordered him to fly to dry land at the Oceana, Virginia, Naval Air Station.

Javier flew back to the *Stennis* the next day in another Hornet. Its HUD worked fine. But by then Javier was too frazzled. He had psyched himself out. He kept boltering and boltering when he tried to land at night until finally the landing signal officers gave up and ordered him to return to Oceana. Javier had failed a second time.

He was angry and frustrated again. He felt helpless. What would he do now? They would surely kick him out of the cockpit. Should he go back to ground-pounding in the infantry? Should he try to transfer to the Air Force, which didn't require its pilots to land on ships at night? The fatalism that set in when Javier's son had died returned. What would be would be. He had given it his best try. Landing on a carrier at night wasn't easy. If it was, everybody could do it. Maybe he couldn't.

But why? Javier sat for hours asking himself that question. He had scored in the top of his class in every other aspect of the Hornet's operation—aerial bombing, dogfighting, even practice carrier landings at Cecil Field. Why couldn't he land on the boat at night?

Maybe it was Geronimo's death, his wife, Beverly, suggested. The only down he had ever received in flight school was after his son died, when Javier tried his first carrier landings in a T-45 jet trainer. Maybe he was suffering from a mental block. Maybe the carrier symbolized in his subconscious the tragedy of Geronimo's death. Maybe he would never overcome that barrier.

But Javier decided he would not quit. He would not drop out of flight training. They would have to throw him out.

Then it dawned on him. The meatball on the Fresnel lens. As his jet neared the stern of the carrier at night, Javier had difficulty picking up the ever-so-slight movements of the Fresnel lens's amber light up and down. He kept reacting to the shifting light instead of anticipating its movement and keeping his jet on the proper glideslope to the carrier. Were his eyes the problem? Did he need glasses?

Javier marched in Commander Gortney's office and begged to be given another chance. "I'm not seeing the deviations," he insisted to Gortney. The training squadron commander sent him to the base ophthalmologist.

Javier, it turned out, was slightly farsighted. If he had been a civilian, the problem would not have been severe enough to even require reading glasses. But landing a jet at night, Javier's eyes had to dart quickly from the instrument readings in his cockpit to the carrier's meatball outside. Not being able to focus perfectly on the instrument readings up close in the cockpit put a slight strain on his eyes when he looked back outside at the amber light—enough of a strain, the ophthalmologist told him, so that he might miss the meatball's slightest movements. Javier was allowed a third try at night carrier landings, this time aboard the *Eisenhower*, which was steaming off the Atlantic coast the next month, and this time wearing reading glasses.

The reading glasses worked. He felt he picked up the meatball's movement better. Or perhaps it was still all psychological. Perhaps the glasses, which didn't change his vision much, were a lucky charm. Javier didn't care. All he knew was that he never boltered on his third try. He scored "average" or "above average" on all his night landings. He would never slam down on a carrier deck at night again without those reading glasses.

It was past midnight when Javier finished his second set of night landings aboard the *Eisenhower* and filled out paperwork in the carrier's maintenance room. The two dozen maintenance technicians from his training squadron, who had been with him on his two previous cruises and had watched him struggle to land at night, now crowded into the room where he wrote on the forms. Javier had made friends with them during the months he had been with the squadron. They had inspected and reinspected his jet carefully this time to make sure no glitches would mar this critical flight. When the last sailor squeezed into the room, the technicians began clapping. Javier had finally qualified. One of the technicians presented him with three trophies to honor his achievement: a can of cold Pepsi, a bag of M&Ms, and the C-shaped hook from the end of his jet's tailhook that had caught the arresting cables.

Javier fought back tears. A half hour later, he had talked the carrier's communications center into patching a phone call to Beverly in Florida. "Pack your bags, we're moving to Beaufort," Javier said when they were finally connected. In Beaufort, South Carolina, was VMFA-115, the Marines' Silver Eagles F/A-18 squadron, which had been waiting for four months for Captain Javier Ball to arrive.

Phillip Clay didn't make top hook in the training squadron. But he did score high enough on the second day of landings aboard the *JFK* to easily pass the night carrier qualification. He became a pilot with the VFA-131 Wildcats, a Hornet squadron based at Cecil Field.

SOURCE NOTES

INTRODUCTION

Psychological profiles of Navy pilots are drawn from "The Outstanding Jet Pilot," Captain Roger F. Reinhardt, *American Journal of Psychiatry*, 127:6, December 1970; "Pilot Personality Testing and the Emperor's New Clothes," Robert O. Besco, *Ergonomics in Design*, January 1994; and "Gender and Performance in Naval Training," Annette G. Baisden, Proceedings: Psychology in the Defense Department, 13th Symposium, USAF Academy Department of Behavioral Sciences and Leadership, April 15–17, 1992. Interviews also were conducted with Commander Lawrence Frank, Commander Jeff Moore, and Lieutenant Commander Jennifer Berg of the Operational Psychology Department at the Naval Aerospace and Operational Medical Institute, Naval Air Station, Pensacola.

The sections on cultural changes in the Navy are drawn from author interviews of Naval officers, including Lieutenant Paula Coughlin, during the Tailhook scandal from May to July 1992. Also consulted: *Tailspin: Women at War in the Wake of Tailhook*, Jean Zimmerman (New York: Doubleday, 1995); *Fall from Glory: The Men Who Sank the U.S. Navy*, Gregory L. Vistica (New York: Simon & Schuster, 1995); *Feet Wet: Reflections of a Carrier Pilot*, Paul T. Gilchrist (New York: Pocket Books, 1990); *The Naval Aviation Guide*, Richard C. Knott (Annapolis, Md.: Naval Institute Press, 1985); *Super Carrier*, George C. Wilson (New York: Berkley, 1988); "A Storm Called Tailhook," Kerry Derochi and Phyllis W. Jordan, *Virginian-Pilot*, Feb. 13, 1994, p. A1; "Deepening Shame," Douglas Waller et al., *Newsweek*, Aug. 10, 1992, pp. 30–36; "'Women Can't Fly Jets' and Other Myths," Douglas Waller, Newsweek, Aug. 10, 1992, p. 36; "Tailhook: What Happened, Why and What's to Be Learned," W. Hays Parks, *Naval Institute Proceedings*, Vol. 120/9/1099, September 1994, pp. 89–103.

PROLOGUE

The account of flight operations aboard the USS *Eisenhower* is based on interviews the author conducted with Lieutenants John Gadzinski, Kristin Dryfuse, and other pilots and crew members aboard the ship in June 1994 and March 1995. Also consulted were: "A Thirty-Year Cruise," James L. Burke, *Naval Institute Proceedings*, Vol. 121/1/1103, January 1995, pp. 32–35; "Welcoming Aboard Female Aviators," Clifford A. Skelton, *Naval Institute Proceedings*, Vol. 120/7/1097, July 1994, pp. 68–70; "Trust Us," Robert E. Norris, *Naval Institute Proceedings*, Vol. 120/11/1101, November 1994, pp. 58–59; "U.S. Pilots Target Nasty Nicknames," Christine Hauser, Reuters, Jan. 26, 1994; "Navy Ouster of Chopper Pilot Called 'PC,'" Rowan Scarborough, *Washington Times*, March 3, 1995, p. 6; "Navy to Seek Out More Black Pilots," Ronald W. Powell, *San Diego Union-Tribune*, Dec. 3, 1993, p. A1.

CHAPTER ONE

Biographical material is based on interviews with Ensign Charles G. McKinney II, and students in Aviation Pre-Flight Indoctrination Class 3195, May 22–June 1, 1995.

The passages dealing with hypoxia and the hyperbaric chambers come from author's observation of the flight physiology and hyperbaric chamber instructions Class 3195 received. Additional interviews of students undergoing flight physiology and hyperbaric chamber training were conducted at Pensacola Naval Air Station on July 14, 1994. Also consulted were: Naval Aerospace and Operational Medical Institute Lesson Plan NP 1.1, Naval Aviation Physiology Training Program Indoctrination; NP 1.1 Oxygen Systems Student Handout; "Way, Way Off in the Wild Blue Yonder," Mark Thompson, *Time*, May 29, 1995, pp. 32–33; and "Navy: Jet Crew Removed Oxygen Masks Before Crash," Dana Priest, *Washington Post*, June 24, 1995, p. A4.

Historical material was drawn from: *The Cradle: Naval Air Station, Pensacola* (Pensacola, Fl.: Pensacola Engraving Co., 1989); *Wings of Gold*, Robert F. Dorr and Robert D. Ketchell (Osceola, Wis.: Motorbooks International, 1990); *Flying the Edge: The Making of Navy Test Pilots*, George C. Wilson (Annapolis, Md.: Naval Institute Press, 1992); and "'Women Can't Fly Jets' and Other Myths," Douglas Waller, *Newsweek*, Aug. 10, 1992, p. 36.

CHAPTER TWO

Biographical material was based on interviews with First Lieutenant Mary Margaret Kenyon as well as interviews with students in Aviation Pre-Flight Indoctrination Class 3195, May 22–June 1, 1995. The account of the Multi-Station Disorientation Demonstrator comes from the author's observation of Class 3195 being trained in the device and interviews with MSDD instructors. Additional interviews and observation of MSDD training occurred on July 13, 1994.

Also consulted were: "Unaided Night Vision and Visual Problems, Spacial Disorientation," Naval Aviation Physiology Training Program Instructor Guide, B-322-0043, Naval Aerospace and Operational Medical Institute; "MSDD Post Brief Indoc Ride," Naval Aerospace and Operational Medical Institute; "Unaided Night Vision Training Guide, 1991," Naval Aerospace and Operational Medical Institute; "NAMI: Naval Aerospace and Operational Medical Institute," *Wings of Gold*, Winter 1993, pp. 17–22; and Captain Matthew Waack, head of physical examinations, Naval Aerospace and Operational Medical Institute.

CHAPTER THREE

Biographical material was based on interviews with Lieutenant J. G. Stacie Fain, Ensigns Brian Hamling, Bill Perkins, Dan Smellick, and Doug Spencer. Material on academic competition, flight rules and regulations, gouge, and the grading system was based on interviews with instructors, Captain Michael Ott and Lieutenant Francis Scott, and with students in Aviation Pre-Flight Indoctrination Class 3195, May 22–June 1, 1995.

Also consulted were: "Flight Rules and Regulations Workbook," Naval Air Training Command, CNAT P-909 (Rev. 10-94) PAT, 1994; Aviation Preflight Indoctrination Training Schedule, Naval Aviation Schools Command, NAS Pensacola, October 20, 1994; Firehouse Print Shoppe booklet, "API"; " 'Gouging' the Honor System," Lincoln Caplan, *Newsweek*, June 6, 1994, p. 33; and "Awash in Scandal, Naval Academy Debates Its Course; Its Honor Code vs. Loyalty as 111 Middies Await Fate," Fern Shen and Paul W. Valentine, *Washington Post*, March 31, 1994, p. A1.

CHAPTER FOUR

This chapter was based on interviews with First Lieutenants Bret Hines and Danny Johnson as well as students in Aviation Pre-Flight Indoctrination Class 3195. Also interviewed were Lieutenant Reed Dunne and Scott Mickley with the Florida Department of Health and Rehabilitation Services.

Also consulted were: "A Better Navy for Women, Minorities," Patrick Pexton, *Navy Times*, Feb. 13, 1995, p. 16; "Sailor Says She Was Beaten While Training in Orlando in '92," Kerry Derochi and Jack Dorsey, *Orlando Sentinel*, Dec. 13, 1994, p. A1; "Sex-Case Accuser Demands Pilot Slot, Rejects Pentagon Job," Rowan Scarborough, *Washington Times*, June 29, 1994, p. 3; "Sex Harassment at Academies," *Washington Post*, April 5, 1995, p. 17; "Harassment Issue Won't Go Away for the Navy," Barbara Slavin, *Los Angeles Times*, March 1, 1995, p. B10; "Who's to Blame When Women Don't Measure Up?," Ellen B. Hamblet, *Naval Institute Proceedings*, Vol. 120/4/1094, April 1994, p. 101; "Red Light/Green Light: Next Time the Navy Plans to Play by the Book," Eric Schmitt, *New York Times*, April 17, 1994; "Tailhook Helped Steer Navy to New Course on Sex Abuse," *Los Angeles Times*, March 16, 1994, p. A5; "Navy Chief Vows No Repeat of Tailhook," John F.

Harris, *Washington Post*, May 5, 1994; "Navy Chief Trying to Make Amends in Harassment Case," Eric Schmitt, *New York Times*, June 1, 1994, p. A1; "Navy Officer Accused of 'Unduly Personal Relationship,'" Susanne M. Schafer, *Washington Post*, May 9, 1995, p. A3; "Are Women OK on Combatants? Not Really, Says Navy Survey," Becky Garrison, *Navy Times*, Sept. 4, 1995, p. 14; "Enough Scandal: Time for a Stand-Down," Becky Garrison, *Navy Times*, Nov. 20, 1995, p. 4; "The Hard Lessons of Adm. Macke," Becky Garrison, *Navy Times*, Dec. 4, 1995, p. 4; "Admiral Fired for Affair," Ernest Blazar, *Navy Times*, Dec. 18, 1995, p. 4; "Female Sailor Details Assault During Airline Flight," Sharon Waxman, *Washington Post*, Dec. 30, 1995, p. A10; "Trouble Surfaced in All Ranks," Becky Garrison, *Navy Times*, Jan. 1, 1996, pp. 18–19; and "Car Theft Ring Case Broadsides the Naval Academy," Robert L. Jackson, *Los Angeles Times*, April 12, 1996, p. A1.

CHAPTER FIVE

This chapter was based on interviews with Ensigns Bill Mallory, Ross Niswanger, and Alwin Wessner, as well as with First Lieutenant Mary Margaret Kenyon and Lieutenant Junior Grade Stacie Fain. Also interviewed were Carol Ross and retired Captains David W. Davis and Philip Ryan, formerly assigned to the U.S. Naval Academy. The accounts of the Dilbert Dunker and Helo Dunker were based on the author's observations of students from Class 3195 training in the devices.

CHAPTERS SIX THROUGH NINE

These four chapters were based on the author observing the VT-2 squadron at Whiting Field, Florida, October 16–27, 1995. The students interviewed included Ensigns Jenna Hausvik, Magnus Leslie, and David Perrin, as well as Amy Leslie. The instructors interviewed included Commander David Jenkins, Commander Bill Yeager, Major Loren Barney, Captain Tony Greco, Lieutenant Tony Chatham, Lieutenant Mike Consoletti, Lieutenant Mike Hicks, Lieutenant Dave Morey, Lieutenant John Prickett, Lieutenant Rob Hoehl, Lieutenant Steve Schutt, Lieutenant Gary Moe, Captain Guy Vilardi, and Lieutenant Rich Jackson.

Historical material on Milton, Florida, Whiting Field, and Naval aviation was drawn from the Santa Rosa County (Florida) Historical Society, Naval Air Station Whiting Field booklet, and Hill Goodspeed of the National Museum of Naval Aviation.

Also consulted were: NATOPS Flight Manual, Navy Model T-34C Aircraft, Updated June 28, 1994; NATOPS Pilot's Pocket Checklist, NAVAIR 01-T34AAC-1B, April 15, 1992; T-34C Flight Crew Checklists; "Flight Training Instruction: Familiarization, T-34C," Naval Air Training Command, CNATRA P-356 (Rev. 02-94) PAT, 1994; "Aircraft Engines and Systems," Naval Air Training Command, Naval Air Schools Command, Aviation Pre-Flight Indoctrination, 1994; "Master Curriculum Guide: Primary Flight Training, T-34C," Naval Air Training Command, CNATRA P-361 (Rev.

01-94) PAT, 1994; "Flight Training Instruction: Visual Navigation, T-34C," Naval Air Training Command, CNATRA P-359 (Rev. 01-94) PAT, 1994; "Flight Training Instruction: Basic Instruments, T-34C," Naval Air Training Command, CNATRA P-355 (Rev. 01-94) PAT, 1994; "Flight Training Instruction: Precision Aerobatics and Landings Primary and Intermediate, T-34C," Naval Air Training Command, CNATRA P-365 (Rev. 12-94) PAT, 1994; "Flight Training Instruction: Radio Instruments, T-34C," Naval Air Training Command, CNATRA P-360 (Rev. 02-94) PAT, 1994; "Flight Training Instruction: Formation, T-34C," Naval Air Training Command, CNATRA P-357 (Rev. 05-93) PAT, 1993; "Introduction to Basic Aerodynamics: Student Guide," Naval Air Training Command, CNATRA P-202 (Rev. 07-94) PAT, 1994; Naval Air Station Whiting Field Fact Sheet; T-34C Fact Sheet; Welcome to Whiting Field Fact Sheet, Feb. 22, 1994; and "Training Flight Kills Navy Lieutenant," *Navy Times*, Feb. 27, 1995, p. 2.

CHAPTER TEN

This chapter was drawn from interviews with instructors and students of VT-21, April 8–12, 1996, at the Naval Air Facility El Centro and at the Naval Air Station Kingsville. Interviewed were: Navy Commander Bill Shewchuck; Navy Lieutenants Eric Scheulin, Edward Miller, Greg Schuster, Felton Elders, W. D. Agerton, and Dan Doherty; Marine Captains Todd C. Vaupel, S. J. Sinner, Scott F. Andersen, and Eric W. Hildebrandt; Navy Lieutenants Junior Grade Jonathan Wise, Maria Grauerholz, Trey Sisson, Lance Luksik, and Brian Douglass; Marine First Lieutenants Mike Sobkowski, Robert W. Schroder, and Matthew Ward.

Also consulted were: "Flight Training Instruction: Weapons Delivery, T-45TS and T-45 (Advanced)," Naval Air Training Command, CNATRA P-1219 (Rev. 11-94) PAT, 1994; "Flight Training Instruction: Familiarization, T-45TS and T-45 (Advanced)," Naval Air Training Command, CNATRA P-1212 (Rev. 04-95) PAT, 1995; "Flight Training Instruction: Master Curriculum Guide: T-45TS Strike Training," Naval Air Training Command, CNATRA P-1201 (Rev. 07-95) PAT, 1995; "Flight Training Instruction: Operational Navigation, T-45TS, ADV, and IUT," Naval Air Training Command, CNATRA P-1217 (Rev. 06-94) PAT, 1994; "Flight Training Instruction: Instruments, T-45TS and T-45 (Advanced)," Naval Air Training Command, CNATRA P-1215 (Rev. 04-95) PAT, 1995; "Preliminary NATOPS Flight Manual, Navy Model T-45A and Up Aircraft," Naval Air Systems Command, A1-T45AB-NFM-000, 15 October 1994; and "Are Annapolis's Problems Systemic?" Michael Janofsky, *New York Times*, April 12, 1996, p. 20.

CHAPTER ELEVEN

This chapter was drawn from interviews with instructors and students from VT-21, April 8–12, 1996, at the Naval Air Facility El Centro: Navy Commander Steve Ross and Navy Lieutenants Bill Sigler and Kevin Nibblelink.

Also consulted were: *Fighter Combat: Tactics and Maneuvering,* Robert L. Shaw (Annapolis, Md.: Naval Institute Press, 1985); "Flight Training Instruction: Air Combat Maneuvering, T-45TS and T-45 (Advanced)," Naval Air Training Command, CNATRA P-1210 (Rev. 06-94) PAT, 1994; "Flight Training Instruction: Gunnery, T-45TS, ADV and IUT," Naval Air Training Command, CNATRA P-1214 (Rev. 09-95) PAT, 1995; "Flight Training Instruction: Out-of-Control Flight, T-45TS and T-45 (Advanced)," Naval Air Training Command, CNATRA P-1216 (Rev. 03-95) PAT, 1995; "Flight Training Instruction: Formation, T-45TS and T-45ADV," Naval Air Training Command, CNATRA P-1213 (Rev. 06-95) PAT, 1995; "Flight Training Instruction: Tactical Formation, T-45TS and T-45 (Advanced)," Naval Air Training Command, CNATRA P-1218 (Rev. 10-94), PAT, 1994; "Preliminary NATOPS Flight Manual, Navy Model T-45A and Up Aircraft," Naval Air Systems Command, A1-T45AB-NFM-000, Oct. 15, 1994; "Pilot Error, Engine Fire Downed F-14s," Otto Kreisher, *San Diego Union-Tribune,* April 17, 1996, p. A2; "Navy Tightens Rules for Pilots," James W. Crawley, *San Diego Union-Tribune,* April 13, 1996, p. A1; "F-14 Crash Probe Reveals Weakness in Rating Aviators," James W. Crawley, *San Diego Union-Tribune,* April 14, 1996, p. A1; and "String of F-14 Crashes Continues," James W. Crawley, *San Diego Union-Tribune,* April 18, 1996, p. A1.

CHAPTER TWELVE

This chapter is based on interviews with Lieutenants Junior Grade Jonathan Wise and Maria Grauerholz.

CHAPTER THIRTEEN

This chapter is drawn from interviews with instructors and students from VT-21 at Naval Air Station Kingsville, Naval Base Key West, and aboard the USS *John C. Stennis,* May 22–28, 1996. Interviewed were: Navy Captain Charles Nesby; Navy Commander Steve Ross; Navy Lieutenants Mike Carr and Rusty Wolfard; Marine Majors Scott Evans and Joel Paulsen; Navy Lieutenants Junior Grade Robert T. Dunn, Brian Burke, Paul Rasmussen, Brian Bronk, and Trey Sisson.

Also consulted were: "Flight Training Instruction: Carrier Qualification, T-45TS and T-45 (Advanced)," Naval Air Training Command, CNATRA P-1211 (Rev. 04-94) PAT, 1994; "Preliminary NATOPS Flight Manual, Navy Model T-45A and Up Aircraft," Naval Air Systems Command, A1-T45AB-NFM-000, Oct. 15, 1994; "Ens. Steven Pontell, 23, Student Pilot in Crash," Jonathan Higuera, *Washington Times,* Nov. 2, 1989, p. B4; and "It Was Like a Battle Zone," Bill Dipaolo, Gannett News Service, Oct. 30, 1989.

CHAPTER FOURTEEN

This chapter is based on interviews with Lieutenants Michael Carr, Rusty Wolfard, and Bill Sigler; and Lieutenants Junior Grade Robert T. Dunn, Paul Rasmussen, and Brian Burke.

CHAPTERS FIFTEEN AND SIXTEEN

These chapters are based on interviews conducted September 1–13, 1996, at the El Centro Naval Air Facility with Lieutenant Junior Grade Richard Whiteley, Lieutenant Curtis Carroll, Lieutenant Joe Gelardi, Captain Javier Ball, Lieutenant Paul Miller, Lieutenant Junior Grade Scott Troyer, Lieutenant Joe Evans, Lieutenant Tim Taylor, Lieutenant John Klas, Lieutenant Kieron O'Connor, Commander Bill Gortney, Lieutenant Ryan Delong, and Commander Jack Holt.

Also consulted were: "Tactical Aircraft Modernization Issues for Congress," Bert H. Cooper, CRS Issue Brief IB92115, Congressional Research Service, Dec. 1, 1994, Washington, D.C.; "F/A-18 E/F Aircraft Program," Bert H. Cooper, CRS Issue Brief IB92035, Congressional Research Service, Dec. 1, 1994, Washington, D.C.; "Mismanagement, Budget Cuts, Doubts Over Role Have Navy Sailing Against the Wind in Congress," Andy Pastor, *Wall Street Journal*, June 4, 1991, p. 20; "Anatomy of Decline," Department of Defense Briefing slides by Franklin C. Spinney, March 1, 1995, Washington, D.C.; "Naval Aviation and Stealth," Dennis Krieger, *Naval Institute Proceedings*, September 1994, pp. 59–63; "Carrier Aviation Is Doing More with Less," Jon B. Anderson, *Navy Times*, Nov. 18, 1996, pp. 12–14; *The Pentagon Paradox: The Development of the F-18 Hornet*, James P. Stevenson (Annapolis, Md.: Naval Institute Press, 1993); "F/A-18: Stores Management and Mission Programming Workbook," 28.7-WB, ASTK 005, VFA-106, Sept. 28, 1994; "F/A-18: LAT Phase Manual," VFA-106, Jan. 22, 1996; "F/A-18A/B/C/D 161353 And Up Aircraft," NATOPS Flight Manual A1-F18AC-NFM-000, Jan. 15, 1994, Naval Air Systems Command; "Radar Bombing with the Hornet," monograph by Charles Van Gorden Jr.; and "F/A-18 Close Air Support Workbook," 31.1.1-WB, ASTK 029, VFA-106, Jan. 3, 1996.

Post–Cold War Navy roles and missions are drawn from interviews with Naval historian Norman Polmar, William Kaufman of the Brookings Institution, Scott Truver of Techmatics Inc., and Ron O'Rourke of the Congressional Research Service. Also consulted: "Handbook of U.S. Aircraft Carrier Programs," Scott Truvar, Techmatics Inc.; "Navy Carrier-Based Fighter and Attack Aircraft: Modernization Options for Congress," CRS Report for Congress, 93-868 F, Congressional Research Service, Oct. 1, 1993; "Aviation Forces," Secretary of Defense Les Aspin, Annual Report to the President and the Congress, Department of Defense, January 1995; "Is There a Doctrine in the House?," Scott A. Hastings, *Naval Institute Proceedings*, Vol. 120/4/1094, April 1994, pp. 34–38; "Is a Change in Doctrine Enough?," Seth Cropsy, *Naval Institute Proceedings*, Vol. 121/2/1104, February 1995, p. 9; "Bosnia, Tanks, and '. . . From the Sea,'" Daniel E. Moore Jr., *Naval Institute Proceedings*, Vol. 120/12/1102, December 1994, pp. 42–45; "Forward . . . From the Sea," John H. Dalton et al., *Naval Institute Proceedings*, Vol. 120/12/1102, December 1994, pp. 46–49; "Sea Power Is Grand Strategy," Don DeYoung, *Naval Institute Proceedings*, Vol. 120/11/1101, November 1994, pp. 73–77; "Staying the Course," James A. Winnefield, *Na-*

val Institute Proceedings, Vol. 120/5/1095, May 1994, pp. 32–39; "Presence: Forward, Ready, Engaged," Philip A. Dur, *Naval Institute Proceedings*, Vol. 120/6/1095, June 1994, pp. 41–44; "The Navy and the Nation," Michael Vlahos, *Naval Institute Proceedings*, Vol. 120/5/1094, May 1994, pp. 56–63; and "Admiral Sees End of Large Deck Carriers After CVN-76," Kerry Gildea, *Defense Daily*, Apr. 13, 1995, p. 59.

CHAPTER SEVENTEEN

This chapter is based on interviews conducted September 9–October 4, 1996, with Lieutenant Phillip Clay, Lieutenant John Brotemarkle, Lieutenant Glenn M. Crabbe, Lieutenant Bryan D. Williams, Lieutenant Junior Grade Chris Brown, Captain Stan D. Hester, Lieutenant Alexander M. Howell, Lieutenant Mike Rourke, Commander Bill Gortney, Lieutenant Commander Mike Wettlaufer, Lieutenant Commander Tom Ganse, Lieutenant Nick Mongillo, Commander Jack Holt, and Captain John Stufflebeem.

Also consulted were: "F/A-18 CQ Class Workbook," Student Workbook, Nov. 2, 1994, VFA-106, NAS Cecil Field; "CV NATOPS Manual," NAVAIR 00-80T-105, Nov. 1, 1995, Washington, D.C.; NATOPS Flight Manual A1-F18AC-NFM-000, Jan. 15, 1994, Naval Air Systems Command; "CAS III Procedures," Workbook, VFA-106, NAS Cecil Field; "USS *John F. Kennedy*," Brochure; "Naval Aviation Physiology Training Program (NAPTP)" Initial Training, B-322-0043, Oct. 7, 1993, Naval Aerospace and Operational Medical Institute, Pensacola, Florida; and "F/A-18 LSO Calls Workbook," 41.0.1-WB, May 6, 1993, VFA-106, NAS Cecil Field.

INDEX